Malware Diffusion Models for Modern Complex Networks

Theory and Applications

Malware Diffusion Models for Modern Complex Networks

Theory and Applications

Vasileios Karyotis

M.H.R. Khouzani

AMSTERDAM • BOSTON • HEIDELBERG • LONDON
NEW YORK • OXFORD • PARIS • SAN DIEGO
SAN FRANCISCO • SINGAPORE • SYDNEY • TOKYO

Morgan Kaufmann Publishers is an Imprint of Elsevier

Acquiring Editor: Brian Romer
Editorial Project Manager: Amy Invernizzi
Project Manager: Priya Kumaraguruparan
Designer: Mark Rogers

Morgan Kaufmann is an imprint of Elsevier
50 Hampshire Street, Cambridge, MA 02139, USA

Library of Congress Cataloging-in-Publication Data
A catalog record for this book is available from the Library of Congress

British Library Cataloging-in-Publication Data
A catalogue record for this book is available from the British Library.

ISBN: 978-0-12-802714-1

For information on all Morgan Kaufmann publications
visit our website at www.mkp.com

Working together
to grow libraries in
developing countries

www.elsevier.com • www.bookaid.org

Contents

PART 2 STATE-OF-THE-ART MALWARE MODELING FRAMEWORKS

PART 4   APPENDICES

Preface

Malicious software (malware) has become a serious concern for all types of communications networks and their users, from the laymen to the more experienced administrators. The proliferation of sophisticated portable devices, especially smartphones and tablets, and their increased capabilities, have propelled the intensity of malware dissemination and increased its consequences in social life and the global economy. This book is concerned with the theoretical aspects of such malware dissemination, generically denoted as *malware diffusion*, and presents modeling approaches that describe the behavior and dynamics of malware diffusion in various types of complex communications networks and especially wireless ones.

The main objective of this book is to classify and present in adequate detail and analysis, families of state-of-the-art mathematical methodologies that can be used for modeling generically malware diffusion, especially in wireless complex networks. However, with minor and straightforward adaptations, these techniques can be further extended and applied in other types of complex networks as well.

In addition, the book covers holistically the mathematical modeling of malware diffusion, starting from the early emergence of such attempts, up to the latest, advanced and cross-discipline based frameworks that combine diverse analytic tools. Starting from the basic epidemics models that are based on systems of ordinary differential equations, the content proceeds to more exotic analytic tools founded on queuing systems theory, Markov Random Fields, optimal control and game theoretic formulations, respectively. Numerical and simulation results are provided, in order to validate each framework and demonstrate its potentials, along with system behavior studies. The book also provides a summary of the required mathematical background, which can be useful for the novice reader. Furthermore, it provides guidelines and directions for extending the corresponding approaches in other application domains, demonstrating such possibility by using application models in information dissemination scenarios.

Consequently, this book aspires to stimulate inter-disciplinary research and analysis in the broader area of modeling information diffusion in complex networking environments. It mainly focuses on the diffusion of malicious information (software) over wireless complex networks, however, as will become evident, most of the results can be easily extended and adapted for other types of networks and application domains.

Intended Audience

The content of this book is presented in a fashion aiming mainly at first-year graduate audiences, postdoctoral researchers, professors and the more experienced/interested professional engineers that are involved in computer security research and development. Most of them are assumed already familiar with the practical topics included in the broader research area and the book provides for them a solid quantitative background on the available mathematical malware modeling approaches in a more systematic manner than the works available nowadays (essentially scattered journal/conference papers and surveys), i.e. with formal definitions, references to the mathematical methods and analysis of the advanced techniques. The text presents and analyzes the latest mathematical tools that can be of use in the research and development activities of the above audiences. However, despite its semi-advanced nature, students in their last undergraduate year can also benefit from such a specialized treatment and involved methodologies, by obtaining a solid background of the corresponding area.

The book focuses on the mathematical modeling of malware diffusion dynamics, and as such, some familiarity on basic mathematical techniques, such as probability theory, queuing theory, ordinary differential equations, optimal control and game theory is needed. The required quantitative level will be no higher than that of the first graduate year. Consequently, the book is ideal for graduate students at the beginning of their programs, both for coursework level (graduate textbook) and as a companion in their own research endeavors. Basic elements of the required mathematical tools are presented in the three appendices, providing quick background reference for those not familiar with the corresponding fields.

The main discipline for which this book was developed for is computer science and system engineering. It has been specifically written for those involved in computer and system security. Academics from these fields can use the book in their research and graduate classrooms. The material provided offers a complete set of existing state-of-the art methodologies accompanied by an extensive bibliography and application examples. It provides a coherent perspective of the area of malware diffusion and security, and guidelines for developing and broadening one's knowledge and research skills in the corresponding areas.

Regarding the application content of the book, the main audience is expected to be scientists and engineers active in the field of communications/computer networks, namely the broader community of computer scientists and electrical engineers, and more specifically, computer and systems security are expected to form the main audience. However, at the same time, a number of researchers and professionals working in other disciplines that study problems sharing several characteristics with the problems emerging in malware diffusion can be also accommodated by the contents of the book, at least partially. Network Science is the most prominent such area that has already brought together disciplines as diverse as sociology, biology, finance, computer science and electrical engineering, in order to jointly study problems and share methods and results. Malware diffusion may be considered in a more generic fashion as information diffusion and professionals from all the aforementioned disciplines studying information dissemination problems are expected to have potential interest. The generic form of the presentation and especially the applications of the presented techniques into practical and diverse problems, such as information dissemination dynamics is suitable for diverse professionals as social scientists, epidemiologists and marketing professionals, as well.

Consequently, the level of the book accommodates practically all levels of expertise, with more emphasis on the intermediate to advanced. The applications are relevant mainly to engineers and scientists in the field of communications and computer science, but also relevant to inter-disciplinary scientists and professionals from the information-related disciplines and Network Science. The book has attempted to balance both depth (technical level) and breadth (application domains) of the included methodologies, originally presented for malware diffusion.

Scope and Outline of the Book

Scope

The topics addressed regarding malware diffusion, are treated in this book from an inter-disciplinary Network Science perspective, and are currently rapidly evolving at rates that other research areas have been enjoying for many years now. Within such framework, some fields of Network Science have already been well-shaped and advanced to a desired degree, e.g. social network analysis (SNA) [125, 164], while others still consist of fragmented contributions and scattered results.

Malware diffusion in computer networks in general, and wireless ones in particular, qualifies as one of the latter fields. Until recently, most of the proposed approaches for modeling the dynamics of malicious software dissemination followed more or less the same practices and they were essentially based on some restrictive assumptions. Most of them required the diffusion process to take place first, in order to later develop/fit accurate models based on the observed data afterwards, lacking predictive power for generic anticipated attacks. Thus, it was not possible to holistically capture the behavior of dynamics and predict the outcomes of attacks before they actually take place.

However, in the last decade, several advanced modeling methodologies were presented, which are capable of describing more accurately malicious software diffusion over diverse types of networks, and more intelligent attack strategies as well. Generic models have been presented, and when necessary they can be adapted to describe accurately the observed behaviors in other types of networks. Such approaches utilize different mathematical tools for their purposes and capture properly the most important aspects of malicious software diffusion dynamics.

Still, the literature is missing a systematic classification, presentation and analysis of all these advanced methodologies and obtained results, in a manner compatible to the broader scope of the discipline of Network Science and with reference to key legacy approaches as well. This book aspires to fill this gap, by methodically presenting the topic of malware diffusion in complex communications networks. More specifically, the book will focus on malware diffusion modeling techniques especially designed for wireless complex networks. However the presented methodologies are applicable for other types of complex communications and social networks and the wireless network paradigm will be employed mainly for demonstration purposes. The mathematical methodologies that will be presented, due to their generic analytical nature can be easily adapted and used in other types of complex networks, even non-technological ones. Thus, the book will not only present and analyze malicious software modeling methods for wireless complex networks, but also demonstrate how these methods can be extended and applied in other settings as well, e.g. generic information dissemination over complex networks of any type such as human, financial, etc.

In short, this book aspires to become a cornerstone for a systematic organization and mathematical modeling of malicious software and information diffusion modeling within the broader framework of Network Science and complex networks. Furthermore, it aspires to provide long-term reference to the required background for studying in-depth and extending the corresponding field of research.

Outline

This book is organized in three main parts and a set of auxiliary appendices with respect to the core mathematical areas required in order to understand the main contents of the book. The introductory Part 1 consists of Chapters 1–3, and constitutes a thorough introduction to the general malware diffusion modeling framework we consider in this book. Part 2, which includes Chapters 4–8, presents state-of-the-art malware diffusion modeling mathematical methodologies and corresponds to the main and unique contribution of this book in the literature. It presents, while also explaining in detail, malware diffusion modeling mathematical methodologies utilizing alternative, yet powerful analytical tools. Part 3 summarizes the key points of the presented methodologies and presents directions for potential future research. It also sets the presented theoretical knowledge into a broader application perspective, which can be exploited in other disciplines as well. Finally, the appendices contain brief, but complete reviews of the basic mathematical tools employed in this book, namely elements of ordinary differential equations, elements of queuing theory and elements of optimal control theory, which can be very helpful for the non-familiar reader, in order to quickly obtain a solid understanding of the mathematical tools required to understand the presented models and approaches.

In more detail, Chapter 1 serves as a concise introduction to the topics addressed in the book, introducing complex communication networks, malware diffusion, as well as some historical elements of the evolution of networks and malware.

Chapter 2 defines the malware diffusion problem, along with the node infection models that emerge in the literature. It also collects and presents characteristic examples of computer network attacks which are of interest in the study of malware diffusion in the framework of the book.

Chapter 3 provides a concise presentation and quick reference analysis of the malware modeling methods, with respect to the emerging incidents in the early days of modeling malicious software propagation dynamics and by focusing on the wireless scenarios. The content of this chapter will serve as background for some of the state-of-the-art approaches presented later in Part 2.

The following chapters in Part 2 present advanced malware modeling techniques, each dedicated to a family of approaches distinguished by the rest according to the employed mathematical tools. Thus, the first chapter of Part 2, namely Chapter 4, presents approaches modeling malware diffusion by means of queuing theory, and especially queuing networks. The basic idea is that the time spent by each node in a state of an infection model[1] can be mapped to the waiting time of a customer in a pure queuing system. Due to the superposition of node behaviors in a network, the corresponding queuing system will be a network of queues for modeling the behavior of malware over the network.

Chapter 5 in its turn presents and analyzes malware modeling approaches that exploit the notion of Markov Random Fields (MRFs). MRFs are sets of random variables that can cumulatively describe the overall state of a system, where in this case, the system is an attacked network. By exploiting several properties of MRFs, it is possible to obtain solutions in a simple manner, without sacrificing important detail, for diverse types of complex networks.

Chapter 6 covers malware modeling approaches that are based on stochastic epidemics and optimal control. Such approaches allow analyzing the robustness potentials of networks and attacks and obtain optimal or semi-optimal policies for dealing with attacks and their outcomes.

Chapter 7 builds on the previous and presents, analyzes and demonstrates malware modeling approaches that exploit principles from game theory to model epidemics. It casts the problems in an interactive framework and combines them with optimal control strategies.

Finally, Chapter 8 provides a qualitative comparison of all the previously (Chapters 4–7) presented approaches with the ulterior goal to reveal the distinct features of each approach in a comparative fashion, allowing selecting the most appropriate one for different applications.

In Part 3, Chapter 9 presents other application areas where the presented models may be applied successfully, thus, exhibiting their potential for creating more holistic information diffusion frameworks. Chapter 10 summarizes the lessons learned, explains the ground covered until now and provides potential directions for future work in the specific topic of malware diffusion modeling and the broader vision of information diffusion. Finally, Chapter 11 concludes this book, highlighting the most important aspects of malware diffusion, in particular, and information dissemination in general.

Appendix A provides background on differential equations, Appendix B on queuing systems theory, and Appendix C on optimal control theory and Hamiltonians, for the interested readers.

---

[1] The infection model will be explained in Chapter 2 and it describes how nodes of a network change states with respect to malware and their own behavior.

List of Figures

List of Tables

Malware diffusion modeling framework

Fundamentals of complex communications networks

1.1 INTRODUCTION TO COMMUNICATIONS NETWORKS AND MALICIOUS SOFTWARE

In complex networks [7, 164, 165] and the broader area of Network Science[1] [125, 155], modern analysis methodologies developed lately have identified multiple and diverse types of interactions between and among peer entities. Such interactions regarding humans, computer devices, cells, animals, and in general, whatever one might think of, vary in their degree of criticality. Peer interactions have been holistically modeled by various research disciplines, e.g. in engineering, social sciences, biology, and financial sciences and lately systematically within the framework of Network Science, as different types of network structures, i.e. communications, social, biological, and financial networks. These network structures bear distinct and characteristic properties of broader interest for science and daily human lives. The key feature across all such different networks is the flow of information, which typically takes place spontaneously, e.g. in biological types of networks, or in specific cases in an on-demand manner, e.g. in communications networks. The information dissemination processes over networks are usually controlled, and typically they are of useful nature for all peers participating in the corresponding network. However, frequently, and especially in the prospect of potential financial benefit, information dissemination over networks can take a malicious form, either for the entities of the network individually or the whole network cumulatively.

In order to explain the latter better, nowadays, it is often observed that the disseminated information can be harmful, or it could be controlled by malicious peers, not the legitimate information owners/producers/consumers. Especially in communication networks, users experience almost on a daily basis several types of malicious software (malware), usually suffering personal, industrial, and/or financial consequences. Similarly, in biological networks, viruses can be transferring malicious signals through various blood cells or nerve networks of a living organism, leading eventually to diseases with sometimes lethal consequences, e.g. extreme cases of the flu virus and malaria. Also, this is especially evident in classic cases of virus

---

[1]Both concepts of Network Science and complex networks will be explained in detail later in this chapter.

spreading between humans, from the simplest seasonal flu scenarios to the more serious scenarios of, e.g. HIV and malaria. [87, 99, 160].

Especially for biological networks, their robustness against the aforementioned threats is very critical for sustaining all forms of life, while for science, such a feature is very fascinating with respect to the sustainability that these networks exhibit to the various forms of threats throughout so many years of evolution and virus spreads. Similarly, the study and analysis of malware behavior in communication networks are rather important for maintaining the coherency of modern information-based societies and the efficiency of the underlying networking infrastructures. The most frequent consequences of such malware infections render computer hosts at least dysfunctional, thus preventing the execution of routine or important tasks, while in more serious situations, the incurred cost may be much higher and diverse. Frequently, the targets of malicious attacks are public utility networks, e.g. water and electricity grids, or social networks, e.g. social network (facebook, twitter, instagram, linkedin, etc.) accounts and email accounts. For all these examples, the underlying computer/communications network operations are implicitly or explicitly targeted by the malicious attacks.

Motivated by the aforementioned observations, the main objective of this book is to present, classify, analyze, and compare the state-of-the-art methods for modeling malware diffusion in complex communications networks and especially wireless ones. The term *malware diffusion* cumulatively refers to all types of malicious software disseminating in various types of networks and could also be extended to characterize cumulatively all types of malicious information dissemination in complex networks, as will be explained in the following section. On the other hand, the term *complex network* characterizes generically the potential structure that a network might have and in this book we will present and analyze modeling frameworks for malware diffusion that are applicable to multiple types of diverse network structures. Thus, all of the presented approaches could be used to model malware or information dissemination in multiple and diverse types of networks, e.g. communications, social, and biological.

The main focus and application domain of the book will be focused on wireless complex networks, a term which includes all types of wireless networks cumulatively. Wireless complex networks can be characterized by the presence or absence of central infrastructure, e.g. cellular [168], *ad hoc* [39], sensor, mesh, and vehicular networks [5], in most of which nodes operate in a peer fashion, acting as both routers and relays [5]. The presented methodologies are also applicable to networks with centralized organization, e.g. wired types of network topologies, via straightforward extension of the corresponding approaches involving distributed network operations. Similarly to the scope of this book, for these types of networks, rather diverse modeling approaches have emerged lately aiming at modeling malware diffusion specially in wireless decentralized networks. Such approaches yield similar results with respect to the trends of malware diffusion dynamics, but more restricted in terms of generality or control potential compared to the results provided by the approaches that will be described in this book.

The book will focus on wireless complex networks primarily for demonstration purposes and in order to better facilitate the practical explanation of the concepts. Extrapolations of the presented methodologies in other types of networks and other types of application contexts, e.g. diffusion of information dissemination over communications networks or even social networks, will be provided across the book and especially in a dedicated chapter, namely, Chapter 9. Such extension will be straightforward and when more complicated extensions are required, the appropriate directions are pointed out and details on the required steps are provided as well.

In the main part (Part 2) of this book, we classify and present state-of-the-art techniques for malware modeling according to the type of mathematical framework employed for the modeling and analysis of the corresponding malware diffusion problems. In a comparative manner, we highlight the strengths and weaknesses of each methodology, thus enabling the interested researcher and professional engineer to select the most appropriate framework for a specific problem/application. Furthermore, we provide a concrete presentation of each different mathematical methodology, which will allow the reader to grasp the salient features and technical details that govern malware (and in general information) diffusion dynamics. Finally, we compare the complexity to obtain analytical results and implementation of solutions of the presented approaches. Within this framework and for those approaches addressing similar or comparable objectives under similar modeling settings, e.g. epidemic and optimally controlled epidemic approaches, we evaluate the obtained results with respect to factors that would be important in an operational environment. Thus, we qualitatively assess whether each framework is accurate and simple, tractable and scalable, etc., for dense or sparse regimes and other topological variations of each network.

Before we proceed with the main topics of the book, we next take a small detour and provide some background on the evolution of networks and malicious software, enabling a better understanding of the systems and applications over which malicious information typically spreads. This will also help understand better the dynamics of information spreading later in Part 3. In the next subsection, we will start with a brief history of networks, from the first academic interconnected systems to today's complex commercial networks.

1.2 A BRIEF HISTORY OF COMMUNICATIONS NETWORKS AND MALICIOUS SOFTWARE

1.2.1 FROM COMPUTER TO COMMUNICATIONS NETWORKS

The emergence of the first network structures in nature took place in the form of messaging pathways in chemical bonds and biological elements [182]. Contrary to these naturally formed networks, humans have developed social interactions that led to various networked developments, such as friendships, smaller of larger

communities, nations, and open markets. Eventually, other artificial and technological networks made their appearance, mainly aiming at making life easier, either
through the transfer of commodities and resources, e.g. electricity grid/water pipe
networks, or via transferring bits of information, i.e. computer and communications
networks. Even though natural and social networks appeared much earlier than
technological ones, the first networks to be systematically studied, analyzed, and controlled were the computer and communications networks. In fact, such networks were
first conceived and designed and then developed. On the contrary, social networks
have been long established before the first mathematical treatises of their structure,
properties, and dynamics emerged. In addition, computer networks, and long before
them telephone networks, were the first networks that were initially designed/studied
by means of analytical methods before actually being developed/implemented.

The first data network, ARPANET, was built in 1973 in order to interconnect
and promote research among the US American universities and various US research
centers. The ARPANET was mainly based on the infamous TCP/Internet protocol
(IP) stack suite, which was originally developed exactly to serve this network, i.e.
provide a layered and modular substrate for developing the required mechanisms
that will ensure the reliable, transparent, and efficient transfer of data between
endpoints physically located in distinct places. For more information on the operation
of TCP/IP, which is even today the protocol suite of choice for the majority of
networks and its use in the ARPANET and later networks until today, there exists
a vast amount of literature available, e.g. [55, 148, 208] and many relevant references
provided in them. The evolution of the infrastructure of ARPANET, as well as the
TCP/IP stack, followed the accumulated knowledge by their analysis-driven design
and implementation. Ever since, computer networks have gone into a development
loop, where progress is dictated either from technology or theory. Once technology
develops due to deeper knowledge, it immediately spurs a research frantic, which in
turn leads technology to even higher complexities and benefits.

In 1980–1981, two other networking projects, BITNET and CSNET, were initiated. BITNET adopted the IBM RSCS protocol suite and featured direct leased line
connections between participating sites. Most of the original BITNET connections
linked IBM mainframes to university data centers. This rapidly changed as protocol
implementations became available for other machines. From the beginning, BITNET
had been multidisciplinary in nature with users in all academic areas. It had also
provided a number of unique services to its users (e.g. LISTSERV). BITNET and
its parallel networks in other parts of the world (e.g. EARN in Europe) had several
thousand participating sites. In its final years, BITNET had established a backbone
that used the TCP/IP protocol suite with RSCS-based applications running above
TCP. As of 2007, BITNET has essentially ceased operation.

CSNET was initially funded by the National Science Foundation (NSF) to
provide networking for university, industry, and government computer science
research groups. CSNET used the Phonenet MMDF protocol for telephone-based
electronic mail relaying and, in addition, pioneered the first use of TCP/IP over
X.25 using commercial public data networks. The CSNET name server provided

an early example of a white pages directory service. At its peak, CSNET had approximately 200 participating sites and international connections to approximately fifteen countries.

In 1987, BITNET and CSNET merged to form the Corporation for Research and Educational Networking (CREN). In the Fall of 1991, CSNET service was discontinued having fulfilled its important early role in the provision of academic networking service. To help speed the connections, the NSF established five super computing centers in 1986, creating the NSFNET backbone. In 1987, the NSF signed a cooperative agreement to manage the NSFNET backbone with Merit Network, Inc., and by 1990, ARPANET had been phased out. NSFNET continued to grow, and more countries around the world were connected to this Internet backbone.

In 1986, the US NSF initiated the development of the NSFNET, which provided a major backbone communication service for the Internet. The National Aeronautics and Space Administration (NASA) and the US Department of Energy (DoE) contributed additional backbone facilities in the form of the NSINET and ESNET, respectively. This further spurred the development of national network infrastructures for research and experimentation. In a similar fashion, in Europe, major international backbones, such as GEANT, and other national ones, such as GRNET and NORDUNET, provide today connectivity to over millions of computers on a large number of networks. Back in 1986 commercial network providers in the US and Europe began to offer Internet backbone and access support.

The year of 1991 was a big year for the Internet: The National Research and Education Network (NREN) was founded and the World Wide Web was released. At the time, the Internet was still dominated by scientists and other academics, but had begun to attract public interest. With the release of the Mosaic Web browser in 1993 and Netscape in 1994, the interest in the use of the World Wide Web exploded. More and more communities became wired, enabling direct connections to the Internet. In 1995, the US Federal Government relinquished its management role in the Internet and NSFNET reverted back to being a research network. Interconnected network providers were strong enough at the time to support US backbone traffic on the Internet. However, the administration at the time encouraged continued development of the US backbone of the Internet, also known as the National Information Infrastructure (NII)—and, most commonly, as the "Information Superhighway."

Throughout its lifetime, "regional" support for the Internet has been provided by various consortium networks and "local" support was provided through each of the research and educational institutions. Within the United States, much of this support had come from the federal and state governments, but a considerable contribution had been also made by industry. In Europe and elsewhere, support arose from cooperative international efforts and through national research organizations. During the course of its evolution, particularly after 1989, the Internet system began to integrate support for other protocol suites into its basic networking fabric. The present emphasis in the system is on multiprotocol internetworking, and in particular, with the integration of the open systems interconnection (OSI) protocols into the architecture. During the early 1990s, OSI protocol implementations also became available and, by the end of 1991, the Internet had grown to include some 5000 networks in over three dozen countries, serving over 700,000 host computers used by over 4,000,000 people.

Over its short history, the Internet has evolved as a collaboration among cooperating parties. Certain key functions have been critical for its operation. These were originally developed in the DARPA research program that funded ARPANET, but in later years, this work has been undertaken on a wider basis with support from US Government agencies in many countries, industry, and the academic community. The Internet Activities Board (IAB) was created in 1983 to guide the evolution of the TCP/IP suite and to provide research advice to the Internet community. During the course of its existence, the IAB has been reorganized several times. It now has two primary components: the Internet Engineering Task Force and the Internet Research Task Force. The former has primary responsibility for further evolution of the TCP/IP suite, its standardization with the concurrence of the IAB, and the integration of other protocols into Internet operation (e.g. the OSI protocols). The Internet Research Task Force continues to organize and explore advanced concepts in networking under the guidance of the IAB and with support from various government agencies.

The recording of Internet address identifiers, which is critical for translating names to actual addresses of machines requested, is provided by the Internet Assigned Numbers Authority (IANA) who has delegated one part of this responsibility to an Internet registry (IR) which acts as a central repository for Internet information and which provides central allocation of network and autonomous system identifiers, in some cases to subsidiary registries located in various countries. The IR [105] also provides central maintenance of the domain name system (DNS) [55, 148, 208], root database which points to subsidiary distributed DNS servers replicated throughout the Internet. The DNS distributed database is used to associate host and network names with their Internet addresses and it is critical to the operation of the higher level TCP/IP stack including electronic mail.

There are a number of network information centers (NICs) located throughout the Internet to serve its users with documentation, guidance, advice, and assistance. As the Internet continues to grow internationally, the need for high quality NIC functions increases. Although the initial community of users of the Internet was drawn from the ranks of computer science and engineering, its users now comprise a wide range of disciplines in the sciences, arts, letters, business, military, and government administration, and of course primarily and most importantly, private citizens (users) of the Internet.

This subsection provided a brief overview of the evolution of computer networks to communications networks, which are nowadays publicly and massively available. From this concise historical overview, it becomes evident that as networks become of more public use and as they grow (evolve) to serve more users, the initially introduced technology is still in use, or at least with minor adaptations. This means that malicious users have had the chance in the course of time to obtain deep knowledge on the operation of such infrastructures and in addition, to enhance their arsenal with sophisticated tools, which as will be shown later have allowed them to launch massive attacks, aiming at critical large-scale systems and individual users as well, with significant consequences.

There are a lot of references providing more details on the history and function of the Internet and its passage to broader communications networks in general. Some of these presentations can be found with [55, 148, 208] and references therein. In the following, we briefly touch on a special type of communications networks, i.e. wireless networks, which will constitute the main network application domain employed in the main part of the book.

1.2.2 THE EMERGENCE AND PROLIFERATION OF WIRELESS NETWORKS

Wireless communications have now a long history. The first one to discover and produce radio waves was Heinrich Hertz in 1888, while by 1894, Marconi demonstrated the modern way to send a message over telegraph wires. By 1899, Marconi sent a signal nine miles across the Bristol Channel and 31 miles across the English Channel to France. In 1901, he was able to transmit across the Atlantic Ocean.

However, it was during World War II, that the United States Army first used radio signals for data transmission. This inspired a group of researchers in 1971 at the University of Hawaii to create the first packet based radio communications network called ALOHANET [2]. ALOHANET was the very first wireless local area network (WLAN). This first WLAN consisted of seven computers that communicated in a bi-directional star topology and spurred the research for the development of more efficient protocols for wireless medium access.

The first generation of WLAN technology based on the previously mentioned ALOHA protocol used an unlicensed band (902–928 MHz-ISM), which later became very popular eventually becoming interference-crowded from networked small appliances and industrial machinery operating in this band [162]. A spread spectrum was used to minimize this interference, which operated at 500 kilobits per second (kbps). However, such rates were proving unsatisfactory in practice, calling for immediate improvements. The second generation (2G) of WLAN technology was four times faster and operating at 2 Mbps per second. The third generation WLAN technology operates on the same band as the 2G and we still use it today. It is popularly denoted as the IEEE 802.11 family, and with respect to the WLAN standardization, in 1990, the IEEE 802 Executive Committee established the 802.11 Working Group to create a WLAN standard to be widely adopted by professionals in the area. The 802.11 standard specified an operating frequency in the 2.4 GHz ISM band [110]. In 1997, the group approved IEEE 802.11 as the world's first WLAN standard with data rates of 1 and 2 Mbps, now evolved into multiple and diverse variations with speeds that can reach up to 6.75 Gbps (802.11ad—December 2012) [110].

At the same time, mobile phone technology is continuously evolving, seemingly at an accelerating rate of innovation and adoption [103, 227]. Examining the strides taken from 1G to 4G, the technology has both created new usage patterns and learned from unexpected use cases. Compared to the more distributed nature of the previously

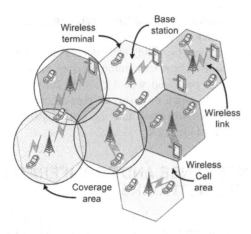

FIGURE 1.1

Simple architecture model of a cellular network and terminology employed (cell, terminal, base station, coverage area).

presented wireless networks, mobile networks operate in a more centralized manner throughout their inception.

In the 1970s, the first generation, colloquially referred to as 1G, of mobile networks was introduced [154]. These systems were referred to as cellular, which was later shortened to "cell," due to the approach employed for the network architecture that covers the network area through cellular base stations (a simple example of such system is shown in Fig. 1.1). Cell phone signals were based on analog system transmissions, and 1G devices were comparatively less heavy and expensive than prior devices, e.g. portable military radios. Some of the most popular standards deployed for 1G systems and implementing the corresponding concepts were advanced mobile phone system (AMPS), total access communication systems (TACS), and Nordic mobile telephone (NMT). The global mobile phone market grew from 30% to 50% annually with the appearance of the 1G network, and the number of subscribers worldwide had reached approximately 20 million by 1990.

In the early 1990s, 2G phones deploying global system for mobile communications (GSM) technology were introduced. GSM is a standard developed by the European Telecommunications Standards Institute (ETSI) to describe protocols for 2G digital cellular networks used by mobile phones, first deployed in Finland in July 1992 [101]. GSM used digital modulation to improve voice quality but the network offered limited data service. Such systems are still partially in use in some countries around the world. As demand drove the uptake of cell phones, 2G carriers continued to improve transmission quality and coverage. The 2G carriers also began to offer additional services, such as paging, faxes, text messages, and voicemail. The limited data services under 2G included WAP, HSCSD, and MLS. An intermediary phase 2.5G was introduced in the late 1990s. It used the GPRS standard [101], which delivered packet-switched data capabilities to existing GSM networks. It allowed users to send graphics-rich data as packets. The importance for packet-switching increased with the rise of the Internet and the IP. The EDGE network is an example of 2.5G mobile technology.

The 3G revolution allowed mobile telephone customers to use audio, graphics, and video applications. Over 3G it is possible to watch streaming video and engage in video telephony, although such activities are severely constrained by network bottlenecks and over-usage. One of the main objectives behind 3G was to standardize on a single global network protocol instead of the different standards adopted previously in Europe, the US, and other regions. 3G phone speeds deliver up to 2 Mbps, but only under the best conditions and in stationary mode. Moving at a high speed can drop 3G bandwidth to a mere 145 Kbps. 3G cellular services, also known as UMTS [101], sustain higher data rates and open the way to Internet style applications. 3G technology supports both packet- and circuit-switched data transmission, and a single set of standards can be used worldwide with compatibility over a variety of mobile devices. UMTS delivers the first possibility of global roaming, with potential access to the Internet from any location.

The currently widely employed generation of mobile telephony, denoted as 4G [175], has been developed with the aim of providing transmission rates up to 20 Mbps while simultaneously accommodating quality of service (QoS) features. QoS will allow the device and the telephone carrier to prioritize traffic according to the type of application using user's bandwidth and adjust between user's different telephone needs at a moment's notice. However, it is only recently that we are we beginning to realize the full potential of 4G applications. A 4G system, in addition to the usual voice and other services of 3G, provides mobile broadband Internet access, for example to laptops with wireless modems, to smartphones, and to other mobile devices. It is also expected to include high-performance streaming of multimedia content. The deployment of 4G networks will also improve video conference service functionality. It is also anticipated that 4G networks will deliver wider bandwidth to vehicles and devices moving at high speeds within the network area.

Nowadays, a very large portion of Internet connectivity occurs over wireless networks, either WLANs or 4G cellular, and it is expected that this tendency will further increase. Thus, the fraction of wireless connectivity is expected to further rise sharply, which is the main reason that relevant malware is anticipated to cause major problems. This is also the main reason that this book focuses and uses wireless complex networks as the main type of network demonstrator. In addition, it is expected that future patterns of wireless access are expected to further diversify. Additional communication and wireless access paradigms, such as machine-to-machine (M2M) communication,[2] are expected to increase more the complexity of

---

[2]M2M [10, 103] refers to technologies that allow both wireless and wired systems to communicate with other devices of the same type. It is a broad term as it does not pinpoint specific wireless or wired networking, information and communications technology. It is considered an integral part of the Internet

the network and allow more flexibility to malicious attackers. Thus, the malware diffusion process in future wireless networks is expected to be more complicated.

The future wireless and mobile networks, denoted as 5G, envisage a more converged environment that includes heterogeneous networks (4G, WLAN, *ad hoc*,[3] cognitive radio networks,[4] etc.) and devices under a holistic management and control framework. Within the 5G networks, performance and flexibility are expected to rise drastically, offering more capabilities and features, but at the same time creating more opportunities for exploiting vulnerabilities and management holes by malicious users and attackers, as will be explained in more detail in the following subsection and chapters of the book.

1.2.3 MALICIOUS SOFTWARE AND THE INTERNET

Malicious software (malware) is not new in computer and communications networks. In fact, it emerged as soon as the first publicly accessible infrastructures were made available. However, in 1986, most viruses were found in universities and their propagation was primarily via infected floppy disks from one machine to another, not through the network. A computer virus is a malware program that, when executed, replicates by inserting copies of itself (possibly modified) into other computer programs, data files, or the boot sector of the hard drive [13]. Viruses often perform some type of harmful activity on infected hosts, such as stealing hard disk space or CPU time, accessing private information, corrupting data, displaying political or humorous messages on the user's screen, spamming user contacts, logging user keystrokes, or even rendering the computer useless. However, not all viruses carry a destructive payload or attempt to hide themselves. The defining characteristic of viruses is that they are self-replicating computer programs that install themselves without user consent. Notable virus instances during the era of their emergence included Brain (1986), Lehigh, Stoned, and Jerusalem (1987), and Michelangelo in 1991 (the first virus to make it to the news headlines) [126].

By the mid-1990s, businesses were equally impacted by malware and its propagation had moved to the network protocol layer, exploiting the flexibility allowed by the TCP/IP stack employed by networks, and thus allowing even more automated

---

of Things (IoT) [103] and brings several benefits to industry and business in general as it has a wide range of applications such as industrial automation, logistics, Smart Grid, Smart Cities, e-health, defense, etc., mostly for monitoring, but also for control purposes.

[3] A wireless ad hoc network (WANET) is a decentralized type of wireless network [212]. The network is *ad hoc* because it does not rely on a pre-existing infrastructure, such as routers in wired networks or access points in managed (infrastructure) wireless networks. Instead, each node participates in routing by forwarding data for other nodes, so the determination of which nodes forward data is made dynamically on the basis of network connectivity. In addition to the classic routing, *ad hoc* networks can use flooding for forwarding data.

[4] A cognitive radio is an intelligent radio that can be programmed and configured dynamically [96, 109]. Its transceiver is designed to use the best wireless channels in its vicinity. Such a radio automatically detects available channels in wireless spectrum, then accordingly changes its transmission or reception parameters to allow more concurrent wireless communications in a given spectrum band at one location. This process is a form of dynamic spectrum management.

propagation capabilities. Notable malware for the period included the Morris worm[5] (1988), i.e. the first instance of network malware, DMV (1994), the first proof-of-concept macrovirus,[6] Cap.A (1997), the first high risk macro virus, and CIH (1998), the first virus to damage hardware [126]. By late 1990s, viruses had begun infecting the machines of home users as well, and virus propagation through email was increasing remarkably. Notable malware included Melissa (the first widespread email worm) and Kak, the first and one of the very few true email viruses, both in 1999 [126].

At the start of the new millennium, Internet and email worms were making headlines across the globe. Notable cases included Loveletter (May 2000), the first high-profile, profit-motivated malware, the Anna Kournikova email worm (February 2001), the March 2001 Magistr, the Sircam email worm (July 2001), which harvested files from the "My Documents" folder of windows operating systems (OSs), the CodeRed Internet worm (August 2001), and Nimda (September 2001), a Web email and network worm [126].

As the decade progressed, malware almost exclusively became a profit-motivated tool. Throughout 2002 and 2003, Web surfers were plagued by out-of-control popups and other Javascript bombs.[7] FriendGreetings ushered in manually driven socially engineered worms in October 2002 and SoBig began surreptitiously installing spam proxies on victim computers [126]. Credit card frauds also took off during the period. Other notable threats included the Blaster and Slammer Internet worms.

In January 2004, an email worm war broke out between the authors of MyDoom, Bagle, and Netsky worms [126]. Ironically, this led to improved email scanning and higher adoption rates of email filtering, which eventually led to a near demise of mass-spreading email worms.

In November 2005, discovery and disclosure of the now infamous Sony rootkit led to the eventual inclusion of rootkits in the most modern day malware. Money mule and lottery scams grew rapidly in 2006. These kinds of attacks aim at novice users, typically through email, and invite them to follow hyperlinks that eventually prove harmful for the user machine. They do so by advertising potential winnings in lotteries and other claims for money-winning-related offers. Though not directly malware-related, such scams were a continuation of the theme of profit-motivated criminal activity launched via the Internet.

Website compromises escalated in 2007, in large part, due to the discovery and disclosure of MPack, a crimeware kit used to deliver exploits via the Web. An exploit is a piece of software, a chunk of data, or a sequence of commands that takes advantage of a bug or vulnerability in order to cause unintended or unanticipated behavior

---

[5]A computer worm is a standalone malware computer program that replicates itself in order to spread to other computers using some type of network infrastructure [200].

[6]A macro virus is a virus that is written in a macro language, i.e. a programming language which is embedded inside a software application (e.g. word processors and spreadsheet applications).

[7]Javascript bombs are malicious programs developed in javascript, a dynamic programming language, most commonly used as part of web browsers, whose implementations allow client-side scripts to interact with the user, control the browser, communicate asynchronously, and alter the document content that is displayed, thus allowing for great flexibility for the attacker.

to occur on computer software, hardware, or something electronic (usually computerized). Such behavior frequently includes things like gaining control of a computer system, allowing privilege escalation, or a denial-of-service (DoS) attack. Notable compromises of this type included the Miami Dolphins website, Tomshardware.com, TheSun, MySpace, Bebo, Photobucket, and The India Times websites [126].

By the end of 2007, SQL injection attacks[8] had begun to increase, netting victim sites, including world-famous company websites. In a 2012 study, security company Imperva observed that the average web application received four attack campaigns per month, and retailers received twice as many attacks as other industries. Following the evolution of malware, by January 2008, Web attackers were employing stolen FTP credentials and leveraging weak configurations to inject tens of thousands of pop style websites. In June 2008, the Asprox botnet facilitated automated SQL injection attacks, claiming famous commercial websites among its victims. A botnet[9] is a number of Internet-connected computers communicating with other similar machines in an effort to complete repetitive tasks and objectives. This can be as mundane as keeping control of an Internet relay chat (IRC) channel, or it could be used to send spam email or participate in distributed DoS attacks.

Advanced persistent threats emerged during this same period as attackers began segregating victim computers and delivering custom configuration files to those of highest interest. In early 2009, Gumblar, the first dual botnet, emerged. Gumblar not only dropped a backdoor on infected PCs and used it to steal FTP credentials, it used those credentials to hide a backdoor on compromised websites as well. In a computer system (or cryptosystem or algorithm), a backdoor is a method of bypassing normal authentication, securing unauthorized remote access to a computer, obtaining access to plaintext, and so on, while attempting to remain undetected. The backdoor may take the form of a hidden part of a program, a separate program may subvert the system through a rootkit.[10] This development was quickly adopted by other Web attackers. As a result, today's website compromises no longer measure up to a handful of malicious domain hosts. Instead any of the thousands of compromised sites can interchangeably play the role of malware host.

The volume of malware is merely a byproduct of distribution and purpose. This can be best seen by tracking the number of known samples based on the era in which it occurred. For example, during the late 1980s, most malware were simple boot sector and file infectors spread via floppy disk. With limited distribution and less focused purpose, unique malware samples recorded in 1990 by AV-Test.org numbered just

---

[8]SQL injection is a code injection technique, used to attack data-driven applications, in which malicious SQL statements are inserted into an entry field for execution (e.g. to dump the database contents to the attacker). SQL injection must exploit a security vulnerability in an application's software, for example, when user input is either incorrectly filtered for string literal escape characters embedded in SQL statements or user input is not strongly typed and unexpectedly executed. SQL injection is mostly known as an attack vector for websites but can be used to attack any type of SQL database.
[9]The word botnet is a combination of the words robot and network.
[10]A rootkit is a stealthy type of software, typically malicious, designed to hide the existence of certain processes or programs from normal methods of detection and enable continued privileged access to a computer.

9044. As computer network adoption and expansion continued through the first half of the 1990s, distribution of malware became easier and malware volume increased. In 1994, AV-Test.org reported 28,613 unique malware samples [149].

As technologies were standardized and their operational details became more specific and easy to obtain and study, certain types of malware were able to gain ground. Macroviruses not only achieved greater distribution by using the email service but they also gained a distribution boost as the email penetration increased in society (nowadays it is almost holistically adopted among connected users). In 1999, AV-Test.org recorded 98,428 unique malware samples [149].

As broadband Internet adoption increased, Internet worms became more viable. Distribution was further accelerated by increased use of the Web and the adoption of the so-called Web 2.0 technologies that fostered a more favorable malware environment. In 2005, AV-Test.org recorded 333,425 unique malware samples [149]. Increased awareness in Web-based exploit kits led to an explosion of Web-delivered malware throughout the later part of the millennium's first decade. In 2006, the year MPack was discovered, AV-Test.org recorded 972,606 unique malware samples. As automated SQL injection and other forms of mass website compromises increased distribution capabilities in 2007, malware volume made its most dramatic jump, with 5,490,960 unique samples recorded by AV-Test.org in that year. Since 2007, the number of unique malware has sustained exponential growth, doubling or more each year. Currently, vendors' estimates of new malware samples range from 30k to over 50k per day [149]. For comparison purposes, this scale is such that the current monthly volume of new malware samples is greater than the total volume of all malware before 2006 cumulatively.

Lately, following the proliferation of smart portable devices, e.g. smartphones, tablets, and handhelds, another emerging trend is observed regarding malware propagation. Malware has progressively moved to the wireless part of the infrastructures, where the main victims are plain users, with far less technical technological involvement than the average computer user, but the stakes are higher due to the current size of the wireless market and the sensitivity of the data now exchanged via smart handhelds. As will be presented in more detail in the next chapter, nowadays wireless malware propagation is a reality similar to the one presented above for the traditional (wired) broadband Internet.

This book will focus more on the malware dissemination over wireless complex network cases, presenting mathematical frameworks that allow the modeling of such malware spreading. Nevertheless, the presented approaches are possible to be easily adapted and extended in cases of other communications networks, as it will become more evident in the sequel. Before delving into such detail, in the next subsection, we present and explain the networking substrate considered.

1.3 COMPLEX NETWORKS AND NETWORK SCIENCE

In this section, we will introduce the notion of complex networks and Network Science research area, which cumulatively describe the network environments considered in this book. They describe not only the physical networks used for information exchange but also various mechanisms, functionalities, and emerging problems developing over these physical, or more abstract concepts of networks. From this perspective, the overall content of this book focusing on the propagation of malware over complex communications networks may be also considered as part of the Network Science and complex networks research fields. The following subsections set the stage for this purpose.

1.3.1 COMPLEX NETWORKS

In brief, and in order to facilitate the following discussion of complex networks toward Network Science, the *"network"* might be thought of as a set of entities (nodes) with a set of pairs from those entities denoted as links which interconnect pairs of nodes, representing some kind of interaction or association between them. In that sense, the network might be seen as an alternative term for the concept of graph [61, 94] in mathematics. However, apart from the physical/systemic interpretation it has in, e.g. communications networks and power grids, it also refers to abstract notions of node-entities and their interactions, e.g. humans and their affiliations and diffusion of information. Thus, within complex network theory, the network is considered either as an abstract representation of potential entity interconnections, and/or a system of actual physical associations, depending on the application context.

The research and industrial interest for network functions and dynamics has increased substantially in the last decade [125, 155]. Various forms of networked structures (systems) are nowadays omnipresent, e.g. public utility and communications, while many of them have been proven crucial in their operation. Computer and mobile networks have enabled pervasive communication across continents, in diverse conditions and situations of varying importance, e.g. search & rescue and monitoring. Other forms of abstract networks have been also developed, e.g. affiliations in the business world, researcher affiliations, and information distribution networks. In this sense, modern societies are nowadays characterized as *connected*, *interconnected*, and *interdependent*. They are connected due to the existence of interactions between various entities, interconnected since they can exchange various types of resources, most prominently information, and finally, they are interdependent in the sense that even though they act independently, they rely on their directly connected counterparts for sustaining their progress, e.g. connected open markets and unified smart grids.

A key observation for networks emerging across different disciplines is that the complexity of most interconnected systems forming networks is not in the behavior/operation of a single unit or larger component, but rather in the cumulative behavior/operation exhibited by their interconnection and exchange of information. Traditionally, the main research efforts were centered in the understanding and analysis of the behavior of individual entities (the nodes of the network), of the clusters they may form, or the interactions among them. However, lately, the interdependent behavior of such basic modules gains in interest and importance, and

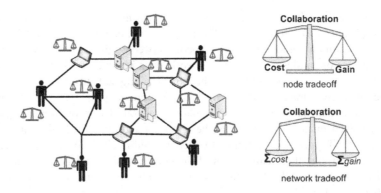

FIGURE 1.2

Network formation tradeoff: cost versus benefit of collaboration. For the network tradeoff, the total cost and total gain, summed over all entities are considered.

exploiting it yields progressively the desired level of control and flexibility over these networked structures.

A new term has been employed lately to refer cumulatively to all relevant newly observed and emerging behaviors, mechanisms, and dynamics regarding the network-oriented research in diverse disciplines, namely, *complex networks* [7, 125, 155, 164].

Definition 1.1 *(Complex Network Behavior). A complex network is a network that exhibits emergent behaviors that cannot be predicted a priori from known properties of the individual entities constituting the network.*

The above definition does not explain the notion of a network, in general, but mainly asserts that complex networks exhibit cumulative behaviors that can be rather diverse and possibly different from each other, even within the same application context. For example, in social networks, different types of networks may form even when the network refers to the same set of social entities, e.g. the same group of people may yield a different network structure when the interactions refer to friendly affiliations or business associations. At the same time, networks emerging across different contexts can also exhibit unexpected similarities, e.g. malware propagation dynamics in wireless multihop networks resembling the virus propagation in animal species or humans.

The notion of network can be explained in its part via the generic network formation mechanism, which is observed in any type of application framework and diverse operation where network structures emerge.

Network formation: The main reason for the formation of any network observed in any aspect of nature or human society is an underlying tradeoff of gain versus cost of collaboration, either for each of the individual entities constituting the network or cumulatively for the whole network (Fig. 1.2).

The above "definition" exemplifies two aspects of network formation, in general, and complex networks, in particular. The first is some inherent type of collaboration, e.g. some common communication protocol, between pairs of network entities or among groups of entities forming in a network, even in the case where each of them unilaterally seeks to maximize its own benefit by acting selfishly, e.g. maximize the number of received/transmitted information in the previous case. The second aspect regards the dynamics of the network under formation. The evolution and mechanisms involved in such dynamics can be observed and quantified with measurable indicators in all types of emerging networks and within all application perspectives they appear through the identification of the corresponding underlying benefit-cost of collaboration tradeoffs. For instance, some indicators, such as the number of neighbors and their obtained utility, can be used/measured if properly defined and observed in the formation process. Thus, it should be emphasized that with every networking structure observed, regardless of whether it was spontaneously formed or artificially developed, there is one or more associated quantifiable benefit(s) (gain)-cost tradeoff(s) involving each of the network entities individually, or the whole network cumulatively.

The field of complex networks covers, in general, a very broad span of network types and their emerging features, developing problems and applied mechanisms of broader scientific interest. However, regarding the current information-based and network-dependent societies, two prominent features arise, namely, the diversity of emerging network structures first, and second, that networks consist of participating entities of varying intelligence and potentials.

Complex network analysis (CNA) [7, 125, 155, 164] refers to the modeling and analysis of different networks and the behavior of the entities constituting such networks, with respect to their computational capabilities, properties, operations, etc. CNA is usually tightly combined with social network analysis (SNA). Examples of approaches combining CNA with SNA can be found in [201, 202]. In CNA, a complex network represents as a generic model, diverse types of networks, thus providing a holistic framework for studying their properties, behaviors, etc., in a unified manner. In order to consider emerging variations, for instance, studying routing jointly over heterogeneous networks consisting of *ad hoc* and cellular components, a generic "complex" network type is assumed. The term "complex" in this occasion signifies the various and diverse properties exhibited by the different topologies, which need to be taken into account in the generic study of a process, i.e. packet routing in communication network, malware diffusion, and information pathways in cell networks. In the second case, the term "complex" refers to the actual nodes of a network, which can also vary in their application and scope, and characterizes the multiplicity/complexity of their features. More specifically, complex network nodes may vary in intelligence and processing capabilities. However, if nodes are capable of executing some type of computation, simple or more advanced, such as a numerical or algorithmic operation, then the cumulative behavior could exhibit various degrees of complexity and/or diversity, and thus, complex behavior may be observed in natural or engineered network cases.

Table 1.1 Examples of Complex Network Classes Based on the Origin of their Formation

Natural	Human-Initiated	Artificial
Biological	Social	Computer
Brain/cortex	Online social	Mobile
Genetic	Open markets	Sensor
Transcriptional	Corporate	Delay-tolerant
Immuno-suppressive	Production	Mesh
Neuron	GDP flow	Transportation
Ecologic	Scientific	Roadmap
Protein	Affiliation	Air-traffic
Food webs	Linguistic	Public utility
Virus/diseases	Malware	Artificial neural networks
Crystal structures	Newsfeeds	Electronic circuits

From the above discussion, it becomes apparent that complex networks may be significantly different to each other. Thus, many possible classifications are possible with respect to different criteria. One of the most useful ones takes into account the origin of network formation and operation, namely, whether the network was formed spontaneously or artificially and whether its operation is dictated by natural or artificial factors as well. According to this parameter, complex networks may be characterized as *natural*, *human-initiated*, and *artificial*, as shown in Table 1.1 for various complex network examples. A more detailed analysis of various possible complex network classifications and features of the corresponding taxonomies can be found in [125] and references therein.

Perhaps the most practical classification is the one considering the structural nature of networks, i.e. the interactions developed between the entities of a network and the properties/features of their interactions cumulatively. This structural classification of networks is based on the notions of connectivity, node degree, and degree distribution of the underlying graph, i.e. the graph corresponding to the structure of the network under discussion. A graph (network) is connected when there is a path between every pair of nodes [61, 94]. The *degree* (or valency) of a node of a graph (network) is the number of links incident to the node, namely, the number of neighboring nodes incident to the specific node. In the study of graphs and networks, the *degree distribution* is the probability distribution of node degrees over the whole network. That is, the degree distribution $P(k)$ gives the probability that a node randomly chosen in the network has k neighbors. Additional definitions on network structure and its properties are provided in Section 1.3.3. Table 1.2 contains such classification of complex networks (first column), as well as a nonexhaustive list of

Table 1.2 Complex Network Classification Based on Topology Structure

Complex Network Type	Main Features	Examples of Networks
Regular	Uniform degree distribution	Lattices, Grids Crystals Optical ring networks Cellular phones (hexagonal grid) Cloud infrastructures Sensor
Random	Normal degree distribution	Peer-to-peer Gas molecules (in equilibrium) Email viruses Online social networks Immunization networks
Mesh	Arbitrary degree distribution	Sensor Delay-tolerant networks Optical networks Cognitive radio networks Zigbee/Bluetooth LTE-Advanced (4G) WiFi (802.11x networks)
Power-law	Power-law degree distribution	Metabolic Population of cities Word frequencies Affiliation networks Neurons (nerve cells)
Scale-free	Power-law degree distribution	(Mobile) Social networks WWW Internet (AS/DNS routers) Protein interaction networks Inter-bank payments Airline connections
Multihop	Arbitrary degree distribution	Tactical networks TETRA Packet radio networks (CSMA/CA) Sensor Vehicular Roadmaps (inter-city highways) LTE-A (4G) networks Cognitive radio networks

examples for each complex network type (second column). In this book, we will be especially concerned with all types of wireless complex networks in the first column of Table 1.2, at various capacities and with respect to the malware diffusion dynamics developing over them. In each case, the features of each wireless complex network will be explicitly mentioned and related to the corresponding malware diffusion processes and their properties.

It should be mentioned that some networks at large, e.g. sensor networks, belong to more than one class of complex network structure depending on the corresponding applications, e.g. some sensor nodes may have grid topologies, while others mesh or multihop layouts. More details on the structure and general features of the complex network types in Table 1.2, are given in [7, 125, 155, 164] and their references.

The behavior of complex networks of any type cannot be currently predicted and/or controlled because the scientific basis for analyzing, building, and evaluating complex networks is yet immature [125, 155]. Thus, getting a grip of the fundamental science of networks in terms of structure, dynamics, functionalities, and evolution is a topic of immense interest and significant value for the benefit and progress of human societies and activities, which has lately attracted the attention of the research community. Several books and monographs have been devoted to such topics from a general perspective, e.g. [7, 125, 155, 164]. However, in this book, we shall focus on more targeted aspects, and specifically on the dissemination dynamics of malware over wireless complex networks. Furthermore, the methodologies presented in this book can be extrapolated to the broader topic of information diffusion in complex networks in general. In the next subsection, we briefly cover the current state of research in CNA, in order to set the methodologies presented in Part 2 into a more practical perspective and position them better in the corresponding roadmap for current and future network research.

1.3.2 NETWORK SCIENCE

Nowadays there exists a cumulative effort to develop generic models for various types of network structures, irrespective of their application context and practical use. In many cases, it suffices to accurately identify the type of a complex network, then identify its features, and finally, employ more general methodologies developed to solve easier several network-related problems. However, as in all other scientific fields, for complex networks too, it is desired to develop a broader framework, where it will be possible to combine theory with application and create a strong bond between modeling and practice. The term "Network Science" has been used to denote this broader and more ambitious effort for a proper scientific field devoted solely to the study, analysis, and applications of networks, wherever and whenever they emerge [125, 155].

Definition 1.2 *(Network Science). Network Science is the study of the abstract (generic) networking properties of systems appearing in different and diverse domains, by means of formal scientific methods.*

Network systems are meant in the sense of a set of entities and their pairwise interconnections (network graph). The above definition segregates explicitly the scientific from the technological part of network study. Graph theory, probability, differential equations, stochastic, and other forms of optimization are only some of the formal scientific methods implied in the above definitions for CNA.

Taking as an example information networks, the components of modern communication and information networks are the results of technologies, which are based on fundamental knowledge emanating from physics, mathematics, and circuits and systems in various capacities. Especially for communications networks, several advances and novelties have enabled the design of modules critical for their operation/performance. However, the assembly of all such novelties into the development of networks is based extensively on empirical knowledge rather than on a deep understanding of the principles of network behaviors underlying the science of networks. For example, regarding the emergence of the protocol layering concept and the infamous TCP/IP stack used extensively in modern networks [55, 148, 208], the technology was first developed within the industry and it is only lately that it was possible to consider the whole of the protocol stack across layers through a uniform mathematical methodology (network utility maximization—NUM), and thus, essentially reverse engineer the whole stack [49].

Considering the field from a holistic perspective, it seems that practitioners in each major application area of Network Science have their own local nomenclatures to describe network models of the phenomena in which they are interested and their own notions of the content of Network Science. For example, spatially distributed networks, e.g. wireless sensor networks and street map networks, emerge both in communications networks and highway traffic engineering (multihop and roadmap networks, respectively). Scale-free (SF) networks emerge frequently in biology and the Internet. However, different terms have been traditionally employed for the same concepts and varying mechanisms/computational tools have been employed to solve the same problems. Consequently, Network Science is emerging as a cross-disciplinary network investigation, aspiring to reach the maturity level of other advanced scientific fields, such as fluid mechanics.

Until today, there is no complete theory offering the required fundamental knowledge for analyzing, designing, and controlling large (or any scale) complex networks. Network Science is currently lively evolving, yet there is still restricted concrete understanding of its potential long-term outcomes. However, as with so many other scientific disciplines in the past, such as medicine, where applications based on empirical knowledge were massively prevailing before the corresponding knowledge established, it is expected that this will be the case with Network Science. In fact, with all the aforementioned fields, it was their formal shaping that helped in better understanding the underlying phenomena and achieve more progress. This is also anticipated with Network Science. The fragmentation in Network Science is evident at large in that not all the involved disciplines scale at the same order, namely, some are progressing well, e.g. communications networks, while others at a much slower pace, e.g. financial networks. Network Science aspires to become a cross-disciplinary study of network representations of physical, logical, and social interactions, leading to predictive models of these phenomena and relations. This book will contribute toward this direction, by attempting a cross-disciplinary transfer of knowledge on malware diffusion modeling (and more generally of information dissemination modeling) from computer networks and more specifically wireless complex ones, to more general complex networks.

The communities from which Network Science is expected to emerge encompass many different and diverse disciplines of research or application areas. Among others, a characteristic example is the field of biology, which provides the most diverse observations of complex network structures of arbitrary order and capabilities, all of them working usually in conjunction to the rest and efficiently on their part. Computer networks have employed substantial mathematical tools for their analysis and development and could provide solid background for quantitative methods for many other disciplines in Network Science and complex networks. In fact, the first studies on computer viruses were inspired from epidemiological models developed by biologists or social scientists. Other disciplines expected to contribute in the development of Network Science include mathematics and physics, statistics, sociology, etc. This book will adopt a Network Science based approach and attempt to expand it within the field of information dissemination, thus, paving the way for similar attempts in other application frameworks.

1.3.3 NETWORK GRAPHS PRIMER

A complex network is typically represented and analyzed through the mathematical notion of a graph. Thus, in this subsection, we provide a short primer on the involved notation, functions, and properties of the network graphs of interest.

A graph is an ordered pair $G = (V, E)$ where V is the set of vertices (nodes) with cardinality $n = |V|$ and E is the set of edges (links) with cardinality $|E|$. The edges of a graph are two element subsets of V. An edge between two vertices i, j is typically represented with (i, j) (ordered or unordered for directed/undirected graphs, respectively). The neighborhood of a node $x \in V$, denoted by $\mathcal{N}_x$, is the set of all nodes of G adjacent to x. The degree of a node i, k_i in an undirected graph is the number of edges having as one of their endpoints the vertex i. In directed graphs, each vertex is characterized by two degrees, the in-degree k_i^{in} which counts all edges pointing to node i and the out-degree k_i^{out} counting all vertices starting from node i. The degree distribution $P_i(k)$ provides the probability that node i has degree k, namely, k direct neighbors. In both the directed and the undirected cases, we denote by $A = [a_{ij}]$ the adjacency matrix, where $a_{ij} = 1$ if there is a link from i to j, otherwise $a_{ij} = 0$. Based on this matrix representation, spectral graph theory has emerged [53], offering several convenient computations and interpretations of graphs. A *clique* in an undirected graph is a subset of its vertices such that every two vertices in the subset are connected by an edge.

A graph (directed or undirected) is weighted, if a measurable quantity (referred to as weight and usually denoted by w) is assigned to each edge: $w : E \rightarrow \mathbb{R}$

[19, 24, 169]. Similarly with the adjacency matrix, the weight matrix $W = [w_{ij}]$ can be defined, where w_{ij} is the weight of the link (i, j). A joint metric of both node degree and adjacent link weights might be defined, denoted by strength of each node (or the in-strength s_i^{in} and the out-strength s_i^{out} correspondingly for digraphs). Node strength expresses the total amount of weight that reaches or leaves node i correspondingly [213]. Thus, $s_i^{out} = \sum_{j=1}^{n} w_{ij}$ and $s_i^{in} = \sum_{j=1}^{n} w_{ji}$. In the case of undirected graph-network, if s_i is the strength of node i, $s_i^{out} = s_i^{in} = s_i = \sum_{j=1}^{n} w_{ij}$.

Connectivity is a fundamental notion in graphs and networks [144]. The connectivity $\kappa = \kappa(G)$ of a graph G is the minimum number of nodes whose removal results in a disconnected or trivial graph. Two nodes are said to be connected whenever it is possible to find a sequence of edges belonging to the graph (path) from one node to another. Then, the average path length of a graph is a connectivity indicative metric [78, 144]. The *clustering coefficient* is an important metric for complex networks [171, 192], used to characterize the structure of a social network both locally, i.e. at the node level, and globally, i.e. at the network level. It computes the cliquishness of the network and more details on computing it can be found in [69, 125, 171]. In various network types, it has been required/desired to characterize the importance of network nodes. *Centrality* has been conceived as an evaluation metric for characterizing such aspect of networks [76, 77, 170]. Typically, the focus on the importance of nodes or connection links, but other features of the graph structure under consideration may also be considered. Such metric is, in principle, subjective, depending on numerous aspects of a network, such as the structure, the network objectives, network operation, and even other more context-oriented factors characterizing a network [75, 77, 150, 181]. For these reasons, various centrality definitions have been established and employed in social and complex communication networks [150]. More detailed definitions of the above notions, computational methods, as well as other quantities and properties of graphs can be found in dedicated works [61, 94].

The structure of a network is represented by the corresponding network graph, the structure of which is determined in turn by the connections of each node, namely, by the node degree. Thus, the total degree distribution $P(k)$ of the network that provides the probability for each node to have k neighbors distinguishes different network structures. This was used in the classification of Table 1.2 (first and second columns). In the remaining of this chapter, we briefly overview the topologies of interest included in Table 1.2 and their properties.

A *d-regular graph* is an undirected graph where each node has the same degree equal to d. A *complete graph* on n vertices, denoted by K_n, is an undirected graph where all vertices are connected with all other vertices and thus is a $(n - 1)$-regular graph. Regular networks emerge oftentimes in nature and occasionally in engineered applications as well, such as communications and power networks [23].

In mathematics, a *random graph* is the general term to refer to probability distributions over graphs. Random graphs may be described simply by a probability distribution, or by a random process which generates them [67]. In general, one encounters two basic and closely related models of random graphs in the literature

[34]. The probability space in each case consists of graphs on a fixed set of n distinguishable vertices $V = [n] = \{1, 2, \ldots, n\}$. If $M = |E|$, then for $0 \leq M \leq N$, the space $\mathcal{G}(n, M)$ consists of all $\binom{N}{M}$ subgraphs of K_n with M edges. Thus, the $\mathcal{G}(n, M)$ model describes graphs which could be obtained by the subspace of K_n that contains only graphs with M edges. All elements (graphs) of this space are assumed equiprobable, and due to the assignment of a probability measure to its elements, the space becomes a probability space. It is customary to write $G_M = G_{n,M}$ for a random graph in the space $\mathcal{G}(n, M)$. The probability that G_M is precisely a fixed graph H on $[n]$ with M edges is $\mathbb{P}_M(G_M = H) = \binom{N}{M}^{-1}$. The space $\mathcal{G}(n, p)$ ($\mathcal{G}(n, \mathbb{P}(edge) = p)$) is defined for probability $0 \leq p \leq 1$. A random element of this space corresponds to selecting edges independently with probability p, for all possible existing edges in a graph of n nodes. This means again that the potential probability space includes all possible subgraphs of the space of K_n graphs. In this case however, only those corresponding to a selection process where each edge is selected with probability p are selected. Similarly to the $\mathcal{G}(n, M)$ model, the probability of a fixed graph H on $[n]$ with m edges is $p^m (1-p)^{N-m}$, where each of the m edges of H has to be selected, and none of the $N - m$ is allowed to be selected.

A *small-world* (SW) network is a type of mathematical graph in which most nodes are not neighbors of one another, but most nodes can be reached from every other by a small number of hops or steps [147, 224]. Formally, a SW network is defined as a network where the typical distance L between two randomly chosen nodes grows proportionally to the logarithm of the number of nodes N in the network, i.e. $L \propto \log N$ [125, 164, 225]. Practically, this means that nodes of the networks are linked by a small number of local neighbors; however, the average distance between nodes remains small. The most popular model of SW networks is the Watts-Strogatz (WS) model, which is a constructive process for obtaining SW and some other types of random graphs, beginning with a regular lattice. The WS model starts from a clustered structure (regular lattice) and adds random edges connecting nodes that are otherwise far apart in terms of hop distance [141]. These random "long" edges will be denoted as "shortcuts" in this book. The initial clustered structure ensures high clustering coefficient for the final network, while a suitable number of added shortcuts can further reduce the average path length, up to a sufficient level, so that the created graph may be characterized as SW. One significant question is whether various types of networks can be turned into SW via some evolutionary process. Some considerations of this issue have been presented in [45, 51, 65, 202].

A *scale-free* network is a network whose degree distribution follows a power-law, at least asymptotically. Formally, the fraction $P(k)$ of nodes in the network having k connections to other nodes follows for large values of k: $P(k)\ k^{-\gamma}$, where γ is a parameter whose value is typically in the range $2 < \gamma < 3$, although occasionally it may lie outside these bounds [1, 15, 16]. In SF networks, different node groups exhibit differences in scaling of their node degree, interpreted as scale difference in connectivity and neighborhood relations. The highest-degree nodes are often called "hubs," and typically they serve specific purposes in their networks. SF networks exhibit two key features. The first one is "growth" [20, 21, 62, 222]. The way a network evolves indicates that new nodes tend to link to existing ones. The second is what is most popularly known as *preferential attachment* [1, 15]. The fact that when nodes form new connections, they tend to connect to other nodes with probability proportional to the popularity of the existing one. Since not all nodes are equally popular, some of them are more desirable than others.

The SF property strongly correlates with the network's robustness to failure. If failures occur at random and the vast majority of nodes are those with small degree, the likelihood that a hub would be affected is almost negligible. Even if a hub-failure occurs, the network will generally not lose its connectedness, due to the remaining hubs. On the other hand, if one chooses a few major hubs and take them out of the network, the network is turned into a set of rather isolated graphs. Thus, hubs are both a strength and a weakness of SF networks. Another important characteristic of SF networks is the clustering coefficient distribution, which decreases as the node degree increases. This distribution also follows a power-law. This implies that the low-degree nodes belong to very dense subgraphs and those subgraphs are connected to each other through hubs. A final characteristic concerns the average distance between two vertices in a network. As with most disordered networks, such as the SW network model, this distance is very small relative to a highly ordered network such as a lattice graph. Notably, an uncorrelated power-law graph having $2 < \gamma < 3$ will have ultrasmall diameter L_d $lnlnN$ where N is the number of nodes in the network, while the diameter of a growing SF network might be considered almost constant in practice.

We should note that a SF network mentioned in Table 1.2 is essentially a network whose node degree distribution follows a power-law, at least asymptotically. Thus, for the rest of this book, we will employ the term "power-law" to denote networks following explicitly a power-law degree distribution and "scale-free" to denote those that follow power-law degree distribution asymptotically only.

Finally, a *random geometric graph* (RGG) is the simplest spatial network, namely, an undirected graph constructed by randomly placing N nodes in some topological space (according to a specified probability distribution) and connecting two nodes by a link if their distance (according to some metric) is in a given range, e.g. smaller than a certain neighborhood radius, R [180]. It is customarily employed to model distributed multihop networks, e.g. *ad hoc*, sensor, mesh, and others.

Malware diffusion in wired and wireless complex networks

2

2.1 DIFFUSION PROCESSES AND MALWARE DIFFUSION

This section will first introduce the notion of diffusion processes in general, then the more specific process of malware diffusion and finally, it will be involved with more specific types and examples of malware diffusion over wired and wireless communication networks.

2.1.1 GENERAL DIFFUSION PROCESSES

Malware diffusion is a general term used to describe various forms of malware dissemination processes encountered in today's wired and wireless communication networks. In order to better analyze malware diffusion, the "diffusion" needs to be understood. This subsection will provide a concise overview of the fundamentals of diffusion processes toward enabling such understanding.

The broader notion of diffusion processes is used in various disciplines such as physics, chemical engineering, and material science [145]. Diffusion processes characterize a type of transfer phenomena observed frequently in nature or society, and it can emerge naturally (i.e. spontaneously) or artificially (i.e. in a controlled manner). A distinguishing feature of diffusion compared to other transfer processes is that it results in the mixing of, or the massive spread of large interacting entities, e.g. atomic particles, humans, liquids, or gas molecules, without requiring a bulk transfer. A characteristic example that one may encounter in his/her daily routine is the mixing of two liquids, e.g. pouring red color in a water tank. The random process taking place from the start of the mixing until the whole mixture becomes homogeneous is an instance of diffusion. Thus, diffusion is a gradual random process taking place progressively, as opposed to convection or advection which are rapid transfer processes [57]. Of course, the rate of transfer for diffusion may be relatively fast, but even in such cases the transfer should not take place in bulk form, i.e. if particle diffusion is under discussion, particles can be transferred one by one at very high speeds, but not in clustered groups.

Intuitively, diffusion could be thought of as a process of spreading out, as the Latin origin of the word diffusion suggests.[1] In physics, it is defined as the process by

---

[1] The word diffusion is derived from the Latin word, "diffundere," which means "to spread out," i.e. move from an area of high concentration to an area of low concentration.

which there is a net flow of matter from a region of high concentration to one of low concentration. Thus, when diffusion refers to malware, it describes the process where malware starts from a source node-user, or groups of nodes-users (high malware-concentration area), and progressively spreads or propagates to other users (low malware-concentration areas). Apart from physics (e.g. particle diffusion), chemistry (e.g. diffusion in gases and liquids), and material science (e.g. atomic diffusion in solids) mentioned above, diffusion processes emerge in biology (between cells and chemical substances) in sociology, economics, and finance (as diffusion of people, ideas, and price values) as well.

In mathematics, diffusions appear in two forms. The first is usually encountered in processes that behave like fluids, in which cases, the underlying mathematical modeling is based on systems of partial differential equations obtained by the laws of fluid dynamics [57]. In the second case, diffusion can be identified whenever the concept of random walk[2] in ensembles of individuals can be applied [145]. For instance, in physics, particle diffusion that is widely expressed through the stochastic process of Brownian motion [143] is essentially an ensemble of atoms performing random walks in space. So is an ensemble of people exchanging news on a social network such as facebook. In the latter example, news disseminate among humans in a random and gradual contact-based fashion, analogous to physical particles moving from one location to another.

2.1.2 DIFFUSION OF MALWARE IN COMMUNICATION NETWORKS

For the purposes of this book, it suffices to consider diffusion as a transfer process of some form of "particles," which will correspond to malware bearers. These "particles" can be packets of a network flow, whole pieces of malware, people, etc., and in general bits of information. The transfer of malware should take place as a type of nonbulk process, and it will be characterized by certain laws, achieving an average transfer rate. In particular, the diffusion of malware may be defined as follows:

Definition 2.1 (*Malware Diffusion*). *Malware diffusion describes generically and holistically the process by which any type of malicious software, from simple viruses and local outbreaks to massive worm spreads that can possibly emerge in a communication network, transfer from one user-attacker (or groups of attackers) to another user-node (or groups of users-nodes), eventually contaminating larger components of the network.*

Thus, the diffusion process describing malware transfer essentially represents a transfer process of malicious software between nodes belonging to two major interdependent groups, namely, attackers/malicious nodes and legitimate nodes of

---

[2]A random walk is a mathematical formalization of a path that consists of a succession of random steps. For example, the path traced by a molecule as it travels in a liquid or a gas, the search path of a foraging animal, the price of a fluctuating stock and the financial status of a gambler can all be modeled mathematically as random walks.

the network.[3] Attackers can be considered as "areas" of high malware concentration, while legitimate nodes as "areas" of no malware concentration, signifying a controlled (artificial) malware flux potential from the high concentration areas to the areas of no malware concentration. Once a node leaves one group, e.g. the group of legitimate nodes, it enters the other one, in this case the group of malicious nodes. Such interdependence dictates the size of the population of each group as a function over time under the assumption of a closed population, while all the rest of decision-making behaviors of nodes are independent. For instance, nodes might be considered to follow their own security policies, which is in contrast to the potential centralized policies followed in major corporal or public infrastructure/utility networks and this could affect the possibility of becoming contaminated by some diffused malware component. Once a node becomes contaminated, the number of legitimate noninfected nodes will decrease by one and the number of infected nodes will increase by one at the same time (interdependence), irrespective of different policies followed by the nodes.

Malware diffusion can be dissected into two major categories, namely, *spreading* and *propagation*, denoting two operationally different mechanisms of malware transfer between nodes (Table 2.1). The "spreading" mechanism models scenarios where the transfer of malware takes place only between two specific and distinct groups of nodes, namely, from attack (malicious) nodes to legitimate noninfected nodes. On the contrary, malware propagation describes the cases where infected legitimate nodes are able to contaminate other noninfected legitimate nodes, in addition to the (original) attack nodes. Thus, both malicious and legitimate infected nodes control the dynamics of malware diffusion against the currently noninfected legitimate nodes. This can occur once these legitimate nodes become infected and can last as long as they remain in the infected state. When infected legitimate nodes cannot infect their peers, we will refer to such networks as *nonpropagative*. In contrast, when a newly infected legitimate node is able to infect other legitimate nodes, we call such networks *propagative*. In the rest of this book, we stand firm with this convention: we will refer to malware diffusion in nonpropagative networks as malware "spreading," whereas we denote malware diffusion in propagative networks by malware "propagation."

We also distinguish cases where malware is transferring between pairs of nodes and where the transfer is in bulk between groups of nodes, which can reflect the nature of the underlying communication protocol in use, e.g. unicast *versus* multicast (Table 2.1). In this book, we will mainly focus on the case of unicast transmissions, and thus malware transfer between pairs of nodes. When transfers occur in bulk, the process cannot be characterized as diffusion in general. Such cases of concurrent group malware spreading or propagation have not been extensively studied, despite the fact that some incidents of such malware outbreak have been observed sporadically. We will return to this issue as a potential future area of research in the later chapters of the book.

---

[3]Depending on the structure and objective of a complex communication network, more types of nodes may be defined with respect to malware diffusion, as we will be discussed later in the chapter.

Table 2.1 Malware Diffusion Categories and their Coverage in this Book. Symbols '+, -, *' Mean the Corresponding Category is Addressed, Not Addressed, Only Touched Upon in the Book, Respectively

Features	Contamination Type		
	Peer-to-peer	Hybrid	Group
Spreading (nonpropagative networks)	+	-	+
Propagation (propagative networks)	+	-	*

2.2 TYPES OF MALWARE OUTBREAKS IN COMPLEX NETWORKS

In the history of computer and communication networks, numerous malware outbreak incidents have been observed and documented. Some of the emerged outbreaks became popular even among the laymen of the field, due to the considerable impact they had on infrastructures, users' work, and the local or global economies. Several of these incidents have been also analyzed and quantified with good accuracy based on real measurements and evaluations, e.g. CodeRed [235] worm and Love virus. Nowadays, the detection, measurement, and documentation of malware outbreaks have become more systematic. In this section, we present the most representative types of malware that have been observed to date and then we refer to current and expected emerging malware trends.

Table 2.2 provides some of the most representative types of malware and many of their underlying subcategories, e.g. viruses and their variations. It also provides some notable instances of each malware type, as well as a qualitative severity characterization of their operation based on their overall past outcome. The severity of each attack is qualitatively assessed as "low-average-high." It should be noted that Table 2.2 is a nonexhaustive summary of malware type classification, and other works in the literature, e.g. [13, 52, 163, 166], can be consulted for more detailed descriptions and analysis.

The most frequently encountered types of malware tend to be those that are targeting individual users, rather than those that attack large-scale centralized systems and infrastructures, such as banking, military, and public utility systems. Their severity can be minor or grave depending on the scale of attack and the intelligence of malware. Most of these malware sources are derived from exploiting operating system (OS) holes[4] and bugs and rarely due to rather complicated software that is

---

[4]Such operating system holes are usually denoted by the term 'vulnerabilities'.

Table 2.2 A Non-exhaustive Classification of Malware Types with Examples

Malware Type	Notable Attacks	Severity
Worms	Blaster, Welchia	Highest
Botnets	SDBot, RBot, Agobot, Spybot, Mytob	High
Rabbit	Fork bomb	Average
Logic bombs	Medco Health Solutions, Fannie Mae, CSOC	Average
Trojan horse	Netbus, Sub7, Back Orifice, Beast, Zeus	High
Sinkhole/wormhole	Styx EK, SweetOrange EK	High
Spyware	CoolWebSearch, WinTools, Zango, Zlob	Average
Adware	Typhoid	Low
Trapdoors/Backdoors	Sobig, Mydoom, Skynet, MD5	Average
DoS	Teardrop, Smurf, SYN flood, Sockstress	High
Zombies (DDoS)	SPEWS, Blue frog and smartphone attacks	Average
Phishing	AOHell, warez, Heartbleed	High
Viruses (boot-sector, file, macro)	CodeRed, Sasser, Melissa, Conficker	High
WiFi viruses	Chameleon (experimental virus)	High
Bluetooth viruses	Cabir, Ronie, Commwarrior	Average
Smartphone viruses	Cabir, Duts, Skulls, Commwarrior, Ikee	High
Socialnet app viruses	Net-Worm.Win32.Koobface.a/b	High
Hybrid and blended	Storm worm, Klez, Bobax, CIH	High

possible to unlock whole software systems. Among the malware types that attack individual users, the most frequently encountered is the worm type and its variations, affecting both Internet and mobile users. Worms usually hit massively and suddenly as many unprotected machines they can, usually after some OS hole has been discovered by hackers. In fact, most of the malware examples employed in the literature refer to attacks by one or more types of worm malware.

In 2001, the CodeRed worm was released only 25 days after a relevant vulnerability was announced, signifying the order of magnitude of the potential capability of malware authors, even a decade ago. Since then, even though significant awareness has been raised and precautions have been taken to avoid similar cases, there still exist several similar incidents, varying in severity and targets. Just for comparison purposes, in 2006, a Microsoft Windows vulnerability was exploited by a worm in only 5 days after it was revealed.

Malware of the worm type, similarly to trojan horses and spyware software, aims mainly at ordinary individual users, rather than the complex centralized grids. Spyware in their turn is malware types that exploit machine vulnerabilities, and typically they exploit the lack of knowledge that characterizes simple users, in order to install themselves into host machines and monitor user activity, obtain passwords, etc., without the user's consent.

On the other hand, malware of botnet type applies to compromised machines infected by targeted software that the attackers use in order to launch large-scale attacks to important and typically large interconnected systems and networks, frequently of commercial or governmental use, such as public utility and defense networks. A similar purpose is attained by DoS/distributed denial-of-service (DDoS) attacks usually aiming more at the commercial operation of websites in the WWW. A DoS attack is an attempt to make a machine or network resource unavailable to its intended users, such as to temporarily or indefinitely interrupt or suspend services of a host connected to the Internet. A DDoS is where the attack source is more than one and often thousands of unique IP addresses. In such cases, the individual hosts essentially behave as "zombie soldiers" by either infecting even more machines and thus creating a large army of compromised machines, or as soldiers of this army in order to attack large and usually well-guarded network systems. Several documented incidents prove that this has been a very popular and successful practice among the attackers with various undesired results for the legitimate systems and users.

The increase of malware outbreaks in computer and communication networks is nowadays more than ever evident. According to a relevant study by Symantec [195], the last half of 2005, 1896 new outbreaks appeared, corresponding to about 70 new per week. Among those, 50% was characterized as having high severity of their outcome, while about 45% were of medium severity, leaving only a small percentage of 5% exhibiting small severity. In another incident, where the `Welchia` worm "fought" against the `Blaster` worm as a released countermeasure, the unrestricted spread of this "patch" created so much additional traffic in the Internet that almost destabilized even well-provisioned critical subnetworks. Perhaps, what is known as the "highlight of worm outbreaks" was registered in 2004 and denoted as War of the Worms between the `NetSky`, `Bagle`, and `MyDoom` worm variants [207]. This war created complex interactions among worms citing instances of one worm terminating another worm. Lately, in March 2012, significant dysfunctions were caused by a DNS Changer Malware involving Internet users, denoted by a name "Operation Click Ghost." This malware was essentially a bundle of viruses such as TDSS, `Alureon`, `TidServ`, and TDL4 viruses that changed DNS settings. Table 2.3 presents a mapping of specific attack threats to the malware types (spreading/propagation) and contamination types (peer-to-peer/group) presented earlier in Table 2.1.

All these examples illustrate the variability of outcomes that several malware types and their variations have on the systems over which they propagate. These outcomes may include financial, operational, ethical, and life-critical issues emerging for the attacked networks and their users and in most of the cases the outcomes turn out to be more severe than measured or expected.

Table 2.3 Mapping of Malware Threats to Malware Attack Types

Threat	Malware Type	Contamination Type
Worms	Propagation	Point-to-point/group
Botnets	Propagation/spreading	Point-to-point
Trojan horses	Spreading	Point-to-point
Sinkhole/wormhole	Spreading	Point-to-point
Spyware	Spreading	Point-to-point
Trapdoors/Backdoors	Spreading	Point-to-point/group
DoS/DDoS	Spreading	Group/point-to-point
Phishing	Spreading	Point-to-point
WiFi viruses	Propagation	Point-to-point/group
Bluetooth viruses	Propagation	Point-to-point
Smartphone viruses	Spreading/propagation	Point-to-point
Socialnet application viruses	Spreading	Group

Lately, and especially since 2010, the proliferation of smartphones has lead to a paradigm shift in network access services and has also driven several radical changes in the traditional wired infrastructures. It has essentially served as a driver for a paradigm shift in network access from wired to wireless networks and has enabled true mobile access and computing. However, in parallel, an increasing trend in mobile malware emergence has been observed, where the malware modules mainly transfer through mobile devices (tablets and smartphones).

Mobile malware is a malicious software that is specifically built to attack mobile phone or smartphone systems. These types of malware rely on exploits of particular OS and mobile phone software technology and represent a significant portion of malware attacks in today's computing world, where mobile phones are increasingly common.

Within the category of mobile malware, certain kinds of smartphones are targeted more often than others. Industry research shows that an overwhelming majority of mobile malware targets the Android platform, rather than other popular mobile OS systems, like Apple's iOS, mainly due to proprietary platform restrictions, etc. Various types of mobile malware include device data spies that log certain kinds of data and deliver it to hackers. Another type of mobile malware is called root malware and gives hackers certain administrative privileges and file access. There are also other kinds of mobile malware that perform automatic transactions or communications without the device holder's knowledge, signifying a noteworthy variability of mobile malware, similar to the corresponding malware targeted for wired networks.

With mobile malware, essentially a completely new battlefield has emerged, which was previously considered relatively immune due to its low market penetration. In 2012 and according to rough estimates, there were approximately 370 million devices in total, including cellphones, smartphones, and tablets [195]. This indicates a steady paradigm shift in usage toward wireless networking, which also paves the way for a corresponding shift of malware toward wireless networks, and especially those that become more decentralized. It seems that very shortly wireless Internet will have approximately the same volume of malware as the Internet. Regarding smartphones,[5] 35,000 new mobile malware pieces per day are observed. Malware has increased by 46% due to attacks targeting mobile devices specifically [195]. New usage paradigms introduced by social networking via mobile devices create a new malware diffusion medium, where scarcer resources enable more diverse damage, the bandwidth is now common to all users, the batteries are restricted, and the media access more overcrowded than ever. However, the wireless setting enables novel countermeasures as well. As will be seen in the second part of the book, the state-of-the-art methodologies for modeling malware diffusion over wireless complex networks can reveal the key dynamics of malware diffusion, thus enabling designing more efficient countermeasures, even in the worst-case attack scenarios.

2.3 NODE INFECTION MODELS

The behavior of a user (network node) that has received malicious software and has been compromised (transition to a state denoted as infected) varies considerably depending on the level of technical knowledge of the user and the device capabilities. It should be noted that for those nodes that receive malware and this software has no effect on them, e.g. they receive a virus through email but an antivirus software or the user is capable of blocking the virus, behave similarly to users that have not received malware at all; thus, they are not considered as truly infected. The behavior of compromised nodes also depends on the features and capabilities of the malware, as well as the structure and employed management policies of the network. Thus, the users of the network individually might be at different states, signifying diverse behavior for the overall system, e.g. even though a significant number of nodes might be infected, the network could behave in an endemic rather than pandemic fashion (malware diffusion remains but does not dominate the whole network). Transitions between the possible node and system states may thus differ considerably. We refer to the corresponding node transition disciplines as *node infections models* and we describe the most characteristic ones in the rest of this section.

The node infection models are tightly related to the possible states that a legitimate node can be in. Examples of such states with respect to malware diffusion are the susceptible, infected, removed, and dead states. The malware-related states of legitimate nodes considered in this book are presented cumulatively in Table 2.4 with

---

[5]In 2011, 428 million new mobile devices were sold, representing 25% of all mobile devices at the time [195].

Table 2.4 Legitimate Node States in the Considered Node Infection Models and their Interpretation

Node State	Symbol	Interpretation
Susceptible	S	Noninfected node
Infected/infectious	I	Infected node
Removed	R	Recovering node (temporarily removed)
Dead	D	Node not considered anymore (completely removed)
Susceptible-r	Sr	Noninfected node with recharging capabilities
Infected-r	Ir	Infected node with recharging capabilities
Removed-r	Rr	Recovering node with recharging capabilities
Dead-r	Dr	Completely removed node with recharging capabilities

a short description of their interpretation. These states can successfully capture the operational modes of legitimate nodes with respect to malware diffusion, and they will be defined in detail next.

In all cases, a node starts clean of any malware disseminating in the network and this state is denoted as *susceptible* (S). If a legitimate node receives some form of active malware, which affects the device's operation (the impact of malware could vary in type and severity depending on factors explained before), the user is denoted as *infected* (I). Essentially, this user becomes a victim of the disseminating outbreak and the outcome of this "infection" will be determined (behavior-wise) by the corresponding node infection model, i.e. whether a node will sustain this attack or not defines a different infection model and *vice versa*. Sometimes the symbol (I) is used to denote a state called *infectious* in order to better reflect that the corresponding node entering this state is not only infected but also infects other legitimate nodes as well as if it was an attacker. Thus, the infected state is applicable in nonpropagative networks, while the infectious state in propagative ones. If a legitimate user was infected and at the same time it has entered a state where recovery actions take place, the corresponding node state is denoted by *removed* (R). This state implies that as long as the node is "removed" it cannot be reinfected by the outbreak, which however also means that the device is not in a fully operational mode as it is the case in the susceptible state (it is similar to being temporarily removed from the network). Finally, the state where the user is practically considered completely removed from the network, i.e. due to a malware depleting all its resources, or due to malware constituting the device nonoperational, or even because a user/administrator decided it was too dangerous to retain the device in the network and concluded to take it offline completely, is denoted as *dead* (D).

Given all the above, a formal definition of a node infection model can be provided.

Definition 2.2 *(Node Infection Model). A node infection model defines the specific states that legitimate nodes can be in, with respect to malware diffusion. It also describes in a generic manner the transitions of users between their possible states, due to malware-related reasons, for various malware types and network paradigms, structures, operations, etc.*

Table 2.4 also includes states that refer to networks, where nodes may exhaust their energy but have the capability of recharging. This additional feature may have a significant role in malware diffusion as will be shown in the later chapters.

As explained in Definition 2.2, the node infection model practically describes the succession of the aforementioned legitimate user states, aiming at approximating as accurately as possible the actual behavior (operational state) of the nodes in an attacked network. The {S, I, R, D} states in Table 2.4 are the most basic ones and in fact, the simplest node infection models consider only some of them (i.e. {S, I} is the minimum possible subset of states modeling the outcome of an attack). However, in more complicated networks, additional states could be considered or defined, depending on the actual behavior and developing events taking place. Further extending the previous state space and the node infection models that will be presented in the following could be straightforward or very tough, depending on the corresponding behavior to be modeled in each case. In this book, we will only focus on node infection models on the four states with respect to the state space defined above, and only briefly touch upon cases requiring the definition and application of additional states.

Let us now shift the focus to the description of various fundamental and useful node infection models (state transitions). The simplest state transition is the susceptible $\rightarrow$ infected, denoted by SI, which models the scenarios where each legitimate user becomes infected and remains so for the rest of its lifetime. Such model is appropriate for describing the behavior of networks under single or multiple attacks with imminent and fatal outcomes, in which, once a node is infected it remains in this state for the rest of its lifetime. This means that for this model, the infected state is absorbing, namely, once a node gets in this state, stays there for ever. Various examples of such malware types are contained in Table 2.3. The SI model is also suitable for the cases where one is interested in whether each node has received a specific piece of malware or not (this could also apply to the dissemination of a specific piece of information, desired or malicious, and whether the members of a population have received it). Most of the earliest epidemics techniques have assumed this type of user/system evolution, as it will be described in the next chapter.

A more advanced model can be considered by observing that in many cases, the impact of malware infection for a device might be determinant. For instance, several types of malware, shown in Table 2.3, can be harmful for the OS and the corresponding machine will be required to be withdrawn and repaired completely. Battery depletion attacks drain the energy of a device at the highest possible rate, reducing its lifetime at a very fast rate. Depleted devices are equivalent to idle machines, i.e. they do not participate in the dynamics of the system evolution, and this is marked by an additional transition from the infected to the removed state. The state transition is now susceptible → infected → removed, and the corresponding node infection model is denoted by SIR. In the removed state, nodes are practically the same as dead and thus, SID could be an alternative acronym. Typically, the SIR identifier is employed in the literature.

From a macroscopic point of view, in a longer observation time interval, a legitimate node of a network receives multiple instances of malware, e.g. within a year a machine might face 10–20 different threats. A typical desktop will be used for a period of at least 5–6 years, within which multiple and diverse malware eventually reaches it. Once a device receives a malicious piece of software, the medium to advanced user (and even the layman after some elapsed time) will typically initiate some type of recovery actions and eventually, even if the host becomes dead for some short period, it will return to the initial state (regarding the specifically identified malware infection). This is a fundamentally different behavior compared to the previous one-way SI and SIR transitions. The new observed behavior is characterized by recurring state transitions. Nodes eventually oscillate between the SI states depending on the maintenance actions and different types of malware received. The corresponding node infection model is denoted as susceptible-infected-susceptible (SIS) and it is one of the most general models considered to study the cumulative and macroscopic behavior of a network when analyzed in the course of time.

Combining the two basic node infection models, SIR and SIS, one is able to obtain a more general model describing the full range of node state transitions for longer time periods (practically spanning all the lifetime of a node) when attacked by multiple malware threats. The latter corresponds to multiple attackers producing threats of different behaviors and characteristics. The specific model is denoted as SIRS, denoting the susceptible-infected-removed-susceptible state transition.

Another interesting node infection model is the susceptible-infected-removed-dead, denoted by SIRD. This model describes effectively behaviors where an initially intact legitimate node receives malware, thus switching to the infected state. After spending some time in that state, in which it could be dysfunctional, not functional at all, or even worse, infective, acting as a malicious node itself, it will make one of the two possible transitions. Either it will be completely removed from the network in which case the transition will be toward the removed state, or toward the dead state, indicating a different type of removal state (usually due to energy depletion or other technical reasons). The transition to the removed state usually denotes that the node is patched and thus permanently protected against the spreading/propagating malware. The SIRD paradigm is popular in modeling specific types of malware, such as the CodeRed worm and its variations, in wired and wireless networks.

Table 2.5 summarizes the node infection models presented in this subsection that will be also considered later in this book, along with a brief explanation. We also provide explicitly the malware type of each node infection model with respect to the

Table 2.5 Classification of Node Infection Models

Infection Model	Infection Model	Malware Types
SI	Simple epidemic spreading	p2p or group spreading
SIR	Epidemic spreading with patching	p2p or group spreading
SIRD	Epidemic spreading with patching and killing	p2p or group spreading
SIS	Macroscopic epidemic propagation	p2p or group propagation
SISR	Macroscopic epidemic propagation with patching	p2p or group propagation

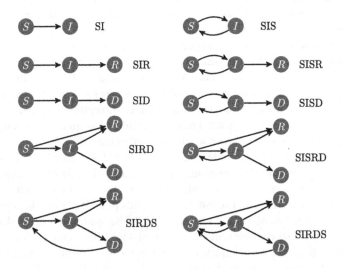

FIGURE 2.1

Examples of node infection models of interest.

classification in Table 2.2, as a quick mapping between node infection models and diffusion dynamics.

Fig. 2.1 presents some examples of node infection models with the corresponding applicable node states that may be encountered in practical scenarios. These models may differ considerably in terms of the system behavior and they can describe radically different application scenarios. In the rest of the book, we will study most of them, explaining in detail the setting in which each of these models emerges, and how effective the modeling of malware diffusion is.

Early malware diffusion modeling methodologies

3.1 INTRODUCTION

Attempts to model malware diffusion have started as early as the first cyber-attacks[1] broke out and their potential financial and technological impact was realized to a pragmatic and accurate extent. These first approaches were mainly based on the earlier methodologies developed in the fields of biology and anthropology describing the spreading of viruses and diseases over living organisms, etc. They utilized elements from the theory of ordinary differential equations (ODEs) [40] (a brief review of the required knowledge from ODE theory is contained in Appendix A), so that the corresponding approaches can be essentially described as deterministic epidemiology modeling [99]. Following suit, various other attempts to extend those initial modeling approaches were employed, by applying stochastic modeling techniques that allowed to incorporate in the developed analytical models various quantitative observations obtained by processing and analysis of real measurement data.

Apart from their historic value, such approaches are quite important for the field of malware modeling, since they are often used as a basis for developing more complicated models. Also, they are successful in revealing the basic emerging trends of malware dynamics observed. In the following, we describe each family of these early modeling approaches separately and then add some additional perspective by illustrating their advantages and limitations, thus paving the way for the more advanced approaches emerging nowadays. The latter are covered in Part 2. We focus on these established analytical models and explain in detail the type of behavior they describe, while also analyzing their relation with realistic malware examples. We also note for each of such examples the degree to which the corresponding model is successful in practical considerations and possible extensions that would be desired in order to be able to cover the existing diffusing attacks even more successfully.

3.2 BASIC EPIDEMICS MODELS

Epidemics models have attracted the research attention for many years, due to the increased interest for modeling and studying lethal attacks, thus eventually protecting humans from the spreading and propagating of viruses of various types and severity.

---

[1]The term cyber-attacks is used oftentimes to characterize computer viruses. In this book we use it in the same sense as malware.

Multiple times in the past, the human species have suffered severe diseases, such as the black plague that incurred a death toll of around 60% of Europe's population in the 1300s. Such cases have motivated scientists, mainly mathematicians, to attempt to describe the diffusion dynamics of the outbreaks, in order to be able to respond properly in future occurrences. Furthermore, observations among dependent species in nature, such as foxes-rabbits, for instance, have lead to several models, e.g. predator-prey Lotka-Volterra equations, most of which describe macroscopically the evolution of such ecosystems that have potential interest and importance for humans [99].

Definition 3.1 *(Epidemics). The term* epidemics *is cumulatively used in epidemiology[2] to characterize instances where documented or new cases of disease occur in a given population of living organisms and during a given period, substantially exceeding what is expected based on the recent or prior experience. Epidemics characterizes the rapidity and magnitude of spread of the corresponding outbreak.*

The above definition refers to the models developed for the case of biological viruses and their spread in different populations. The term "epidemic" characterizes the cases where the spreading outcome grows fast, yielding significant number of infections, either expected or not. Consequently, the term is used in epidemiology to denote either an emergency state (spreading of a disease has gone beyond control) or a state when a massive spreading is expected (even with less severe outcomes), e.g. flu epidemic every winter.

By analogy, the term epidemics in computer and network science (also sometimes referred to as cyber-attacks or malware) characterizes cases of malicious software diffusing (spreading or propagating) in a given user computer/device population for a given period of time, which usually exceeds substantially what is expected in terms of duration, outcome/impact, or any combination thereof. Thus, epidemics (malware) in computer science follows the interaction between users/machines in a communication network via the underlying infrastructure. Even in biology and epidemiology, epidemics follow an underlying network structure formed by the interacting individuals, e.g. an affiliation human network in the case of the flu spreading. In any case of epidemics, the links of the network formed represent the interactions of the corresponding population through which epidemics diffuse. The term also refers to the models developed in order to describe and study such scenarios.

Standard epidemiological models usually consider homogeneous underlying networks. Homogeneity characterizes network topologies where the degree distribution, typically referred to as *connectivity*, is classified as homogeneous. Mathematically, this means that the degree distribution peaks at a value, which is the average network degree (denoted by $\langle k \rangle$), and decays exponentially fast for $k \ll \langle k \rangle$ and $k \gg \langle k \rangle$, where k is the degree of a node in the network. Intuitively, this means that the number of neighbors of each network node is relatively "homogeneous" around a central

---

[2]Epidemiology is the science that studies the patterns, causes, and effects of health and disease conditions in various populations [87].

(average) value, and few nodes deviate from this regime. Characteristic example of deterministic homogeneous networks is the standard hypercubic lattice, while among the random homogeneous networks the random graph model of Erdos-Renyi and the Watts-Strogatz (WS) model of small-world (SW) networks. On the other hand, several other networks, most prominently the router network on the Internet or the WWW and DNS networks, have an inhomogeneous structure characterized by power-law degree distributions. Furthermore, network homogeneity refers also to the OS and protocol suites employed, which is the case for most Internet machines that run on one out of a restricted set of available OS/network platforms. Epidemics models have been also developed for such types of complex networks and in the remaining of this chapter, they will be shortly considered and compared against other models. Additional examples for the networks described above may be found in Table 1.2.

Among various epidemics models that have been developed in epidemiology to describe the spreading dynamics of viruses for herb, animal, human, and other populations, the most important ones that are suitable for modeling malware diffusion are those denoted as simple epidemic model and the Kermack-McKendrick model. These two basic epidemics models will be analyzed in the following two subsections.

3.2.1 SIMPLE (CLASSICAL) EPIDEMIC MODEL—SI MODEL

The simple (or sometimes denoted as classical) epidemic model is a straightforward adaptation of epidemics models used extensively in the sciences of biology and ecology [58, 99, 160]. Traditionally, these models have been used for the study of virus propagations (outbreaks) in closed or interacting living populations, such as the common flu or deadly diseases, e.g. malaria and AIDS in humans, and other viruses in animals, insects, etc.

The main assumption of this approach is that the SI node infection model is considered as the underlying malware propagation model (explained in detail in Section 2.3, Fig. 2.1 and repeated for clarity in Fig. 3.1, which highlights the nature of each state). This is because in the early days of epidemiology, most of these outbreaks, if not all, were lethal, and thus most infected entities would succumb rapidly. A single state transition from the susceptible to the infected state (whichever the outcome of the infected state) was sufficient for an accurate model of the propagation dynamics. The interest of scientists at the time was in estimating the ratio of healthy-to-infected entities as a measure of the infection potential, in order to evaluate the severity of the expected outcome, or predict the potential of a pandemia (state where the infection dominates the population).

Consequently, the purpose of this subsection is to study the simple epidemic model, which is characteristic of the SI malware propagation dynamics. If a piece of malware is identified to follow the SI infection paradigm (or equivalently SI seems to properly describe the infection dynamics of a malware module over a legitimate network), the analytical tools provided in this subsection for studying the SI model

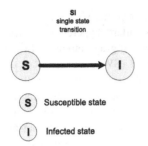

FIGURE 3.1

Simple epidemic model: SI infection paradigm for each member of the population.

can be used directly for analysis of the threat dynamics. In the following, we employ a computer science terminology to describe the analyzed models, i.e. the term "host" or "node" will be used in place of "individual" or "human," commonly used in epidemiology and biology.

In the simple epidemic model, a total number of hosts N are assumed and the number of susceptible host machines in the system is denoted by $S(t)$. Then, since the population is considered closed, the number of currently infected hosts (nodes) at each time is $I(t) = N - S(t)$. We also consider $\alpha(t) = I(t)/N$, as the fraction of infected nodes of the attacked network. Another assumption of the model is that of a homogeneous underlying system, namely, each node has the same probability of contacting a malicious (infected/infectious) node and thus become infected. This means the number of contacts between the two distinct node groups is proportional to the product $S(t)I(t)$. By assuming that β represents the attack infection rate, namely, the number of probes sent out by a malware source per time unit, then the homogeneous simple epidemic model (SI infection dynamics) is given by an ODE, which can be formulated in the following alternative forms:

$$\frac{dI(t)}{dt} = \beta I(t)S(t) = \beta I(t)[N - I(t)], \qquad (3.1a)$$

$$\frac{d\alpha(t)}{dt} = k\alpha(t)[1 - \alpha(t)], \qquad (3.1b)$$

$$\frac{dS(t)}{dt} = -\beta I(t)S(t) = -\beta S(t)[N - S(t)]. \qquad (3.1c)$$

In the above set of ODEs, only one of them is independent, since $N = S(t) + I(t)$ and parameter $k = \beta N$. Eq. (3.1a) with the initial condition that at $t = 0$, $I(0) = I_0$ hosts are infected and thus $S(0) = N - I_0$ are noninfected can be solved straightforwardly.

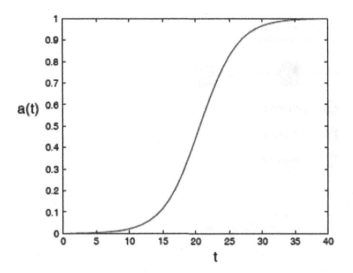

FIGURE 3.2

Simple epidemic model: Percentage of infected hosts as a time function.

The solution of Eq. (3.1a) given the initial condition $I(0) = I_0$ is

$$I(t) = \frac{I_0}{I_0 + (1 - I_0)e^{-\gamma t}},$$ (3.2)

where $\beta = \gamma N$.

Eq. (3.1b) is a normalized version of Eq. (3.1a), where $\alpha(t) = I(t)/N$ is the fraction of the infectious population at time t. Also, Eq. (3.1c) is symmetric with Eq. (3.1a), where $I(t)$ has been replaced with the complementary value $N - S(t)$, denoting the dependence of $S(t)$, $I(t)$ with respect to N given the considered closed population of the legitimate network. The opposite sign of Eq. (3.1c) is due to the fact that in the SI model, the susceptible population is decreasing, while the infected group is increasing. Both of these trends are evident in the set of alternative equations given in Eqs. (3.1).

In general, each of the Eqs. (3.1) constitutes a suitable approximation model for worms and especially for the CodeRed worm [235]. The solution of Eq. (3.1b) is shown in Fig. 3.2, where it is evident that the infection has an exponential growth following a slow-start phase, and finally again leading to a slow-stop phase near the end of the attack. According to the SI dynamics, one expects all hosts to be eventually infected. This is due to the fact that initially, the number of infected nodes is small, and thus the diffusion is restricted. Then, a critical mass of infected nodes is accumulated, in which case the propagation becomes very intense (many malware sources) and the growth of infected nodes becomes exponential. At the final stages,

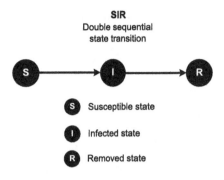

SIR
Double sequential
state transition

S → I → R

S   Susceptible state

I   Infected state

R   Removed state

FIGURE 3.3

Kermack-McKendrick: Underlying infection
model.

few nodes are still susceptible, and thus any incremental increase to the number of infected nodes shown in Fig. 3.2 becomes very rare, yielding the observed saturation.

Finally, it should be also noted that even though the simple epidemic model is inaccurate for many concurrently spreading pieces of malware, or for some specific types of malware, it is a good first model that may be used to build more elaborate and powerful ones. The epidemics models presented next were set in a broader application perspective and have been developed in order to improve the simple epidemic model for more complex malware behaviors.

3.2.2 GENERAL EPIDEMIC MODEL: KERMACK-MCKENDRICK MODEL

The simple epidemic model can be extended to describe the actual behavior of malware spreading in wired computer networks more accurately, by extending it to the Kermack-McKendrick model [235]. The latter is also denoted as a general epidemic model, and in addition to node infection, it considers the removal of infected nodes. Indeed, in real networks, the infected hosts are not expected to stay in this condition forever, since an end-user will notice sooner or later abnormal behavior of the infected device and will act on that. Thus, in addition to the previous simple epidemic model, some of the nodes are removed completely from the network.

Similarly to the simple epidemic model corresponding to the SI infection paradigm, the general (Kermack-McKendrick) epidemic model corresponds to the SIR infection paradigm, as shown in Fig. 2.1 and repeated in Fig. 3.3, highlighting further the considered states of legitimate nodes. The state R is denoted as "removed" and typically represents cases that nodes are either completely damaged (e.g. run out of battery or other resources) by the attack, or eventually recover and they are permanently immunized by the specific malware type. In the latter case, the SIS model would also seem appropriate. However, the SIS paradigm is more accurate for the cases of modeling a system in the long-term (for an extensive period of time, or if the analysis considered multiple possible threats spreading/propagating, where in both cases infections and the susceptible state are recurrent). The SIR paradigm, and thus the Kermack-McKendrick model as well, has been developed for the study of specific malware threats, each potentially diffusing individually across the network. Consequently, if a node recovers from such infection, it will be concurrently patched and protected from future contacts with attack nodes and their malware. Equivalently, if a node becomes completely dysfunctional from the received infection, it will not be able to come in contact with attackers or other legitimate nodes in the future. Thus, the SIR mapping to the Kermack-McKendrick model is more appropriate and the two terms (SIR, Kermack-McKendrick model) will be used interchangeably in the rest of this book when we refer to practical application of SIR epidemics models.

Based on the set of simple epidemics dynamics given in Eqs. (3.1), the general epidemic model (SIR paradigm) dynamics can be obtained by taking into account the node removal process. The Kermack-McKendrick model dynamics are given by the following system of ODEs:

$$\frac{dS(t)}{dt} = -\frac{\beta S(t)I(t)}{N}, \tag{3.3a}$$

$$\frac{dI(t)}{dt} = \frac{\beta S(t)I(t)}{N} - \gamma I(t), \tag{3.3b}$$

$$\frac{dR(t)}{dt} = \gamma I(t), \tag{3.3c}$$

$$N = S(t) + I(t) + R(t), \tag{3.3d}$$

which can be also cast in an alternative form

$$\frac{dJ(t)}{dt} = \beta J(t)[N - J(t)], \tag{3.4a}$$

$$\frac{dR(t)}{dt} = \gamma I(t), \tag{3.4b}$$

$$J(t) = I(t) + R(t) = N - S(t), \tag{3.4c}$$

where β is the infection rate, γ is the rate of removal of infectious hosts from the malware circulation process, and $S(t)$, $I(t)$, and $R(t)$ are the number of susceptible, infected, and removed nodes at time t, respectively. Parameter N is the total size of the population as in the simple epidemic model. $J(t)$ is the equivalent of infected nodes in the simple epidemic model. Just as the group of susceptible nodes reduces and the coupled group of infected increases by the same amount in the simple epidemic model, in the general epidemic model, the groups $S(t)$, $J(t)$ are coupled in the general epidemic model, and thus the simple epidemic model Eq. (3.1a) can be used for modeling the general epidemics behavior where now the term $I(t)$ in the simple epidemic should be replaced by the equivalent $J(t)$ in the general epidemic. Also, a new equation for the removal rate (Eq. (3.4b)) was included. Together with the initial conditions $I(0)$ and $R(0)$, the problem is fully determined via the system of Eqs. (3.4).

One of the most characteristic features of the Kermack-McKendrick model, derived from the solution of Eqs. (3.4), is the emergence of the infamous *epidemic threshold*. The epidemic threshold advocates that a major outbreak occurs if and only if the initial number of susceptible hosts $S(0) > \rho$, where $\rho = \gamma/\beta$. Thus, ρ is called the epidemic threshold. The epidemic threshold quantifies the number of secondary infections caused by a single primary infection. In other words, it determines the number of users infected by contact with a single infected user machine before the death or recovery of the latter.

Furthermore, it can be derived from the system of Eqs. (3.4) that $dI(t)/dt < 0$ if and only if $S(0) < \rho$. The emergence of the epidemic threshold is an instance of threshold phenomena, which can be observed in many complex network processes, e.g. percolation [35, 64, 89]. Another similar and very prominent threshold behavior in some types of wireless complex networks is connectivity and the emergence of a giant cluster of nodes in the network as the transmission radius or network density increases. Such networks are modeled as random geometric graphs and their connectivity exhibits threshold behavior with respect to the scaling of their transmission range radius and their deployment region [180].

3.2.3 TWO-FACTOR MODEL

The previous epidemics models do not take into account the case where an infection of a susceptible node is rapidly detected and the specific node is directly quarantined (i.e. removed from the network, or at least prohibited from communicating with other host machines) from the user or network administrator, in order to prevent it from further propagating the received malware. This is a similar behavior to the one observed in human viruses, e.g. the common flu, where once a person becomes infected she/he is put in a form of isolation avoiding excessive interaction (or even avoiding completely interaction with others) to prevent transmission to other persons, proactively or mandatorily [58].

A basic assumption in the previous epidemics models is that malware and more specifically worms as the ones modeled by the simple and general epidemic models continuously search for available susceptible nodes to infect. However, in reality, worms usually do not continuously spread forever. For instance, the CodeRed worm stopped propagation at 00:00 UTC July 20th, 2001 [235].

In general, one can observe two specific aspects of worm propagation, not taken into account by the simple and Kermack-McKendrick epidemic models:

1. Human countermeasures result in removing both susceptible and infectious computers from circulation. This corresponds to users becoming more aware of the spreading threats and either take some precautions while susceptible, or act rapidly once infected.
2. The infection rate is usually decreasing with time, i.e. $\beta = \beta(t)$ is a decreasing function rather than constant. This could model, for example, the effect of a vast

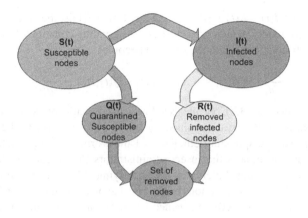

FIGURE 3.4

State transitions in the two-factor spreading model.

worm attack, where Internet routers become bottlenecked and thus the worm
scanning process is slowed down.

Both of these observations have been actually observed in the behavior of CodeRed
worm spreading [235]. The cumulative effect of these two observations is a time-
dependent and decreasing worm infection rate. From the worm's perspective, human
countermeasures remove some hosts from worm spreading circulation, including
both hosts that are infectious and hosts that are still susceptible. In other words, the
removal process consists of two parts: removal of infectious hosts $I(t)$ and removal of
susceptible hosts $S(t)$. Thus, the more suitable infection paradigm model in this case
is an extended SIR one, where the removed hosts come both from S, I populations.
The removed susceptible users are denoted by $Q(t)$ and referred to as quarantined,
since they are usually informed users that take proper precautions. The removed
infected nodes are simply referred to as removed and denoted by $R(t)$. The involved
user groups, $S(t), I(t), Q(t)$, and $R(t)$, are shown in Fig. 3.4 and they are identical
to the dynamic quarantine model assumed states that will be explained in the next
subsection. It should be noted that the category $Q(t)$ of susceptible nodes was not
introduced in Chapter 2; however, a realistic consideration of malware epidemics
calls for such a possible state for susceptible nodes.

The change in the number of susceptible hosts $S(t)$ from time t to time $t + \Delta t$ is
given by

$$S(t + \Delta t) - S(t) = -\beta(t)S(t)I(t)\Delta t - \frac{dQ(t)}{dt}\Delta t. \tag{3.5}$$

Consequently, the basic epidemics equation now becomes

$$\frac{dS(t)}{dt} = -\beta(t)S(t)I(t) - \frac{dQ(t)}{dt}, \tag{3.6}$$

where now $S(t) + I(t) + R(t) + Q(t) = N$ at any time t. Thus, the epidemics equation of the two-factor model (which extends the corresponding equation of the SIR model with the two observations mentioned above) becomes

$$\frac{dI(t)}{dt} = \beta(t)[N - R(t) - I(t) - Q(t)]I(t) - \frac{dR(t)}{dt}. \tag{3.7}$$

In order to solve Eq. (3.7), one needs to know the dynamic properties of $\beta(t)$, $R(t)$, and $Q(t)$. Parameter $\beta(t)$ is determined by the impact of worm traffic on Internet infrastructure, and the spreading efficiency of the worm code. Parameters $R(t)$ and $Q(t)$ involve any awareness that users/administrators might have of the worm, or any patching and filtering difficulties. By specifying their dynamic properties, the complete set of differential equations of the two-factor worm model can be obtained.

To date, obtaining directly the general two-factor worm model in analytical closed-form solutions was not possible. Instead, one may use a numerical model, in which she/he needs to determine the dynamical equations describing $R(t)$, $Q(t)$, and $\beta(t)$ in the model described by Eq. (3.7). For the removal process from infectious hosts, the same assumption as in the Kermack-McKendrick model can be assumed, namely, $dR(t)/dt = \gamma I(t)$. The removal process from the set of susceptible hosts is more complicated. At the beginning of the worm propagation, most people do not know there exists such a kind of worm. Consequently, the number of removed susceptible hosts is small and increases slowly. As more systems become infected, awareness increases. Hence, the speed of immunization (quarantining) increases fast as time passes. The speed decreases as the number of susceptible hosts shrinks and converges to zero when there are no susceptible hosts available. The classical simple epidemic equation can be employed to model such behavior $dQ(t)/dt = \mu S(t)I(t)$. The decreased infection rate can be described as

$$\beta(t) = \beta_0 \left[1 - \frac{I(t)}{N}\right]^n, \tag{3.8}$$

where β_0 is the initial infection rate. The exponent n is used to adjust the infection rate sensitivity to the number of infectious hosts, where $n = 0$ corresponds to a constant infection rate. Using the assumptions above on $Q(t)$, $R(t)$, and $\beta(t)$, the complete differential equations of the two-factor worm model are

$$\frac{dS(t)}{dt} = -\beta(t)S(t)I(t) - \frac{dQ(t)}{dt}, \tag{3.9a}$$

$$\frac{dR(t)}{dt} = \gamma I(t), \tag{3.9b}$$

$$\frac{dQ(t)}{dt} = \mu S(t)I(t), \tag{3.9c}$$

$$\beta(t) = \beta_0 \left[1 - \frac{I(t)}{N}\right]^n, \tag{3.9d}$$

$$N = S(t) + I(t) + R(t) + Q(t), \tag{3.9e}$$

$$I(0) = I_0 \ll N; S(0) = N - I_0; R(0) = Q(0) = 0. \tag{3.9f}$$

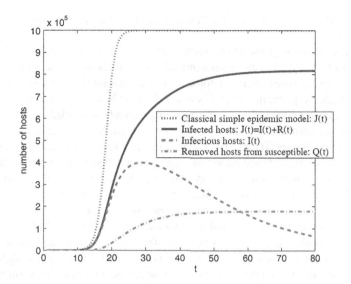

FIGURE 3.5

Two-factor model: numbers of infected and removed hosts.

It can be derived that $dI(t)/dt = \beta(t)S(t)I(t) - dR(t)/dt = [\beta(t)S(t) - \gamma]I(t)$, and the number of susceptible hosts $S(t)$ is a monotonically decreasing function of time. The maximum number of infectious hosts will be reached at time t_c when $S(t_c) = \gamma/\beta(t_c)$. At the same time, $\beta(t)S(t) - \gamma < 0$ for $t > t_c$; thus, $I(t)$ decreases after $t > t_c$.

Fig. 3.5 demonstrates the number of infected and removed hosts as a function of time. The susceptible removed hosts $Q(t)$ are also depicted for comparison purposes. It also compares with the infected nodes value predicted by the simple epidemic model.

3.2.4 DYNAMIC QUARANTINE

Similarly to the two-factor model, the dynamic quarantine model has been developed to take into account cases where a worm infection is rapidly detected and some of the infected machines are quarantined to prevent them from further propagating the received malware [234]. The quarantined node can be considered as a form of removed node from the network, and as shown in Fig. 3.4, these nodes are not further considered in the malware propagation process, similarly to the two-factor propagation model.

With respect to epidemics control in the real world, people usually react under the principle that whenever a person exhibits a symptom slightly similar to an expected one for a disease, he or she will be quarantined immediately. The quarantine will be released after the person passes the disease latent period without showing up further symptoms. This is referred to as the "assume guilty before proven innocent" rule of thumb.

In terms of worm propagation, such a soft quarantine strategy will allow that every host of the system can be quarantined individually when the worm anomaly detection program raises alarm for some host. The quarantine on a host under alarm is released after a quarantine time T, even if the host has not been inspected by the user or an administrator yet. Once the quarantine on a host is released, this host can be quarantined again if the anomaly detection program raises an alarm for this host some time later. The specific means of quarantine may vary considerably, from simple traffic filtering/blocking based packet sniffing and port filtering according to identified malware traffic signatures to more sophisticated ones where constant monitoring of traffic and information exchange is required. All these countermeasures depend on the availability of the specific infrastructure and associated resources.

The dynamic quarantine method has two advantages: first, a falsely quarantined but otherwise healthy host will only be quarantined for a short time; thus, its normal activities will not be impacted too much. Second, since in this case a higher false alarm rate than the normal permanent quarantine can be tolerated, the worm anomaly detection program can be set more sensitive to a worm's activities.

The simplest such quarantine approach considers constant quarantine time and the anomaly detection threshold throughout the spreading period of a worm. On average, an infectious host will be detected in $1/\lambda_1$ units of time after the host becomes infectious, or after it is released from a previous quarantine. The total time corresponds to the propagation time of an infectious host before having been discovered and quarantined. Parameter λ_1 is denoted as the quarantine rate of infectious hosts. A healthy, nonquarantined host will keep its normal activities for $1/\lambda_2$ units of time on average before it is falsely alarmed and quarantined. Thus, λ_2 is the quarantine rate of susceptible hosts and corresponds to the false alarm rate of the anomaly detection program used in the system. The values of λ_1 and λ_2 are determined both by the threshold and by the performance (and sensitivity) of the anomaly detection program employed. A high performance anomaly detection program has higher detection rate and lower false alarm rate, i.e. the detection program has larger λ_1 and smaller λ_2 than a worm detection program with ordinary performance.

Based on the above notation, we obtain the number of infectious hosts removed as

$$R(t) = \int_{t-T}^{t} [I(\tau) - R(\tau)]\lambda_1 d\tau, \tag{3.10}$$

which is correct only for large populations, due to the use of the average value λ_1. To grasp this better, one should keep in mind that in large populations, a great number of interactions are expected to take place. Furthermore, these interactions tend to be more "random" than in smaller populations, since within the large population multiplexing between different behaviors is expected. In other words, the large population is expected to mix better than a small population, in which case, average values are quite representative for all users. In the following, we will study how the dynamic quarantine affects a worm's propagation by extending the simple epidemic model and the Kermack-McKendrick model, respectively.

For the simple epidemic model, a host is either susceptible or infectious. Due to the dynamic quarantine, a host is also either quarantined or not quarantined at any time t. Based on the state dynamics shown in Fig. 3.4, the interactions of the simple epidemics dynamics are now between $[I(t) - R(t)]$ and $[S(t) - Q(t)]$. Therefore, the dynamics of simple epidemic model under dynamic quarantine now follows

$$\frac{dI(t)}{dt} = \beta[I(t) - R(t)][S(t) - Q(t)] = \beta', I(t)[N - I(t)], \qquad (3.11)$$

where $\beta' = (1 - p_1')(1 - p_2')\beta$ is the effective pairwise rate of infection, i.e. the actual rate of infection that would be observed between a randomly selected pair of nodes, $p_1' = \frac{\lambda_1 T}{1 + \lambda_1 T}$ is the effective quarantine probability of infectious hosts, $p_2' = \frac{\lambda_2 T}{1 + \lambda_2 T}$ the effective quarantine probability of susceptible hosts, and T the observation time interval. It should be also noted that $Q(t) = p_2' S(t)$.

The above equation shows that under dynamic quarantine, a worm still propagates according to simple epidemic model but with slower spreading speed. The dynamic quarantine decreases a worm's pairwise rate of infection β by the factor of $(1 - p_1')(1 - p_2')$: the larger the effective quarantine probabilities p_1' and p_2' are, the slower the worm can propagate [234]. Therefore, when one implements the dynamic quarantine, it can provide precious time to take counteractions—patching vulnerable computers and cleaning infected ones—before a worm infects the major part of a network.

Shifting back to the case of the Kermack-McKendrick model, assume that $U(t)$ denotes the number of removed hosts from infectious only. Then, it follows $dU(t)/dt = \gamma I(t)$. Removal of the infectious hosts takes place uniformly from the set of infected nodes $I(t)$, which could also include nodes that are under quarantine when they are selected for removal. Assume that before removal $R(t) = p_1' I(t)$ and $Q(t) = p_2' S(t)$. When the removal process is in effect, $Q(t) = p_2' S(t)$ still holds. However, Eq. (3.10) should be modified to consider the removed hosts from $R(t)$ during time $[(t-T), t]$. Since the removal process uniformly removes infectious hosts from $I(t)$, the removal rate from quarantined $R(t)$ (quarantined infected hosts) should be $\gamma R(t)$ at time t. Therefore, Eq. (3.10) can be extended to

$$R(t) = \int_{t-T}^{t} [I(\tau) - R(\tau)]\lambda_1 d\tau - \int_{t-T}^{t} \gamma R(\tau) d\tau. \qquad (3.12)$$

With the assumption that $R(\tau) \cong R(t)$ and $I(\tau) \cong I(t)$, $\forall \tau \in [t - T, t]$, it is obtained $R(t) = q_1' I(t)$, where $q_1' = \frac{\lambda_1 T}{1 + (\lambda_1 + \gamma)T}$ is the effective quarantine probability of infectious hosts for a worm's propagation with removal process. It should be noted that such an assumption holds when the quarantine time T is small, compared with the model's dynamics (time scale that changes happen in the network). Some relevant discussion and more details are contained at the end of this subsection. For consistency, $q_2' = p_2' = \frac{\lambda_2 T}{1 + (\lambda_2 + \gamma)T}$ as the effective quarantine probability of susceptible hosts for a worm's propagation with removal process.

The propagation dynamics will be now governed by

$$dI(t)/dt = \beta[I(t) - R(t)][S(t) - Q(t)] - \gamma I(t) = \beta'' I(t)S(t) - \gamma I(t), \qquad (3.13)$$

where $\beta'' = (1 - q_1')(1 - q_2')\beta$ is the effective pairwise rate of infection for a worm's propagation with removal process. The worm propagation model given by Eq. (3.13) is the same as the Kermack-McKendrick model, except that the pairwise rate of infection β'' is decreased from β by the factor of $(1 - q_1')(1 - q_2')$. The new dynamic quarantine system will have an epidemic threshold ρ' that is

$$\rho' = \gamma/\beta'' = \frac{1}{(1 - q_1')(1 - q_2')}\rho. \qquad (3.14)$$

If the initial number of susceptible hosts $S(0)$ satisfies the relationships $S(0) > \rho$ and $S(0) < \rho'$, then according to the Kermack-McKendrick epidemic threshold theorem, a worm will spread out in the original system but will not be able to spread out when we implement dynamic quarantine on the system. In other words, the dynamic quarantine method reduces the chance for a worm to form an outbreak, i.e. a holistic spread of the worm in the whole network (case of pandemia).

In the previous model, all infectious hosts have an equal probability to be removed. However, a more realistic scenario is that some authority inspects the hosts that have raised alarm and have been quarantined. This is justified since limited human resources do not permit the full-scale inspection of all hosts, while alarmed hosts are more likely to be infected by a worm. Therefore, in such a dynamic quarantine system, only infectious hosts in the quarantined population (the latter denoted now by $R(t)$) could be removed. In this case, the number of removed hosts from quarantined infectious population $R(t)$ (denoted by $U(t)$) follows $dU(t)/dt = \gamma R(t)$. Eq. (3.12) remains valid and the propagation dynamics become

$$\frac{dI(t)}{dt} = \beta[I(t) - R(t)][S(t) - Q(t)] - \gamma R(t) = \beta'' I(t)S(t) - \gamma' I(t), \qquad (3.15)$$

where $\gamma' = q_1'\gamma$ is the effective removal rate. This model has the same form as the Kermack-McKendrick; therefore, all outcomes regarding the Kermack-McKendrick model remain valid. The epidemic threshold becomes

$$\rho'' = \gamma'/\beta'' = \frac{q_1'}{(1 - q_1')(1 - q_2')}\rho, \qquad (3.16)$$

where $\rho = \frac{\gamma}{\beta}$ as denoted above. The epidemic threshold theorem states that if $S(0) < \rho''$, a worm will not form an outbreak in this dynamic quarantine system.

On a final note, it should be stated that all the previous analyses are based on two assumptions. First, the quarantine time T is small such that

$$\begin{cases} R(\tau) \simeq R(t), \\ I(\tau) \simeq I(t) \quad \forall \tau \in [t - T, t], \\ S(\tau) \simeq S(t). \end{cases} \qquad (3.17)$$

Second, Eqs. (3.10) and (3.12) rely on the law of large number since they both use the mean values of λ_1 and λ_2 without considering stochastic effects. These two equations are accurate only when $I(t) - R(t)$ is large (the formula of $Q(t)$ is correct only when $S(t) - Q(t)$ is large).

3.3 OTHER EPIDEMICS MODELS

The previously presented epidemics models have been mainly developed for wired networks and especially worms and viruses spreading over the Internet and exploiting elements from TCP/IP, e.g. port scanning and email attachment attacks [233]. However, epidemics models have also emerged in the literature for other types of networks, which can be physical (actual) networks of computers, e.g. local area proprietary networks, or utility grids (e.g. water pipes or electricity poles), overlay/application networks, e.g. user network of online social networks, or any other type of network representation at any intermediate protocol layer.

In the following, we present concisely some of these models for the respective network types and application scenarios. Further details for each of these approaches can be found in the corresponding bibliography provided.

3.3.1 EPIDEMICS MODEL IN SCALE-FREE NETWORKS

One of the major outcomes of epidemics is that the corresponding models, and of course the actual propagation dynamics they describe, are heavily affected by the connectivity patterns emerging in the population over which malware spreads/propagates. Simply put, connectivity affects the outcome of the propagation. Various works have identified this fact and have studied epidemics models specifically over particular network topologies. In [176], the corresponding models are studied over SF networks, which were presented in Chapter 2. The assumption is that SF networks of interest exhibit a power-law connectivity distribution $P(k) \sim k^{-\gamma}$, $2 < \gamma \leq 3$, which implies that each node-member of the population has a statistically significant probability of having a very large number of connections compared to the average connectivity $\langle k \rangle$ of the network. The employed node infection model is the SIS introduced in Section 2.3. Characteristic application scenarios considered are the Internet and maps of human sexual contacts, which in turn are characterized by SF connectivity properties. In any case, the epidemics are first developed for homogeneous networks, in terms of protocols and mixing of nodes. Homogeneous mixing means that nodes do not show special preference when interacting with each other, but rather their interactions have a more "random" nature. Then modeling is generalized to SF networks.

In one of the first analytical studies extending the SIS model over SF networks in [176], the analysis is undertaken in terms of a dynamical mean field (MF) theory [113], where average values and cumulative behavior are considered. For homogeneous networks, in which the connectivity fluctuations are very small, the MF theory is approached by means of a reaction equation for the total prevalence

$\rho(t)$, defined as the density of infected nodes present at time t. This reaction equation, describing the dynamics of malware diffusion, is as follows:

$$\frac{d\rho(t)}{dt} = -\rho(t) + \lambda\langle k\rangle\rho(t)(1 - \rho(t)). \qquad (3.18)$$

In the above equation, $\lambda = \frac{\nu}{\delta}$ is the effective spreading rate, where ν is the probability with which each susceptible node is infected and δ denotes the cure rate of infected nodes. Furthermore, the homogeneous nature of the considered network allows to use just the average connectivity $\langle k\rangle$ (as most of the probability mass for the node degrees is centered tightly around this value). By solving this equation, the main prediction of the SIS model in homogeneous networks is the presence of a positive epidemic threshold, proportional to the inverse of the average number of neighbors of every node, $\langle k\rangle$, below which the epidemics always dies out and endemic states are impossible. Endemic states, in general, characterize cases where the malware is present in a population at consistent levels and periods of time, i.e. it does not die out, nor diffuses over the whole population. Thus, in the preceding discussion, the fact that the epidemics always dies out does not allow endemic states to emerge. The second term in the right-hand side of the above equation is denoted as "creation term."

However, relaxing the homogeneity assumption seems oneway for modeling more accurately real networks and malware dynamics spreading over them. For this reason, the relative density $\rho_k(t)$ of infected nodes with given node degree k is employed. The dynamical mean-field equations in this case can be written as

$$\frac{d\rho_k(t)}{dt} = -\rho_k(t) + \lambda\langle k\rangle(1 - \rho_k(t))\Theta[\{\rho_k(t)\}]. \qquad (3.19)$$

The second term in the right-hand side of the above (creation term) considers the probability that a node with k links is healthy $(1 - \rho_k(t))$ and gets the infection via a connected node. The probability of this last event is proportional to the infection rate λ, the real number of connections k, and the probability $\Theta[\{\rho_k(t)\}]$ that any given link points to an infected node. We make the assumption that Θ is a function of the partial densities of infected nodes $\{\rho_k(t)\}$. In the steady (endemic) state, the ρ_k are functions of λ. Thus, the probability Θ becomes also an implicit function of the spreading rate, and by imposing the stationarity condition $\frac{\partial\rho_k(t)}{\partial t} = 0$, one obtains

$$\rho_k = \frac{k\lambda\Theta(\lambda)}{1 + k\Theta(\lambda)}. \qquad (3.20)$$

With more detailed analysis that can be found in [176], an average probability of a link pointing to an infected node,

$$\Theta(\lambda) = \frac{1}{\langle k\rangle} \sum_k kP(k)\rho_k. \qquad (3.21)$$

The analytical solutions for the Barabasi-Albert model of SF networks [1, 15] can be obtained by computing the stationary points of $d\rho(t)/dt = 0$ (Eq. (3.18)), which holds for the steady-state observation period of the system (not for the initial transient interval), yielding

$$\rho[-1 + \lambda\langle k\rangle(1 - \rho)] = 0,$$

which defines an epidemic threshold $\lambda_c = \langle k\rangle^{-1}$ and yields

- $\rho = 0$ if $\lambda < \lambda_c$,
- $\rho = \frac{\lambda - \lambda_c}{\lambda}$ if $\lambda \geq \lambda_c$.

Details on the analysis of real data on this model may be found in Section 1.4 of [176].

3.3.2 GENERALIZED EPIDEMICS-ENDEMICS MODELS

The SIR epidemics model presented in Section 3.2.2 and Section 2.3 can be generalized in epidemiology to include other states that typically emerge in the cases of human virus spreading. Such states could have a notable interpretation for malware diffusion as well. In this subsection, we briefly discuss this general epidemics (or endemics) model.

The *generalized epidemic model* is shown in Fig. 3.6 and due to the defined succession of states, it is denoted as immunity-susceptible-exposed-infected-removed (MSEIR). The M symbol represents the state of *passive immunity*, namely, when a member of the population is born with antibodies, e.g. inherited by the mother or in cases of computers when the machine comes already with a patch (or service pack) preinstalled for a specific threat. Once the maternal antibodies disappear or the threat changes its parameters and the patch is not valid anymore, the corresponding member of the population transitions to the *susceptible* state (S), which is as described previously. Some members of the population, e.g. machines that have not preinstalled antivirus software, start from the S state. The *exposed* state (denoted by E) denotes cases where there exists an adequate contact of a susceptible with an infective, so that a transmission occurs between them. The following transition is to the *infective* state where members are capable of infecting others, and the final transition occurs toward the *recovered* state, where members of the population have permanent infection-acquired immunity.

In malware diffusion terminology, the passive immunity state M corresponds to machines that have been already properly patched against spreading threats. After some time, when new threats not known to manufacturers emerge, a transition to the susceptible state takes place. The exposed state E is an intermediate state that can be used to denote the state of a machine before it becomes infected. The infective and recovered states are identical to the ones defined for malware models before. This is a generalization of the node infection models presented earlier, and as mentioned already, the choice of states to include in a model depends on the characteristics of the particular malware threat or set of threats being modeled and the purpose of the model. Of course, the passively immune state M and the latent period (exposed) state E are often omitted, because they are not crucial for the SI node pair interaction, as explained in Section 2.3 in macroscopic analysis of the complex networks of interest.

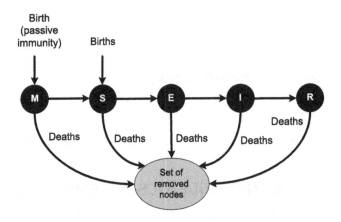

FIGURE 3.6

General epidemics infection model—state transition diagram.

One of the major features of the traditional epidemiology models is that the models for the dynamics of diffusion include both time t and age a as independent variables, because age groups mix heterogeneously, the recovered fraction usually increases with age, risks from an infection may be related to age, vaccination programs often focus on specific ages, and epidemiologic data are often age specific.

A common assumption is that the movements out of the M, E, and I states and into the next possible operational state (not the set of dead members of the population) are governed by exponentially distributed waiting times in each of the states. This assumption will be crucial in many of the more advanced techniques that will be presented in Part 2 of the book, especially in Chapter 4.

Within this framework, there are three key quantities of interest for epidemiologists that can be of particular interest to computer scientists and network engineers for the analysis and control of malware epidemics. The basic reproduction number R_0, also known as basic reproduction ratio or basic reproductive rate, is the average number of secondary infections (secondary infections refer to infections that take place by already infected legitimate nodes, not attackers themselves—primary infections) that occur when one infective is introduced into a completely susceptible host population [99]. In this definition, it is implicitly assumed that the infected outsider is in the host population for the entire infectious period and mixes with the host population in exactly the same way that a population native would mix. The contact number σ is defined as the average number of adequate contacts of a typical infective during the infectious period [99]. An adequate contact is one that is sufficient for transmission, if the individual contacted by the susceptible is an infective. The replacement number R_p is defined to be the average number of secondary infections produced by a typical infective during the entire period of infectiousness [99]. All these three quantities are equal at the beginning of the spread of an infectious disease when the entire population (except the infective invader) is susceptible. R_0 is only defined at the time of invasion, σ and R_p are defined at all times.

In epidemiology literature, the basic reproduction number R_0 is often used as the threshold quantity that determines whether a disease can dominate over a given population or not [99]. For most models, the contact number σ remains constant as the infection spreads, so it is always equal to the basic reproduction number R_0. For these cases, σ and R_0 can be used interchangeably. However, for other cases that new groups of infective nodes with lower infection capability appear, the contact number σ for these models becomes less than the basic reproduction number R_0. The replacement number R_p is the actual number of secondary cases from a typical infective, so that after the infection has dominated a population and no one is in the susceptible state, R_p is always less than the basic reproduction number R_0. After that, the susceptible fraction is less than 1, so that not all adequate contacts result in a new infection instance. Consequently, the replacement number R_p is always less than the contact number σ after the invasion. Combining these results leads to $R_0 \geq \sigma \geq R_p$ with equality at the beginning of the attack ($t = 0, S(t) = 1$, with $S(t)$ the fraction of the population at the susceptible state). In general, $R_0 \geq \sigma$ for most models and $\sigma \geq R_p$ after the beginning of the attack for most models.

Other epidemics models that can be derived from the aforementioned general epidemic model and have notable interest due to the various features and applications they exhibit are the following:

- **The MSEIR model with exponentially changing size (churn)**—The M-S-E-I-R model takes into account the fact that the population of a network changes dynamically (denoted by the term node churn), with users entering and leaving the network due to a multitude of reasons, the most prominent being the virus attack. More details and the formulation of the corresponding epidemics equations can be found in Section 3.1 of [99].
- **The MSEIR model with continuous age structure**—This endemic model takes into account the demographic model with continuous age (presented in Section 4.1 of [99]) and the formulation of the corresponding set of differential equations and its solution to obtain the basic reproduction number and stability properties are presented in detail in Section 5 of [99]. This model essentially takes into account the impact of aging on the members of the population and the spreading of diseases and can be extended to model the impact of user device aging patterns on malware dissemination dynamics.
- **The MSEIR model with vaccination at a specific age** — Such model extends the previous to the cases where vaccination (quarantine in malware terms) takes place for each population member (user). More details are provided in Section 5.5 of [99].
- **The susceptible-exposed-infected-removed (SEIR) model with age groups**— The S-E-I-R model utilizes the demographic model with age groups described in

Section 4.2 of [99] and involves separate age groups. The formulation and solution for the basic reproduction number can be found in Section 6.1 of [99].

3.4 MISCELLANEOUS MALWARE MODELING MODELS

Apart from the epidemics models of various types presented in the previous subsections, many other approaches modeling the spread of worms and other similar types of viruses, popular in wired computer networks, have been proposed in the literature. In the following, we summarize some of these models, in order to provide a more complete picture of the existing references and their potentials.

A mean-field (MF) approach utilizing Markov theory and spectral radius from algebraic graph theory has been presented in [216], with the objective to understand the influence of graph characteristics on epidemics spreading. The model is centered around the adjacency matrix of the network of legitimate nodes A, and its spectral radius $\lambda_{\max}(A)$, defined as the largest eigenvalue of the adjacency matrix A of the network. The adjacency matrix A represents the network structure developed between legitimate nodes, and for each element $a_{ij}, A = [a_{ij}]$, where $a_{ij} = 1$ denotes a link between nodes i, j, while $a_{ij} = 0$ denotes that no link exists between nodes i, j. The major intriguing result of MF models is that the epidemic threshold, mentioned in the previous sections, is related to the spectral radius in a simple manner,

$$\tau_c = \frac{1}{\lambda_{\max}(A)}. \tag{3.22}$$

Extending this, an exact Markov chain model is developed (containing all 2^N possible states for a legitimate population of N users through an N-intertwined model). It can be obtained that the exact Markov chain can provide insight into the virus spread process (with respect to the time of convergence to the absorbing state). The N-intertwined model relates network topology parameters to the spreading process (through the largest eigenvalue and degrees of the nodes). Several upper bounds for the steady-state infection probabilities have been obtained in [216], while under specific simplification assumptions it was deduced that the N-intertwined model reduces for regular graphs to the other basic (e.g. Kephart and White [130]) epidemiological models after additional simplifications. It was also possible to explore phase transition phenomena and show that for fixed graph, the epidemic threshold τ_c is a very characteristic metric, dictating malware diffusion dynamics. It also presented a relation between the spreading rate and convergence time toward the extinction of epidemics for two extreme cases (full mesh and line graph). This is especially important for smaller epidemics in general, where the spreading rate is close to the epidemic threshold and where the lifetime of an epidemic varies significantly. The most prominent result, however, is that the largest eigenvalue of the adjacency matrix of the graph is rigorously shown to define an epidemic threshold of the N-intertwined model (as well as of other MF models).

3.5 SCOPE AND ACHIEVEMENTS OF EPIDEMICS

The previous brief presentation of some of the most popular epidemics models for malware propagation and some other miscellaneous malware modeling approaches allows observing and documenting characteristic facts and emerging trends that describe the more traditional malware modeling cumulatively. We summarize these general trends in the following and note that in most cases, such facets of traditional malware diffusion modeling, especially epidemics, have motivated the more advanced approaches presented in the subsequent sections of this book.

Strictly speaking, worm propagation is a discrete event process. The evolution of the process is according to the specific events taking place and indicating broader system state transitions. For instance, an infection of a specific susceptible node leads to a +1/-1 type of variation of nodes in the infected/susceptible groups, respectively. However, the presented epidemics models and all others based on them, which is the majority of models, treat the worm propagation as a continuous process and use the continuous differential equations models provided earlier to describe it [130]. Such an approximation is accurate for large-scale systems and it is widely used in epidemics modeling [58, 74], Internet traffic fluid modeling [159], etc. The propagation of Internet worms is inherent in cases of large-scale problems, since they take place in the global Internet, so the presented models are suitable to use the continuous differential equations for modeling purposes.

Most of the earliest malware modeling approaches, presented above, have been based on epidemics models and have yielded several confirmations of the designed models through numerical validations and observed data [130]. In fact, one of the emerging facts is that the epidemics models were evolved (from the simple epidemic to general epidemic, the dynamic quarantine, and two-factor models) according to the observed real data that were collected from actual attacks. Utilizing these observations, it was possible to extend previous differential equations to model more complex dynamics of malware spreading. Thus, all these models are very accurate for the specific attacks they have been designed for. Unfortunately, this has two consequences. The first one is that a specific model cannot be used cumulatively for all observed attacks, but rather only for the specific one it was developed for. The second is that these models cannot guarantee accuracy in case of new attacks, not even variations of them. The emergent threats can be considerably modified, casting the already established epidemics models outdated (where not even some parameter modification can alleviate the involved discrepancy). Both consequences reduce the broader value of epidemics models in the field of malware diffusion and their impact on further driving the evolution in the corresponding research.

On the other hand, these models have been proven invaluable at the early stages of malware modeling, when the first Internet connected networks emerged and shortly the first malware threats were introduced. Their simplicity aided in understanding their behavior and design the first countermeasures, which later were used for more advanced defense systems. In fact, if the time dependence is discarded for a moment, the epidemics process of a single threat resembles that of a *random walk* [183] on a line, visiting nodes sequentially. Multiple threats diffusing resemble multiple concurrent random walks. As it will become evident in the subsequent sections, epidemics models have formed a solid basis for more advanced malware diffusion models.

In summary, the earlier and even some of the latest epidemics techniques have the disadvantage of not being able to model the behavior of generic malware propagating attacks, but only special cases of them. The latter is meant in the sense that models developed for specific attacks, e.g. `CodeRed` worm propagation, are not capable to model the behavior and dynamics of any type of worm or other computer virus. Nevertheless, significant targeted results have been obtained and were initially used to design the first countermeasures that enabled also assessing the severity and status of malware attacks. These approaches have constituted a solid basis for more advanced studies and developing the more intelligent and accurate methodologies for malware diffusion modeling, e.g. study of periodicity and stability [100].

Some of the interesting open problems in epidemiology with potential applications in malware diffusion modeling and analysis for communications and computer complex networks include the following:

- As with many real viruses, diffusion might happen over diverse populations. Thus, the new epidemics models must consider heterogeneous populations, divided into subpopulations or groups given the similarities of the corresponding users, and then extend the aforementioned models for obtaining accurate descriptions thereof.
- Another issue to consider is that of proportionate mixing with multiple interacting groups of nodes/users. This refers to the case where the mixing (interactions) between nodes/groups of nodes is not uniformly homogeneous, but rather depends on the size of the interacting groups. In this case, the basic reproduction number is the contact number, namely, the weighted average of the contact numbers of the groups.

In the main part of the book, some of these open problems are tackled, while some others are not. We discuss the latter problems in more detail in Part 3 of the book in Chapter 10, where potential approaches to tackle these problems are identified.

PART

State-of-the-art malware modeling frameworks

2

CHAPTER 4

Queuing-based malware diffusion modeling

4.1 INTRODUCTION

As mentioned in Chapter 1, the wireless communications market has expanded massively in the last decade, constituting nowadays the most preferred way for Internet access by users around the world. Following suit, wireless services and applications, software and wireless devices' technologies have also proliferated, orders of magnitude compared to the ones available ten years ago. Unfortunately, since the emergence of the first computer virus and the corresponding malicious software (malware) targeting mobiles (cabir bluetooth virus in 2004 [221], also see Section 2.1.2 and 2.2 for more examples and details), malware spreading in wireless networks has exhibited exponential growth (see also [179] and references therein).

Motivated by this observation, and the general trend of increasing malware incidents, the last decade witnessed a broad attempt to obtain alternative modeling approaches compared to those presented in Part 1 of the book. The new attempts have strived to provide more generic malware diffusion modeling frameworks that are capable to holistically describe various aspects of malware diffusion, e.g. propagation and spreading, dynamic and static legitimate networks involved, and for diverse application domains. More importantly, these alternatives attempted to capture the stochastic nature of malware diffusion and possible attack strategies employed by malicious sources, as the approaches presented in Part 1 (Chapter 3) were not capable of doing so.

In this chapter, and in general in Part 2 of the book, we present several of these alternative and state-of-the-art modeling frameworks devoted to holistically modeling the important aspects of malware diffusion in various types of complex networks. We will focus especially on those developed recently (within the last decade), which are also founded on solid mathematical grounds, not employed before for modeling malware diffusion. In this chapter, we focus on those approaches exploiting elements of queuing theory and more specifically, elements from the theory of closed queuing networks. For those readers not familiar with queuing theory, Appendix B presents a quick reference to the general theory of Markov processes and queuing, while Appendix B.6 a more focused treatment of the theory of closed queuing networks, applicable to the technical content of this chapter.

The approaches presented in this chapter allow modeling both spreading and propagation malware dynamics, in wireless distributed (and thus centralized as well)

63

networks, including both static and dynamic topologies with network (node/edge) churn (dynamic networks where nodes enter/leave the network and/or links may be deleted/added). In the following, we start with the introduction of the basic concept behind this family of malware diffusion modeling approaches and then present the generic modeling framework that is based on queuing theory. Subsequently, we present analytic and numerical results demonstrating the use of the framework.

4.2 MALWARE DIFFUSION BEHAVIOR AND MODELING VIA QUEUING TECHNIQUES

4.2.1 BASIC ASSUMPTIONS

An important key observation regarding malware diffusion dynamics, in general, is that the time spent by attacked and infected legitimate nodes in the states defined by the corresponding malware infection model, i.e. susceptible (S), infected (I), recovering (R), dead (D) state, is inherently characterized by a stochastic nature, because it is influenced by random events. Indeed, legitimate users remain in each state until the arrival of a certain event triggers a transition to another state. For instance, an infection signifies a transition from the susceptible to the infected state and a recovery a transition from the recovering to the susceptible state again. This observation motivates the use of the theory of stochastic processes in general [187], and of queuing theory [25, 229] in particular, for quantifying and describing this behavior of the network and malware diffusion dynamics.

The waiting in each state and the state-transition triggering events themselves are usually of stochastic nature, thus constituting the overall malware diffusion process probabilistic as well. This observation regarding the waiting times spent in states of an infection model is one of the basic reasons that stochastic modeling of malware diffusion (presented in this and the next chapter) seems more suitable than the plain probabilistic approaches presented in Part 1 (Chapter 3). In the latter, only the probabilistic interaction between malware-legitimate nodes is captured through the contact probability (probability that a legitimate user is infected when in contact with a malicious node) or the overall link infection probability. In the techniques presented in this chapter, the stochastic waiting time of legitimate nodes is captured completely allowing for more holistic results and extending the study significantly to analysis of attack strategies, countermeasures, etc.

The basic concept will be illustrated by application. A network of legitimate nodes is considered, following the SIS infection paradigm [178], explained in Section 2.3. Under this general regime, it is initially assumed that legitimates node can be in one of the two states, namely, noninfected (susceptible) or infected. Other intermediate states, e.g. recovering, dead, and recharging, can be considered in general, but for the sake of simplicity they will not be regarded in the initial introduction of the framework.

Once infected, a legitimate node will go through a process of recovery and then eventually return to the noninfected state. Due to the existence of multiple mobile attackers (bearing probably different types of malware), a recovered node is prone to become infected again, by a different or possibly the same (e.g. when not patched properly) malware threat. As already explained in Chapter 2 (Section 2.3), such assumption is suitable for studying the long-term behavior and evolution of such a network, in which numerous and different malware pieces are diffused and legitimate nodes can potentially become infected by various threats in the course of their lifetime.

In order to study this macroscopic behavior, for the case of distributed wireless networks it will be initially assumed that no energy restrictions apply for the duration of the corresponding study. Then the presented framework will be extended to cover the cases of dynamic topologies as well (where energy or other factors lead to topology variations — nodes/edges added/deleted from the network). This assumption might seem counter-intuitive to many attacks that aim at exactly depleting the energy resources of wireless nodes. However, we note that such behavior can be taken into account when required, through the incorporation of additional states, such as the dead state of nodes. Furthermore, the consideration of dynamic topologies later in the chapter takes into account this scenario as well. In the basic model governing the macroscopic characteristics of malware propagation under the SIS paradigm, auxiliary states and their effect have been all cumulatively taken into account in a combined "infected" superstate for each legitimate node. Besides this simplification, initially not considering energy constraints has additional merits, since this corresponds to removing the impact that purely networking functions would have on the lifetime of nodes individually and the network cumulatively. Thus, such approach enables studying the malware diffusion problem focusing purely on that, removing any "networking" biases.

The network is assumed to consist of N legitimate nodes and M attackers, $1 \leq M, N$, both types initially having a common transmission radius R. In practical situations, it is understood that the number of attackers is much smaller than the number of legitimate nodes, leading to the additional constraint $M \ll N$. With respect to node deployment, it is assumed that initially all nodes are randomly and uniformly spread over the network region. In terms of mobility, both legitimate and attack nodes are assumed to move around the network area following the random walk mobility model with wrapping [37, 38], which maintains the uniformity of node distribution.

Depending on the transmission radius and mobility model, the link connectivity of a decentralized wireless complex network varies significantly over time, affecting its topology. In this chapter, it is assumed that two nodes are connected whenever each one lies in the other's transmission radius (corresponding to a specific transmission power). Therefore, the topology of the considered decentralized wireless complex network at each time instant can be completely determined by the relative positions of the nodes and their transmission powers/radii.

The above observations imply that a random geometric graph [180] representation of a wireless network can be employed, which is a common practice in the literature [85, 220]. Topology control capabilities are also considered for both legitimate and

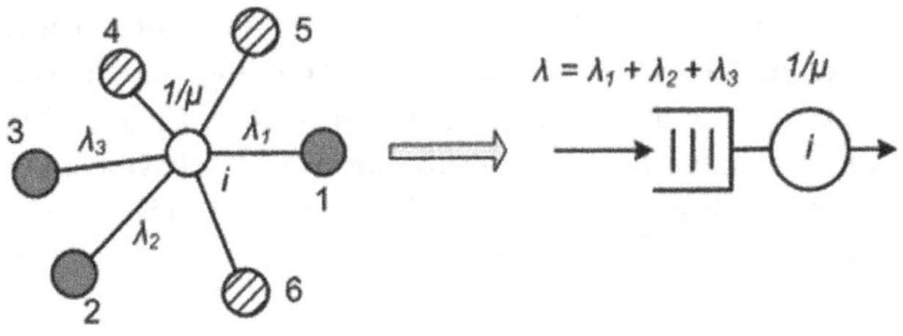

FIGURE 4.1

Mapping of malware diffusion problem to the behavior of a queuing system. The shaded nodes are susceptible legitimate neighbors of node i. The colored nodes are either malicious nodes or legitimate already infected neighbors of i. Node i is considered susceptible at the moment.

attack nodes in the general case and will be utilized as specified in each of the applications that are described in more detail in the sequel.

4.2.2 MAPPING OF MALWARE DIFFUSION TO A QUEUING PROBLEM

A legitimate user/node might have multiple links with other legitimate nodes, but also with infected neighbors and/or the attacker (as shown in the left-hand side of Fig. 4.1). This user is prone to receive infections from any of the links with attack or infected neighbors. For instance in Fig. 4.1, node i is susceptible to receive malware from any of the nodes 1, 2, 3. Moreover, these infections are assumed to occur stochastically, while the recovery process is of similar nature. Following the paradigms employed extensively in the literature [80, 117, 118, 124, 133], it is assumed that the infection arrival process over each link of a legitimate node is Poisson, while the recovery process of each network node takes exponentially distributed time intervals. Symbols λ_i and μ_j are used to denote the infection arrival rate on link with node i and the recovery rate of node j, respectively. Both infections and recoveries are independent of each other and between different instances of the same kind. Thus, in Fig. 4.1, node i is prone to receive malware from nodes 1, 2, 3, with exponentially distributed rates λ_1, λ_2, and λ_3. The total infection process is also exponentially distributed with rate $\lambda = \lambda_1 + \lambda_2 + \lambda_3$, as shown in Fig. 4.1.

The last result is a consequence of a very important feature of the exponential distribution. Let $X_1, ..., X_n$ be independent exponentially distributed random variables with rate parameters $\lambda_1, ..., \lambda_n$. Then $\min\{X_1, ..., X_n\}$ is also exponentially distributed, with parameter $\lambda = \lambda_1 + \cdots + \lambda_n$ [174]. Furthermore, the index of the variable which achieves the minimum is distributed according to the law $\Pr(X_k = \min\{X_1, ..., X_n\}) = \frac{\lambda_k}{\lambda_1 + \cdots + \lambda_n}$. Both of these results have direct applicability in the malware mapping to queuing. The first asserts that the total rate in the queue of Fig. 4.1 is $\lambda = \lambda_1 + \lambda_2 + \lambda_3$, since this rate signifies how fast the earliest infection arrives. The more the neighbors, the higher λ is and thus more infections are expected in a time unit. The second result indicates which infected node (1, 2, or 3) will produce the fastest infection arriving at node i in Fig. 4.1, and thus can be very useful for simulating the behavior of the obtained queuing system and implicitly the behavior of the malware attack.

The model described above (shown in the left-hand side of Fig. 4.1) is a node-specific malware diffusion model. By observing that the evolution of such model is essentially dictated by the time intervals spent by each node in different infection states, this model can be mapped to the operation of an $M/M/1$ queue (shown in the right-hand side of Fig. 4.1), where each infection corresponds to an arrival of a customer that requires service, and the service itself corresponds to the recovery procedure [117–119, 124].

It should be noted that in reality, infection attempts from multiple sources would proceed immediately. However, seen from an infinitesimal time point of view, even multiple infections come at distinct time instants and the above modeling exploits this observation. In addition, once successful, again from an infinitesimal time perspective, all infections are dealt with sequentially, even though at a possibly very fast rate. As long as there are processes awaiting in the queue (corresponding to multiple malware infections) the server is occupied, which expresses in queuing terms the fact that the node is in the infected state. As soon as the queue is exhausted, the last remaining process will move for service in the server. If no additional arrival (infection) takes place, once this process (infection) completes processing, the node becomes idle (it becomes noninfected) and transitions to the susceptible state. In this mapping, both arrivals and departures are independent and exponentially distributed. In the sequel, it is shown how this mapping can be used to model the aggregate behavior of the legitimate network with respect to malware diffusion over decentralized network.

4.3 MALWARE DIFFUSION MODELING IN NONDYNAMIC NETWORKS

Under the employed SIS model, at every time instant, legitimate nodes can be separated into two categories, those that are infected and those that are noninfected (susceptible). Taking into account that each of these legitimate nodes can be represented as an $M/M/1$ queue, as described above, the whole of the legitimate network can be represented by a closed multiqueue network, as shown in Fig. 4.2(a).

Without loss of generality, the queues of the upper part represent the group of noninfected (susceptible) nodes, while the queues of the lower part represent the group of infected nodes. Each node has its own queue in the susceptible and infected node group separately, but at every instant a specific node is actually in only one of the two groups, either the infected or the susceptible, depending on whether the node's state is infected or susceptible, respectively. By this mapping, the group of

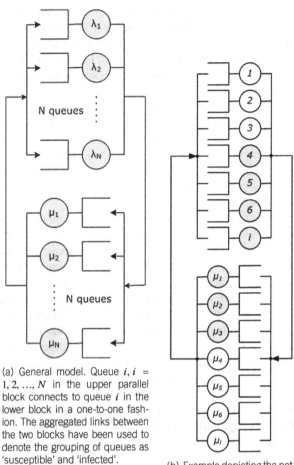

(a) General model. Queue i, $i =$ 1, 2, ..., N in the upper parallel block connects to queue i in the lower block in a one-to-one fashion. The aggregated links between the two blocks have been used to denote the grouping of queues as 'susceptible' and 'infected'.

-Courtesy Karyotis V, Kakalis A, Papavassiliou S. Malware-propagative mobile ad hoc networks: asymptotic behavior analysis. Springer J Comput Sci Technol 2008;23(3):389–99. With kind permission from Springer Science and Business Media.

(b) Example depicting the network for the scenario shown in Fig. 4.1.

FIGURE 4.2

Closed queuing systems modeling malware diffusion over a wireless SIS network.

susceptible nodes and similarly the group of infected nodes, each corresponds to an $M/M/N$ sub-system, as shown in Fig. 4.2(a), where the service rate of each queue is set according to the state of the node (infected or susceptible), the link infection or recovery rate, and topological parameters. For wireless multihop networks, the service rate depends on the node transmission radius and the number of neighboring attackers/infected legitimate nodes for each node separately. Thus, if a node i is noninfected, it will lie in the upper group and its service rate will be equal to the total infection rate for this node, determined by the number of links that the node has with malicious nodes (being equal to the sum of the corresponding link infection rates, as explained in the previous subsection). The corresponding recovery rate for this node, applicable to the bottom "infected" group will be naturally $\mu_i = 0$, because the node is noninfected (i.e. in the upper group). It should be also noted that the $M/M/1$ model described previously refers to the infection-recovery process of a single node, while the closed queuing network of $M/M/N$ queues refers to the aggregated behavior of the network of legitimate nodes. For a more specific example of what the whole network of queues looks like, Fig. 4.2(b) presents the queuing network derived for the simple scenario depicted in Fig. 4.1. The shaded servers indicate which of them are active at the moment and the state in which the corresponding node lies. Thus, nodes $1, 2, 3$ are infected and their queues in the lower block are active, while nodes $4, 5, 6, i$ are susceptible and their queues in the upper block are active.

Obtaining the steady-state distribution of the closed queuing system with the two $M/M/N$ blocks directly is cumbersome and in many cases intractable. The $M/M/N$ queuing model contains the complete information about the state of the system, i.e. how many nodes and which of them specifically are infected, but this comes at the cost of analytical tractability. One approach to deal with this is to sacrifice some of the complete information, e.g. focus only on the cumulative number of infected nodes rather than on which nodes are currently infected, for the favor of analytical simplicity. To achieve this, one can employ the Norton equivalent of the queuing system [152, 194]. The Norton equivalent of a queuing network is an aggregation method that allows the parametric study of complex networks of queues satisfying several conditions. It is an extension of Norton's theorem from circuit theory [44, 112], where each server is considered as a "node" and connections between servers as "wires." The aggregation procedure is the "shorting" of a subsystem or subnetwork (in the same way a part of a circuit is short-circuited in circuit theory), and obtaining the equivalent throughput or other equivalent quantities of this "shorted" subsystem.

When applying the Norton equivalent in the above case, each of the $M/M/N$ subsystems is substituted with an $M/M/1$ subsystem, with total rate equivalent to the rate of the original. In this case, where the queues of each subsystem are parallel and due to the fact that all service rates are exponentially distributed, the rate of the equivalent subsystem will be the sum of the corresponding rates. Consequently, the Norton equivalent of the original system will have the form shown in Fig. 4.3 [117, 118]. Explicit definition of each queue's service rate, i.e. $\lambda(N - k)$ and $\mu(k)$ for a total of N nodes, depends on the specific model parameters and assumptions and constitutes a necessary step for solving the system explicitly. More details will be provided in the rest of this chapter. The "short-circuit" approach for queuing systems holds only for closed "product-form" networks, which is true for the analyzed network and the corresponding proposed closed queuing network model, as can be

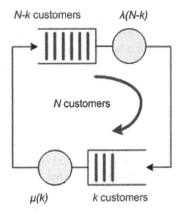

FIGURE 4.3

The Norton equivalent model for malware
propagation in communications networks.
The figure shows the instance where k nodes
are currently infected.

readily seen by the form of expression (4.1) that will be derived in Section 4.3.2. The
detailed methodology for analyzing the Norton equivalent, which in this scenario is
a two-queue closed network, is provided in Appendix B.

As already mentioned, from a macroscopic perspective in the attacked network
legitimate nodes can be segregated in two groups, i.e. susceptible and infected. A
specific node oscillates between the two groups, due to the arrivals of new malware
and corresponding recoveries from the infected state. Thus, each node exhibits a
similar behavior to a customer in closed queuing network as the one depicted in
Fig. 4.3. Thus, one obtains a mapping where the Norton equivalent of the original
queuing system consists of a two-queue closed queuing network, where N customers,
corresponding to legitimate nodes, circulate. Each of these N customers can be in one
of the two distinct system components, either waiting/processing in an upper queue,
or waiting/processing in one queue of the lower $M/M/N$ block. At any instant, if
k nodes are infected, then $N - k$ will be noninfected, since the original network is
considered static. Both service rates are state-dependent according to the number of
customers (nodes) in the corresponding queue. Without loss of generality, we assume
that the lower queue represents the group of infected legitimate nodes and denote it
as "infected," while the upper queue represents the noninfected nodes and we denote
the queue as "noninfected" (susceptible). Furthermore, using the Norton equivalent
retains any topology/mobility information in the network, which will appear in the
analytic expressions for the networks studied in the sequel. The only information loss
suffered is which nodes exactly are infected/susceptible at each time. Rather than

knowing the exact state of each node, we only know how many are in the infected state and subsequently how many remain in the susceptible state.

The closed two-queue model emerges frequently in other application domains too as can be found in the literature, e.g. [25, 115]. The mapping between the N customers of a communications network and their waiting/processing and the legitimate users of a communications network and their susceptible/infected states is now fairly straightforward. Each of the N customers corresponds to a specific user of the legitimate network nodes. When in the upper queues the corresponding customer (legitimate node) is in susceptible state, while it is in the infected when in the lower queues.

Since the number of legitimate nodes in the network remains fixed,[1] the state of one queue is dependent on the state of the other through the total number of customers (i.e. legitimate nodes) in the system and it is sufficient to focus on the one-queue state dynamics for the analysis.

4.3.1 EVALUATION METRICS

Before we present the analysis and study of the steady-state behavior of the previously obtained Norton equivalent, in this subsection we explain some of the evaluation metrics of interest for the analytical models presented. The presented metrics are indicative of the long-term expected behavior of the systems analyzed as will become apparent. They can be used for assessing the overall robustness capabilities of the attacked network, contrary to more specific and time-dependent metrics employed in the literature, e.g. percentage of susceptible or infected nodes.

The network damage caused by malware spreading can have various undesirable forms, such as host unavailability, data corruption, or machine breakdown. In any case, the host node essentially becomes nonoperational, or at least of reduced performance. From a network perspective, we focus on such unavailability and performance degradation, and consider that the imminent result of an infection actually means that an infected node remains nonoperational until it recovers. In that sense, a general attack-evaluation metric that captures the overall impact caused by a group of attackers to the network is a combined metric for the number of infected nodes over the time interval under consideration. The quantity that effectively indicates this impact throughout the system observation period is the average number of infected nodes, since it represents the expected long-term (steady-state) network damage caused by the attackers.

With respect to the Norton equivalent model, the average number of infected nodes in the network equals the average number of customers in the lower queue of Fig. 4.3. The average number of infected nodes will be denoted by $E[L_I]$ (L_I is used to denote the number of customers in the infected queue and L_S will be used to denote the number of customers in the susceptible queue), and will be calculated later in Section 4.3.2 (given by Eq. (4.11)), while the average number of customers in the

---

[1]Energy considerations are not currently taken into account for wireless nodes, i.e. nodes are assumed to have sufficient energy reserves for the duration of the observation, as explained in Section 4.2.1.

noninfected (susceptible) queue can be then obtained as $E[L_S] = N - E[L_I]$, since the sum of the average number of infected and the average number of susceptible nodes remains constant and equal to N.

Furthermore, the throughput of a queue is indicative of the long-term instantaneous service potentials of the system, i.e. the instantaneous service potentials experienced in the steady-state of the system. Similarly, the average throughput represents the average rate of customers served in the steady-state. In terms of the Norton equivalent system of Fig. 4.3, the throughput of the noninfected queue is descriptive of the total infection rate of the malicious nodes experienced in the steady-state. In other words, the throughput of the noninfected queue indicates the long-term capability of an attack strategy to infect legitimate nodes, and therefore the average throughput of the susceptible queue, denoted by $E[\gamma_S]$, obtained in Section 4.3.2 (provided by expression (4.12)), is indicative of the average capability of the group of malicious nodes to infect legitimate network nodes. Here, we use $E[\gamma_S]$ as a second attack-evaluation metric and present relevant results in the following section. It can be observed that by the invariance of input/output rates in each of the two queues of the coupled system one may obtain $E[\gamma_S]$ in terms of $E[L_I]$, as will be shown in the next subsection.

The above metrics will be employed in the following subsections and they are provided for both propagative and nonpropagative networks under attack. Other similar, long-term characterization metrics could have been also employed, depending on the specific application frameworks and the objective of each analysis.

4.3.2 STEADY-STATE BEHAVIOR AND ANALYSIS

In this subsection, we present a detailed analysis of the Norton equivalent system and without harming generality we focus on the infected queue. Its steady-state distribution, denoted by $\pi_I(i)$, coincides with the behavior of the system observed after sufficient operational time, and represents the probability that there are i customers (i.e. legitimate nodes) in this queue (state). Considering the underlying continuous-time Markov chain, the state diagram shown in Fig. 4.4 describes the possible state transitions of the associated discrete (embedded) Markov chain. Using balance equations between state $i-1$ and state i, we can obtain the general expression,

$$\lambda(i-1)\pi(i-1) = \mu(i)\pi(i),$$

yielding $\pi(i) = \frac{\lambda(i-1)}{\mu(i)}\pi(i)$ and eventually

$$\pi(i) = \pi(0) \prod_{j=1}^{i} \frac{\lambda(j-1)}{\mu(j)}.$$

Then, taking into account the state of the system in the scenario of malware diffusion and expressing the infection rate in the numerator in terms of the actual number noninfected nodes in the corresponding state as $\lambda(N - j + 1)$, the accurate expression

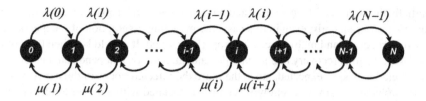

FIGURE 4.4

State diagram for the analysis of the two-queue closed network and for obtaining the expression of the steady-state distribution.

can be substituted in the place of $\lambda(j-1)$ in the product above, yielding the explicit expression for the steady-state distribution

$$\pi_I(i) = \pi_I(0) \cdot \prod_{j=1}^{i} \frac{\lambda(N-i+j)}{\mu(j)}, \tag{4.1}$$

where $\pi_I(0)$ is the probability of no infected nodes in the network. The derivation of the above expression for the general case can be found in Appendix B.6.1. By applying the normalization condition $\sum_{i=0}^{N} \pi_I(i) = 1$, one gets

$$\pi_I(0) \cdot \sum_{i=0}^{N} \prod_{j=1}^{i} \frac{\lambda(N-i+j)}{\mu(j)} = 1. \tag{4.2}$$

For simplicity and without loss of generality, since malicious nodes have been assumed uniformly spread in the network it can be assumed that all legitimate network nodes have the same link infection rate λ, i.e. $\lambda_k = \lambda, \forall k \in \{1, 2, ..., N\}$, and the same recovery rate μ, i.e. $\mu_k = \mu, \forall k \in \{1, 2, ..., N\}$. This is a key assumption for the correctness of the aggregate expressions for the analysis of the Norton equivalent two-queue network in the sequel.

Consequently, the state-dependent service rate in the state having i infected nodes becomes

$$\mu(i) = \sum_{k=1}^{i} \mu_k = i\mu. \tag{4.3}$$

Thus, the total recovery rate is $\mu(i) = i\mu$, and this holds for both nonpropagative and propagative networks since the recovery process is independent of the malware diffusion process, at least for the systems considered in this book. Of course, in some cases, it is possible that the recovery depends on the specific malware type, i.e. when the malware is capable of completely taking over the control of the infected machine. However, even in this case, the user holds physical control of a machine, and sooner or later, the device can be unplugged from the network, thus regaining at least partially the control of the machine, and with that, regain the capability to recover the machine. Consequently, even in this case, the modeling assumption of independence between recovery and malware can be properly justified. It should be noted that from Eq. (4.3), the total recovery rate is state-dependent, namely, it depends on the number of infected nodes, as expected, since the more the infected nodes in the network, the more probable a recovery event will become (associated with a higher total recovery rate).

However, regarding the total infection rate, the situation is not the same and in this case, one needs to distinguish between malware propagative and malware nonpropagative networks. In the latter, the infection rate will depend only on the attack nodes, while in the first the total infection rate will depend on both attack and legitimate nodes that are already infected by attackers. In the following, we describe the two cases separately, starting with the analytically simplest one, namely, nonpropagative networks.

Malware spreading (nonpropagative) networks

In the nonpropagative types of networks, where spreading of malware takes place, malware spreads only through dedicated attack nodes (which cannot be distinguished from legitimate nodes in terms of capabilities in the general case) [118]. This means that infected legitimate nodes are incapable of further propagating the malware they receive. Such approach models the cases where malware is relatively simple and unable to self-propagate by exploiting the infected host. Thus, only infected legitimate nodes suffer from malware and susceptible ones can only receive malware through attackers.

In such cases, the total infection rate $\lambda(N - i)$ in the Norton equivalent model described above depends on the total number of attackers existing in the deployment region of the network. In this case, the total infection rate can be computed as

$$\lambda(j) = \sum_{m=1}^{j} k_m \lambda_m = \lambda \frac{\sum_{m=1}^{j} k_m}{j} j,$$

for j susceptible nodes, where k_m is the number of attackers within the neighborhood of a legitimate node indexed by m. Since the sum of all k_m terms divided by the total number of summands yields the average number $K \equiv \frac{\sum_{m=1}^{j} k_m}{j}$ of malicious attackers perceived by legitimate nodes on average, the above expression becomes $\lambda(j) = \lambda K j$. It should be noted that this average number of malicious attackers K is invariant across the network, due to the assumption of the uniform spread of attackers in the network, and its computation depends only the number of attackers and network topology parameters. The expression in terms of j infected nodes would become $\lambda(j) = \lambda K(N - j)$. Then, the total infection rate, namely, the average total infection rate in the left-hand side of Fig. 4.1, is obtained as

$$\lambda(N - i + j) = \lambda K(N - i + j). \tag{4.4}$$

Using this expression for the total infection rate and after some algebraic manipulations, relation (4.2) becomes

$$\pi_I(0) \cdot \sum_{i=0}^{N} \left[\binom{N}{i} \left(\frac{\lambda K}{\mu} \right)^i \right] = 1. \tag{4.5}$$

Using the binomial identity, $\sum_{k=0}^{n} \binom{n}{k} p^k q^{n-k} = (p + q)^n$ in the above expression, $\pi_I(0)$ is obtained as

$$\pi_I(0) = \frac{1}{\left(1 + \frac{\lambda K}{\mu} \right)^N}. \tag{4.6}$$

Thus, the steady-state distribution in (4.1) can be seen equal to

$$\pi_I(i) = \frac{1}{\left(1 + \frac{\lambda K}{\mu} \right)^N} \binom{N}{i} \left(\frac{\lambda K}{\mu} \right)^i = \pi_S(N - i), \tag{4.7}$$

where $\pi_S(N - i)$ is the steady-state distribution for the noninfected queue. Using Eq. (4.7), the probability of a completely infected network is obtained as

$$\pi_I(N) = \frac{\left(\frac{\lambda K}{\mu} \right)^N}{\left(1 + \frac{\lambda K}{\mu} \right)^N}. \tag{4.8}$$

Expression (4.7) is a binomial distribution with parameters $B\left(N, \frac{\lambda K/\mu}{1+(\lambda K/\mu)} \right)$, clearly indicating the critical parameters that affect the behavior of the system. Assuming a fixed area of the network deployment region, the densities of the legitimate and malicious nodes (K depends on the number of attackers) along with the ratio of the link infection rate to the recovery rate are decisive in the stability and overall performance of the system.

Without loss of generality, considering a square deployment region ($L \times L$), and since the malicious nodes are uniformly and randomly deployed over the network region, the average number of neighboring attackers for each network node is computed as

$$K = \frac{\pi R^2}{L^2} \cdot M = \pi M \left(\frac{R}{L} \right)^2, \tag{4.9}$$

where R is the transmission radius and M the number of attackers. For this particular system model, where a wireless *ad hoc* network is analyzed, K depends also on the transmission radius of the device of a legitimate user, thus making it a critical factor as well.

Based on (4.7), we can calculate the average number of infected nodes (expected number of customers in the queue of infected nodes) $E[L_I] = \sum_{i=1}^{N} i \cdot \pi_I(i)$, as follows:

$$E[L_I] = \frac{\lambda N K}{\mu + \lambda K}. \tag{4.10}$$

Substituting the value of K by Eq. (4.9) in the latter, the average number of infected nodes in an *ad hoc* network of N nodes and M attackers is obtained in the general case

$$E[L_I] = \frac{\lambda \pi M N \left(\frac{R}{L}\right)^2}{\mu + \lambda \pi M \left(\frac{R}{L}\right)^2}. \tag{4.11}$$

Also, for the average throughput of the noninfected queue $E[\gamma_S] = \sum_{i=1}^{N} \lambda(i) \cdot \pi_S(i)$, and by the invariance of input/output rates in each of the two queues of the coupled system one always has $E[\gamma_S] = \sum_{i=1}^{N} \lambda(i)\pi_S(i) = \sum_{i=1}^{N} \mu(i)\pi_I(i)$. So, when $\mu(i) = i\mu$, one obtains $E[\gamma_S] = \mu E[L_I]$. Consequently, the expression for the average throughput of the susceptible queue is

$$E[\gamma_S] = \frac{\lambda \mu \pi M N \left(\frac{R}{L}\right)^2}{\mu + \lambda \pi M \left(\frac{R}{L}\right)^2}, \tag{4.12}$$

as would be verified by analytical computations of $E[\gamma_S] = \sum_{i=1}^{N} \lambda(i) \cdot \pi_S(i)$. The latter expression corresponds to the average cumulative infection rate of nodes in the network. Thus, the average throughput of the susceptible queue is the cumulative average rate at which nodes in the network become infected, represented by a transition from the susceptible to the infected queue. In terms of the Norton equivalent, this corresponds to the average throughput of the susceptible queue, as already mentioned.

Numerical results: Malware spreading

In this subsection, some numerical results that demonstrate the dependence of the average number of infected nodes and the average throughput of the noninfected queue on the various system parameters are presented.

The square deployment region is considered to have a side length of $L = 1500$ m, while the wireless nodes are all assumed to have a common transmission radius of $R = 200$ m. Throughout these evaluation results, the number of legitimate nodes N, the number of attack nodes M, and the infection and recovery rates λ and μ, respectively, are varied.

Initially, the behavior of the system at the two extremes is analyzed, namely, either when no node is infected (probability $\pi_I(0)$) or when all nodes are infected (probability $\pi_I(N)$). Fig. 4.5 shows the trend followed by the probability $\pi_I(0)$ of no infected legitimate nodes. More specifically, Fig. 4.5(a) depicts the dependence on the ratio of link infection and recovery rates λ/μ for three combinations of values of N and M, while Fig. 4.5(b) depicts the dependence on N, M for $\lambda/\mu = 0.5$. In Fig. 4.5(a), the two curves for $N/M = 100/50$ and $200/25$ almost overlap. This is because in the plotted range, $\lambda/\mu \ll 1$, so approximately $\pi_I(0) = \frac{1}{1+NM\pi(R/L)^2}$ and the probability depends (approximately) only through the product NM, which is the same for both curves. The probability of no infected nodes, given by relation (4.6), decreases rapidly with increasing values of λ/μ, since the higher the infection rate, the

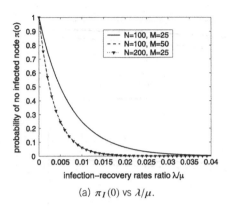

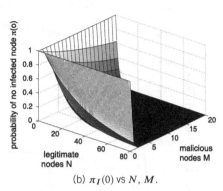

(a) $\pi_I(0)$ vs λ/μ.

(b) $\pi_I(0)$ vs N, M.

FIGURE 4.5

Probability of no infected nodes $\pi_I(0)$.

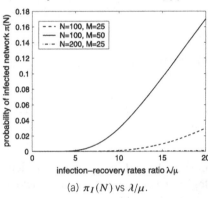

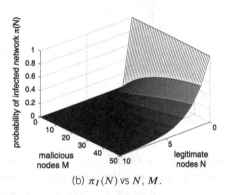

(a) $\pi_I(N)$ vs λ/μ.

(b) $\pi_I(N)$ vs N, M.

FIGURE 4.6

Probability of all nodes infected $\pi_I(N)$.

more likely it is that several nodes in the network will be infected for a period of time and thus $\pi_I(0)$ should be zero. As the number of attackers increases, $\pi_I(0)$ decreases faster, while a similar trend is revealed by Fig. 4.5(b). The probability of no infected nodes depends exponentially on the number of network nodes (for constant number of attackers M) and it has a decreasing behavior for increasing network densities, since in this case, more nodes are prone to become infected and thus the value of $\pi_I(0)$ is further suppressed.

Complementary results, but in the same direction, hold for the probability of a completely infected network $\pi_I(N)$ (Eq. (4.8)), shown in Fig. 4.6. In this case, the probability $\pi_I(N)$ increases nonlinearly with the ratio λ/μ, but only for values in the order of a decade and higher. It can be noticed in Fig. 4.6(a) that the curve corresponding to $N = 200$, $M = 25$ is practically zero and it would lead to nonzero

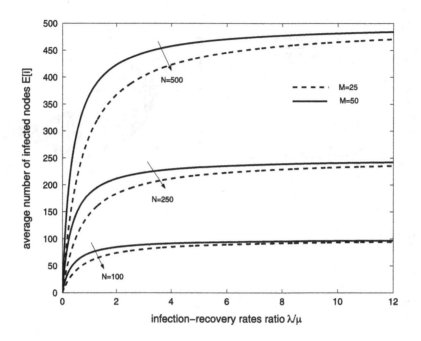

FIGURE 4.7

Average number of infected nodes $E[L_I]$ versus λ/μ.

values only for much larger values of λ/μ. Similarly, the probability of a completely infected network approaches unity for topologies where the density of malicious nodes is much larger than that of the network nodes, (Fig. 4.6(b)).

Fig. 4.7 shows the average number of infected nodes in the network, given by relation (4.11), with respect to the ratio λ/μ of the link infection rate to the recovery rate. The figure depicts two different scenarios with respect to the number of malicious nodes, one with $M = 25$ nodes and one with $M = 50$ nodes, in order to investigate the impact of the density of malicious nodes on the spreading of malware. Also, three different values of legitimate nodes have been used, i.e. $N = 100$, 250, or 500 nodes, representing a sparse, medium, and dense network topology, respectively. Fig. 4.7 reflects the trend of increasing $E[L_I]$ as the ratio of the infection-recovery rates increases. This trend is expected, since the greater the link infection rate is with respect to the recovery rate, the more time a node will spend in the infected state and thus the greater the average number of infected nodes will be. The trend is similar for various densities of attackers; however, the greater the number of attackers, the greater $E[L_I]$, as expected, since larger values of M mean that more attackers lie in the vicinity of a randomly selected node and thus, the probability of infection of the node increases. Nevertheless, the more dense the network is, the greater the significance of the number of attackers.

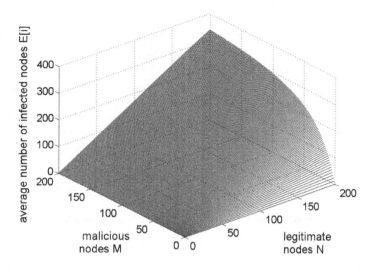

FIGURE 4.8

Average number of infected nodes $E[L_I]$ versus legitimate N and malicious M nodes.

Fig. 4.7 also reveals the trend of the average number of infected nodes to increase rapidly as the ratio λ/μ increases. As the number of infected nodes increases, it approaches asymptotically the total number of legitimate nodes N (upper bound). This means that there exists a value of the ratio λ/μ beyond which the network damage does not increase significantly. Thus, additional effort by the attacking nodes to increase λ (assuming a constant recovery rate) is meaningless.

The dependence of $E[L_I]$ on the sets of legitimate or malicious nodes is evident in relation (4.11) and depicted in Fig. 4.8. By inspection of the figure, the linear dependence of the average number of infected node on the number of legitimate nodes N and the nonlinear dependence on the number of malicious nodes M is evident. This means that the expected network damage will be different for the same increase in node density of legitimate and malicious nodes. Specifically, an increase in the number of attackers would result in greater damage than the damage corresponding to the same increase in legitimate network node density (assuming the density of the other group remains the same). From the attackers' viewpoint, a small increase in the number of malicious nodes should make the network more vulnerable than if legitimate network density increased by the same amount.

Fig. 4.9 shows the dependence of the average throughput of the noninfected queue on the legitimate and malicious nodes (relation (4.12)). It is similar to that of the average number of infected nodes in the network and thus, the same observations apply. Again, the dependence on M (N) is nonlinear (linear). Thus, a greater number of attackers constitute a greater threat for the vulnerability of a denser network of legitimate nodes than other parameters.

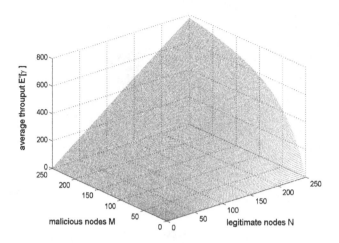

FIGURE 4.9

Average throughput of noninfected queue $E[\gamma_S]$ versus legitimate N and malicious M nodes.

However, as far as the dependence on the infection and recovery rates is concerned, the situation is different. As Eq. (4.12) shows, there is no absolute dependence on the ratio λ/μ. The average throughput will increase nonlinearly with respect to the link infection rate and the recovery rate, but the rate of increase is greater for the link infection rate λ. This can be observed in all scenarios shown in Fig. 4.10, where the curvature of the surface due to λ is greater. Moreover, in cases with greater numbers of attackers, the importance of λ is greater, as can be seen by comparing Fig. 4.10(a) and Fig. 4.10(b). This is not the case for denser networks of legitimate nodes, after comparing Fig. 4.10(a) and Fig. 4.10(c).

In Chapter 9, Section 9.1 results similar to the ones presented above will be exploited in order to study the long-term robustness capabilities of networks against intelligent topology-control [120, 191] based attacks, where attackers can vary their transmission radius via proper transmission power adaptations.

Summarizing the most important findings emerging from the numerical results presented for malware spreading in wireless distributed networks:

- The probability of no infected nodes depends exponentially on the number of network nodes (for constant number of attackers M) and it has a decreasing behavior for increasing network densities.
- The probability of pandemia (a completely infected network) tends to one for very dense networks and a number of attackers in the scale of the number of legitimate nodes.
- A greater number of attackers constitute a greater threat for the vulnerability of a denser network of legitimate nodes than other parameters involved.

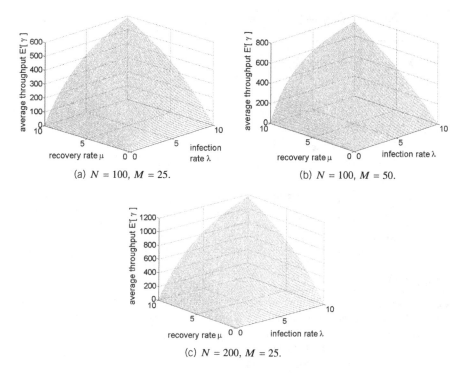

(a) $N = 100$, $M = 25$.   (b) $N = 100$, $M = 50$.

(c) $N = 200$, $M = 25$.

FIGURE 4.10

Average throughput of noninfected queue $E[\gamma_S]$ versus infection λ and recovery μ rate.

- Malware spreading in distributed networks exhibits threshold phenomena, i.e. there exists a value of the ratio λ/μ beyond which the network damage does not increase significantly.

Malware propagation (propagative networks)

As explained in previous subsections, the nodes of the legitimate propagative network can be mapped to customers circulating in a two-queue closed queuing network having the form shown in Fig. 4.3, where N customers, i.e. legitimate nodes circulate. Explicit definition of each queue's service rate, which is a function of the state of each queue ($\lambda(N - i)$ and $\mu(i)$ for i infected nodes), depends on the specific model parameters/assumptions and constitutes a necessary step for solving the system explicitly. As explained earlier, the state-dependent service rate of the noninfected queue depends not only on the number of currently noninfected nodes $N - i$, but the attacker as well, since even when all nodes are noninfected, infections are possible due to the attacker's existence. The notion of state-dependent service rate of the noninfected queue can be extended to the case of multiple attackers, where the service rate will obtain the form $\lambda(N - i + M)$, M being the number of attackers. In this work,

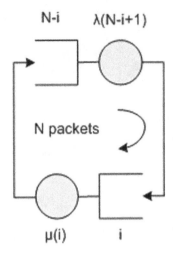

FIGURE 4.11

Norton equivalent of the closed queuing network model for a propagative system. Compared to Fig. 4.3, there is a difference in the
infection rate due to the impact of attacker.

-*Courtesy Karyotis V, Kakalis A, Papavassiliou S.*
Malware-propagative mobile ad hoc networks: asymp-
totic behavior analysis. Springer J Comput Sci Tech-
nol 2008;23(3):389–99. With kind permission from
Springer Science and Business Media.

we consider the case of a single attacker, i.e. $M = 1$, and the corresponding Norton
equivalent is shown in Fig. 4.11.

In this subsection, we present and specify in detail the parameters of the initially
introduced generic queuing model for propagative wireless networks and present
some interesting relevant results. We focus again on the infected queue of the Norton
equivalent. Its steady-state distribution, denoted by $\pi_I(i)$, represents the probability
that there are i customers (nodes) in this queue. Using a similar thread of analysis as
for the spreading case before, the explicit expression for the steady-state distribution
can be obtained as

$$\pi_I(i) = \pi_I(0) \cdot \prod_{j=1}^{i} \frac{\lambda(N + 1 - i + j)}{\mu(j)}, \tag{4.13}$$

where $\pi_I(0)$ is the probability of no infected nodes in the network and the infection
rate $\lambda(N + 1 - i + j)$ has been computed appropriately. Applying the normalization

condition $\sum_{i=0}^{N} \pi_I(i) = 1$ again, we obtain

$$\pi(0) \cdot \sum_{i=0}^{N} \prod_{j=1}^{i} \frac{\lambda(N+1-i+j)}{\mu(j)} = 1. \tag{4.14}$$

As before, all legitimate nodes are assumed to have the same link infection rate λ, and the same recovery rate μ. The total recovery rate of the infected queue will be $\mu(i) = \sum_k \mu_k = i\mu$. The total infection rate of the noninfected queue for i infected nodes is obtained similarly to the spreading case as follows:

$$\lambda(i) = \lambda K(i)(N-i),$$

where $K(i) = \pi\left(\frac{R}{L}\right)^2 (i+1)$ is the average number of infected neighbors for a susceptible node, which depends on the number of currently infected legitimate network nodes i and the one attacker. Note that this time the rate was expressed in terms of infected nodes rather than legitimate ones as it was the case previously for the spreading networks.

Given expression (4.13), the proper value of the total infection rate of the susceptible queue becomes

$$\lambda(N+1-i+j) = \lambda\pi\left(\frac{R}{L}\right)^2 (i-j+1)(N-i+j). \tag{4.15}$$

Using expression (4.15) for the total infection rate into relation (4.14) one obtains

$$\pi_I(0) \sum_{i=0}^{N} \prod_{j=1}^{i} \left[\frac{\lambda\pi}{\mu}\left(\frac{R}{L}\right)^2 \frac{(N-i+j)(i-j+1)}{j} \right] = 1.$$

By setting $\alpha^{-1} = \frac{\lambda\pi}{\mu}\left(\frac{R}{L}\right)^2$, the above expression becomes

$$\pi_I(0) \sum_{i=0}^{N} \prod_{j=1}^{i} \left[\frac{1}{\alpha} \frac{(N-i+j)(i-j+1)}{j} \right] = 1, \tag{4.16}$$

which can be reduced to

$$\pi_I(0) \cdot \sum_{i=0}^{N} \left[\left(\frac{1}{\alpha}\right)^i \frac{N!}{(N-i)!} \right] = 1. \tag{4.17}$$

Setting $N - i = k$ and changing the summation index to k, expression (4.17) becomes

$$\pi_I(0) \sum_{k=0}^{N} \left[\frac{1}{\alpha^{N-k}} \frac{N!}{k!} \right] = 1,$$

or equivalently

$$\frac{N!\pi_I(0)}{\alpha^N} \sum_{k=0}^{N} \frac{\alpha^k}{k!} = 1. \tag{4.18}$$

As the number of legitimate nodes becomes very large, the sum of expression (4.18) converges to the power series expansion of e^α. Thus, the probability of no infected nodes can be approximated as

$$\pi_I(0) \cong \frac{\alpha^N}{N!} \cdot e^{-\alpha}, \tag{4.19}$$

and the steady-state distribution can be obtained as

$$\pi_I(i) = \frac{\alpha^{N-i}}{(N-i)!} e^{-\alpha} = \pi_S(N-i), \tag{4.20}$$

where $\pi_S(N-i)$ is the steady-state distribution for the noninfected queue. From this expression, it can be readily seen that the distribution approximately follows a Poisson form with parameter α. Using relation (4.20), the probability of a completely infected network equals $\pi_I(N) = e^{-\alpha}$. The error incurred by the above approximation is negligible for values of α and N commonly used in practice. The impact of this approximation is more thoroughly investigated in the following subsection.

Expression (4.20) clearly indicates the critical parameters that affect the behavior of the system. Assuming a fixed area of the network deployment region, the number of legitimate nodes (i.e. the density of the network) along with the common transmission radius and the ratio of the link infection rate to the node recovery rate is decisive regarding the overall behavior and stability of the system.

Based on (4.20), the average number of infected nodes (expected number of customers), $E[L_I] = \sum_{i=1}^{N} i\pi_I(i)$, can be calculated as

$$E[L_I] = \sum_{i=1}^{N} \left(i \frac{\alpha^{N-i}}{(N-i)!} e^{-\alpha} \right) = e^{-\alpha} \sum_{k=0}^{N-1} (N-k) \frac{\alpha^k}{k!} = e^{-\alpha} \left(Ne^\alpha - \alpha \sum_{k=1}^{N-1} \frac{\alpha^{k-1}}{(k-1)!} \right),$$

eventually yielding (when N becomes large)

$$E[L_I] = N - \alpha = N - \frac{\frac{\mu}{\lambda\pi}}{\left(\frac{R}{L}\right)^2}. \tag{4.21}$$

This is a result which can be derived more easily since as already mentioned in Eq. (4.20) is approximately a Poisson distribution. Computing directly its expected value yields the same result.

Similarly, to the spreading case and by a stability argument of the two-queue closed Norton equivalent network, $E[\gamma_S] = E[\gamma_I]$. In fact, $E[\gamma_I] = \sum_{i=1}^{N} \mu(i)\pi_I(i) = \mu E[L_I]$. Thus, the average throughput of the susceptible and infected queues can be directly obtained as

$$E[\gamma_I] = E[\gamma_S] = \mu(N-\alpha). \tag{4.22}$$

Additional quantities of interest in the closed queuing networks can be computed and interpreted appropriately.

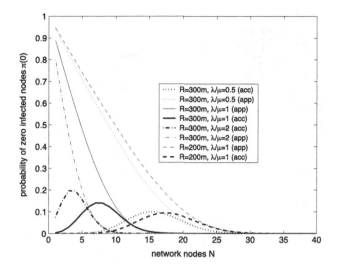

FIGURE 4.12

Probability of zero nodes infected $\pi_I(0)$ (accurate-approximated).

-Courtesy Karyotis V, Kakalis A, Papavassiliou S. Malware-propagative mobile ad hoc networks: asymptotic behavior analysis. Springer J Comput Sci Technol 2008;23(3):389–99. With kind permission from Springer Science and Business Media.

Evaluation of approximation

Throughout the rest of this subsection and without loss of generality it is assumed that the deployment network region is square with side $L = 1500$ m. By fixing the deployment area and varying the number of legitimate nodes, one can study the behavior of the system as the network density increases from values representing a sparse topology to values corresponding to massively dense networks. Varying also the transmission radius of the nodes, the system can be studied under topologies that vary from very strongly connected to relatively weakly connected ones.

As already mentioned, the analytical results presented so far describe the asymptotic behavior of the attacked network in the event of large-scale networks where the number of legitimate nodes increases to infinity, ideally $N \rightarrow \infty$. However, in practice, wireless distributed networks have a finite number of network nodes. In this subsection, in order to provide a realistic idea of the behavior of actual systems with respect to the controllable parameters and in terms of the attacker's impact the results are applied and evaluated on finite networks, investigating the impact of the approximation introduced in Eq. (4.20) and demonstrating that the error incurred by this approximation is negligible for values of the parameters R/L involved in α and values of N commonly used in practice. The results presented in this chapter have considered the values of these parameters within such typical ranges for the evaluation purposes of the study, and more details on the extreme values that these

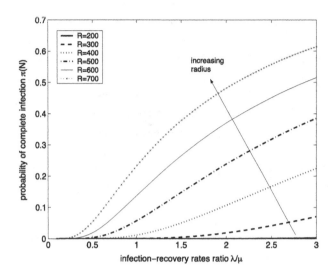

FIGURE 4.13

Probability of all nodes infected $\pi_I(N)$ versus λ/μ.

-Courtesy Karyotis V, Kakalis A, Papavassiliou S. Malware-propagative mobile ad hoc networks: asymptotic behavior analysis. Springer J Comput Sci Technol 2008;23(3):389–99. With kind permission from Springer Science and Business Media.

combinations of parameter ranges can be found in a relatively extensive literature, e.g. [29–32] and their references.

More specifically, in Fig. 4.12, the probability of no infected nodes $\pi_I(0)$ as obtained by relation (4.19) is plotted (denoted with "app" in the figure), as well as its exact value (denoted with "acc" in the figure), obtained by relation (4.18), for different values of the parameter α (with fixed L, different R, λ/μ yield different α). Fig. 4.12 essentially depicts the number of legitimate nodes above which the approximation is valid for given values of α. It is evident that $N = 25$ nodes suffice for all the instances depicted, in order for the approximation to be acceptable. It is interesting to note that for large values of R and λ/μ, $N > 7$ achieves a perfect match between the approximated and the accurate value of $\pi_I(0)$. Furthermore, as R or λ/μ increase, i.e. α decreases, even fewer network nodes suffice. The approximation error for other states of the steady-state distribution follows similar trends, since the probabilities of the rest of the states are expressed in terms of the state with no infected nodes.

Numerical results: Malware propagation

We first focus on the probability of a completely infected network, given by

$$\pi_I(N) = e^{-\frac{\mu}{\lambda\pi\left(\frac{R}{L}\right)^2}} = e^{-\alpha}. \tag{4.23}$$

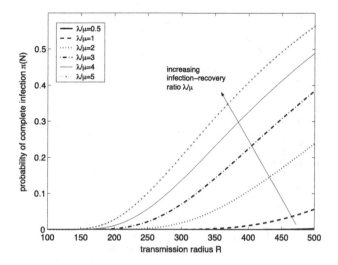

FIGURE 4.14

Probability of all nodes infected $\pi_I(N)$ versus R.

-Courtesy Karyotis V, Kakalis A, Papavassiliou S. Malware-propagative mobile ad hoc networks: asymptotic behavior analysis. Springer J Comput Sci Technol 2008;23(3):389–99. With kind permission from Springer Science and Business Media.

Complete infection of legitimate nodes, denoted as pandemia in epidemics terms, corresponds to an ultimately successful attacker that has achieved its purpose of completely damaging a network. In reality, the network has been damaged severely much earlier than total infection, as its traffic-carrying capability will have been reduced significantly by the time that a large percentage of the legitimate nodes, which depends on the network features, will be infected.

In Fig. 4.13, the probability of total infection with respect to the ratio of link infection to recovery rates is presented for various values of the common transmission radius. Here, it should be noted that as indicated in Eq. (4.23), the probability of pandemia depends only on the transmission range, the link infection rate, and the recovery rate. The number of legitimate nodes does not affect such an event. This is expected, as a denser network does not guarantee that all nodes will be constantly infected. On the contrary, greater N allows more nodes (those placed away of the source of infection) at the early stages of the infection to remain intact. However, a greater link infection-recovery rate ratio means that every node becomes infected easier. Similarly, a greater transmission radius results in more neighbors for a single node and thus more potential infections as the number of transmission links associated with infected nodes increases.

As observed in Fig. 4.13, higher transmission radii lead with higher probability to the pandemic state, even for values of the infection to recovery rates ratio below

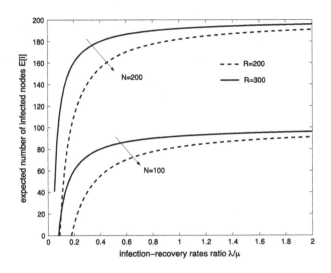

FIGURE 4.15

Average number of infected nodes $E[L_I]$ versus λ/μ.

-Courtesy Karyotis V, Kakalis A, Papavassiliou S. Malware-propagative mobile ad hoc networks: asymptotic behavior analysis. Springer J Comput Sci Technol 2008;23(3):389–99. With kind permission from Springer Science and Business Media.

unity. Note that the form of Eq. (4.23) does not allow the value of $\pi_I(N)$ to become larger than one even for large values of R. In any case, larger values of R increase the speed of converge of $\pi_I(N)$ to unity.

Similar observations hold in Fig. 4.14, where the probability of complete infection is shown with respect to the common transmission radius for various values of the infection to recovery rates ratio. Again, $\pi_I(N)$ becomes significant for values of R and λ/μ larger than those commonly used in practice. However, this time the curves have a smaller rate of convergence to the unit value than the previous case. This means that the behavior of the network is more sensitive to changes in the ratio λ/μ than to changes in the transmission radius R as far as the pandemic state is concerned, since from the form of (4.23), a change in λ/μ has the same effect as a change to $(R/L)^2$, verifying the greater sensitivity against λ/μ than against R.

In Figs. 4.15–4.17, the average number of infected nodes, $E[L_I]$, is plotted against the ratio of link infection rate to the recovery rate, the transmission radius, and the number of legitimate network nodes, respectively. $E[L_I]$ is rather indicative of the aggregated behavior of the attacked network and an overall metric of the attacker's impact on the normal operation of the network. Clearly, higher $E[L_I]$ indicates that the combination of system parameters is more desirable for an attacker.

Specifically, Fig. 4.15 presents $E[L_I]$ against the ratio of the link infection to the recovery rate for two different combinations of the transmission radius and

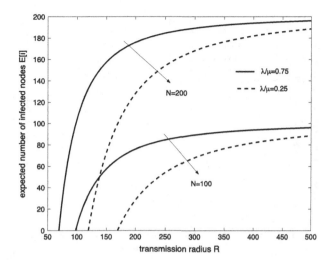

FIGURE 4.16

Average number of infected nodes $E[L_I]$ versus R.

-Courtesy Karyotis V, Kakalis A, Papavassiliou S. Malware-propagative mobile ad hoc networks: asymptotic behavior analysis. Springer J Comput Sci Technol 2008;23(3):389–99. With kind permission from Springer Science and Business Media.

the number of network nodes. It is evident that as the ratio λ/μ increases, $E[L_I]$ approaches the total number of infected nodes for the given network scenario. The difference lies in the rate of increase, which is higher for a larger transmission radius as expected.

Fig. 4.16 presents the behavior of $E[L_I]$ as a function of the transmission radius for two different values of the number of nodes and ratio of the link infection rate to the recovery rate. Again, as the transmission radius increases, $E[L_I]$ approaches the total number of legitimate nodes of the specific network instance. The rate of convergence increases for higher values of λ/μ ratio. Compared with the corresponding results in Fig. 4.15, the rate of increase is faster now, since parameter R is squared in the denominator as opposed to the linear power of λ/μ in the previous case.

The dependence of the average number of infected nodes on the number of legitimate nodes is linear as can be verified by Eq. (4.21) and shown in Fig. 4.17. The combination of larger values of R and λ/μ yields the higher number of $E[L_I]$ for all values of legitimate nodes, while the combination with smaller values of R and λ/μ yields the lower values of $E[L_I]$. An interesting observation is that a higher value of the ratio λ/μ leads to higher $E[L_I]$ as opposed to a higher value of R when the other parameter remains constant. This means that the behavior of the network is more sensitive to changes in the value of λ/μ than variations in R.

As mentioned in Section 4.3.1, an indicative metric of the overall instantaneous capabilities of an attacker to infect network nodes is the average throughput of the

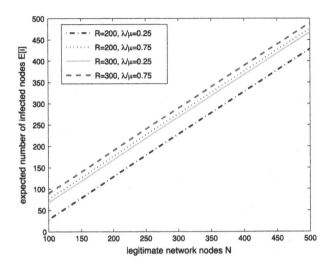

FIGURE 4.17

Average number of infected nodes $E[L_I]$ versus N.

-Courtesy Karyotis V, Kakalis A, Papavassiliou S. Malware-propagative mobile ad hoc networks: asymptotic behavior analysis. Springer J Comput Sci Technol 2008;23(3):389–99. With kind permission from Springer Science and Business Media.

noninfected queue, $E[\gamma_S]$. In Figs. 4.18–4.20, $E[\gamma_S]$ is presented against the link infection rate, the transmission radius, and the number of legitimate nodes. We note that due to the form of $E[\gamma_S]$, given by Eq. (4.22), the dependence on λ and μ is different, making a direct plot of $E[\gamma_S]$ against λ/μ inappropriate. Still, in order to gain some insight about the system behavior with respect to the ratio λ/μ, for the rest of the paper, we fix the recovery rate to $\mu = 1$ and vary only the link infection rate λ.

In Fig. 4.18, the average throughput of the noninfected queue with respect to the link infection rate for various combinations of the transmission radius and the number of legitimate nodes is shown. It is observed that $E[\gamma_S]$ increases with increasing values of the link infection rate λ (and thus the ratio λ/μ), transmission radius R, and number of legitimate nodes N, as expected. Nevertheless, an increase in the number of legitimate nodes (i.e. node density) causes a much greater increase in the achieved throughput than an increase in the transmission radius. It is also interesting to note that for larger node densities (i.e. number of legitimate nodes) the increase in the average throughput $E[\gamma_S]$ due to using a larger transmission radius is greater. This is visible in Fig. 4.18 by comparing the differences between curves corresponding to the same node density for large values of the link infection rate. Similar results hold in Fig. 4.19, where the average throughput of the noninfected queue is plotted against the common transmission radius. Once more, the network density is more influential to $E[\gamma_S]$ than the link infection rate.

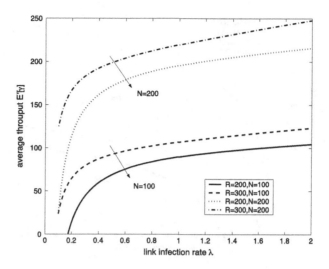

FIGURE 4.18

Average throughput of the noninfected queue $E[\gamma_S]$ versus λ.

-Courtesy Karyotis V, Kakalis A, Papavassiliou S. Malware-propagative mobile ad hoc networks: asymptotic behavior analysis. Springer J Comput Sci Technol 2008;23(3):389–99. With kind permission from Springer Science and Business Media.

Finally, in Fig. 4.20, $E[\gamma_S]$ is depicted with respect to the number of legitimate nodes, i.e. node density. The relation is linear and as expected, the smaller $E[\gamma_S]$ corresponds to the combination of λ and R with smaller values, whereas the higher $E[\gamma_S]$ to the combination with larger values.

Summarizing the most important findings emerging from the numerical results presented for malware propagation in wireless distributed networks:

- The behavior of the network is more sensitive to changes in the ratio λ/μ than to changes in the transmission radius R as far as the pandemic state is concerned.
- The effect of λ/μ on $E[L_I]$ is greater than the effect of R when all the other parameters remain constant.
- The network density emerges again as more influential than the link infection rate for $E[\gamma_S]$.

4.4 MALWARE DIFFUSION MODELING IN DYNAMIC NETWORKS WITH CHURN

As explained extensively in the previous parts of the book, modeling accurately the dynamics of malware spreading is of high research and practical importance with numerous associated benefits. This is even more important for dynamic networks

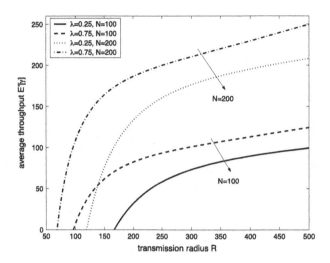

FIGURE 4.19

Average throughput of the noninfected queue $E[\gamma_S]$ versus R.

-Courtesy Karyotis V, Kakalis A, Papavassiliou S. Malware-propagative mobile ad hoc networks: asymptotic behavior analysis. Springer J Comput Sci Technol 2008;23(3):389–99. With kind permission from Springer Science and Business Media.

with churn (node or edge churn) [6, 173], where the impact of malware can be more severe. In this subsection, we focus on touching exactly this aspect of modeling malware dynamics, namely, modeling malware diffusion dynamics in dynamic networks. Such topologies with *network churn* emerge in most of the applications of complex networks and especially wireless multihop topologies, e.g. *ad hoc*, sensor, and vehicular, due to a multitude of reasons [203], such as wireless environment variations, device energy variations, as the outcome of malware attacks, and due to user patterns. The churn rate (sometimes called attrition rate), in its broadest sense, is a measure of the number of individuals or items moving out of a collective group over a specific period of time. In that sense, network churn describes the process of nodes (or edges) of the network entering it or leaving it dynamically in a certain period of time.

In the past, various attempts to model malware have emerged (see [223] and references therein, and Part 1 of this book), each aiming at different objectives and employing different tools [58] (Part 1, Chapter 3). However, all this prior work has not studied explicitly network churn. Lately, some notable effort has been devoted to the macroscopic dynamics of malware propagation, where the term macroscopic modeling is meant in the sense of the aggregate study of malware propagation (interested in the number of nodes in each state, not what the exact state of each node is) for a long time period, where different types of attacks spread and present recurring behavior. Such efforts have especially focused on wireless decentralized

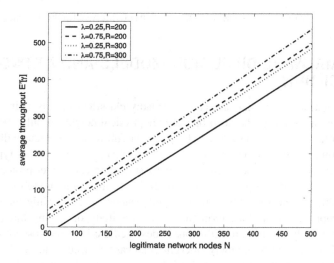

FIGURE 4.20

Average throughput of the noninfected queue $E[\gamma_S]$ versus N.

-Courtesy Karyotis V, Kakalis A, Papavassiliou S. Malware-propagative mobile ad hoc networks: asymptotic behavior analysis. Springer J Comput Sci Technol 2008;23(3):389–99. With kind permission from Springer Science and Business Media.

networks, as partially shown in the previous subsections of this chapter and the various works available in the literature, e.g. [119, 133] and references therein.

This subsection presents and analyzes modeling approaches that extend the queuing-based modeling of malware diffusion presented in the previous subsections of this chapter and analyze the macroscopic modeling of malware propagation for complex communication networks with churn, i.e. wireless *ad hoc*, SF, SW, and random topologies. In the considered framework, legitimate nodes might enter/leave the network due to exhausting their energy/recharging and/or the impact of malware. As in the previous models in this chapter, a queuing network model is again used to capture the transitions of states of legitimate nodes attacked by malicious users. To treat this problem, an open queuing network model is presented for the behavior of legitimate nodes attacked by malicious users, in contrast to the models presented previously where a closed queuing system was sufficient, due to the lack of churn. As before, a product-form solution of the presented model is obtained through the Norton equivalent model, and the dynamics of spreading are studied. The results can be used for assessing the robustness of arbitrary complex networks under diverse operational scenarios.

The results yielded by this approach can be used again for assessing the robustness of the network and can be further exploited in increasing network reliability against the worst possible outbreaks. Given also that very few approaches in the literature have addressed malware diffusion modeling for dynamic networks with churn

[121, 122], the value of this framework is of great interest for both academics and the industry.

4.4.1 MALWARE DIFFUSION MODELS AND NETWORK CHURN

As mentioned in Part 1, malware can be broadly classified in two main types, i.e. direct and indirect, where threats propagate via physical neighbors only [199], or via multihop infections, e.g. email viruses [233]. In the following, the focus will continue being put on propagation via physical neighbor contact, since the other type can be implicitly analyzed as a case of direct malware spreading at a higher protocol layer.

Furthermore, with respect to the SIR and SIS infection models, the first is more suitable for the short-term study of independent threats, while the second is more appropriate for the long-term (macroscopic) study of mixing threats. In the macroscopic study of malware diffusion, nodes oscillate between susceptibility and infection due to recurrent or newly emerging malware. Currently, network churn can be only studied within the framework of the macroscopic malware diffusion study [121, 122], and thus the following analysis will focus on the long-term and steady-state behavior of complex networks. For this reason, the SIS node infection paradigm is adopted here too.

As explained in the previous subsection, in accord with the importance of complex network infrastructures, more generic approaches analyzing malware propagation are required to secure commercial and critical networks. Probabilistic tools have been mostly employed as alternatives to the more popular deterministic models relying on systems of differential equations, e.g. probabilistic models based on interactive Markov chains proposed in [82], or a stochastic optimization framework introduced in [132, 133], which will be described in detail in Chapter 6. Similarly, in the previous subsections of this chapter, a queuing-based framework has been proposed in [124] for wireless multihop networks. This last modeling approach can be exploited in various capacities, as it will be shown in Chapter 9. The approach presented next adopts the same framework but extends it in the more general scenario of dynamic networks with node churn.

The most general behavior of dynamic networks, where legitimate nodes enter/leave the network due to their own operation (e.g. exhausting network energy) and/or the impact of malware, is considered. The approach that will be presented in the following, was developed in [121, 122] and extended the framework of [119] for networks with churn, further applying it in various types of complex communications networks.

4.4.2 OPEN QUEUING NETWORK THEORY FOR MODELING MALWARE SPREADING IN COMPLEX NETWORKS WITH CHURN

In this subsection and the rest of the chapter, we focus on the case of nonpropagative networks where malware spreading takes place under the SIS infection model for

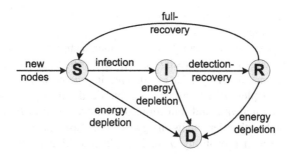

FIGURE 4.21

State-transition diagram for legitimate nodes in a network with churn.

-Courtesy Karyotis V, Papavassiliou S. Evaluation of malware spreading in wireless multihop networks with churn. In: Proceedings of 6th international conference on ad hoc networks (EAI ADHOCNETS). Springer; 2014. With kind permission from Springer Science and Business Media.

describing the macroscopic behavior of users. A static wireless distributed network with churn [104], modeling, e.g. a sensor or *ad hoc* network, is considered.

In a network with churn and under the SIS paradigm, a legitimate node will start susceptible, clear of any malware, denoting the corresponding susceptible state by S (Fig. 4.21). At some point, a susceptible node will become infected by some spreading threat, e.g. virus and worm, and within a long observation period, the node will eventually return to the susceptible state (by removing the malware). The infected state is denoted by I. In the general case of networks with churn, but also for networks without churn, the short-term behavior of nodes and their corresponding state transition may involve other intermediate or terminal states as well, as shown in Fig. 4.21. These transitions may involve an intermediate recovery state (denoted by R) and a terminal state where nodes are considered dead (denoted by D). The dead state cumulatively represents nodes that cease operation due to exhausted energy or due to malware operation. Nodes that complete their recovery, return to the susceptible state, and without loss of generality, it is assumed that new nodes entering the network also begin their lifetime in the susceptible state. Dead nodes are completely removed from the network (potentially reintroduced in the network as new susceptibles after a long time, partially constituting the set of new nodes). Consequently, the overall system follows the SIS paradigm, where it will be possible for susceptible nodes to become infected and eventually recover again to the susceptible state.

Regarding the communication model, we assume that at each time the network has $n = n(t)$ legitimate nodes, each with transmission radius R_{trans}. Whenever there is no danger of ambiguity, we drop the time dependence from the symbol of the total number of network nodes at each time instant, $n(t)$. For simplicity, it is also assumed that node pairs are formed only within the transmission range of devices, as in [117–119, 133]. More general communication models can be incorporated in a straightforward manner.

With respect to Fig. 4.21, it can be observed that each user spends an amount of time in each node state that varies stochastically. Furthermore, given the succession of state transitions depicted in Fig. 4.21, a node entering the network at the susceptible state might either deplete its energy and become dead with probability p or could become infected by malware and transition to the infected state. From the infected state, the node might either deplete its energy as well, cease operation due to malware and become dead with probability q, or it could transition to the recovery state. Finally, from the recovering state, the user either recovers to the susceptible state and the cycle begins again, or the node depletes its energy while recovering and it is removed from the network to the dead state with probability w.

Thus, the behavior of legitimate users can be segregated in two main modes, susceptible and infected-recovering. In the first mode, nodes can be operational (some of which might exhaust their energy and leave the network) or recharging (considered semi-operational). In the infected-recovering mode, nodes become infected and either they move to recovery until they become susceptible again or they are removed due to malware or exhausting their energy.

This behavior can be mapped to the operation of a queuing network as shown in Fig. 4.22(a) [121, 122], where queuing and processing correspond to the time spent by each node in each different state described before. In Fig. 4.22(a), the upper part corresponds to the normal operation of nodes (susceptible operational-susceptible recharging) and the lower part to the infection-recovery mode. The customers of the queuing network correspond to the nodes of the network as they change states due to malware and node churn. It should be noted that the queuing network is open due to node churn, allowing for new customers to enter (corresponding to new susceptible nodes) and customers to leave (corresponding to nodes depleting their energy or becoming dead due to malware).

The depicted input/output rates of the network in Figs. 4.22(a) and 4.22(b) correspond to node churn rates, while the arrival/service rates at each queue correspond to the infection/recovery processes of the legitimate nodes of the actual network under attack. Starting with Fig. 4.22(a), λ is the rate at which new nodes emerge in the network, while p, q, w is the probability for a node to exit the network at the susceptible, infected, and recovering states, respectively. The sum $p + q + w$ corresponds to the total rate at which nodes leave the network. Thus, $\lambda/(p + q + w)$ determines the total node churn rate of the network. In the block of the susceptible queues (upper block) in Fig. 4.22(a), λ_s^o is the effective rate at which nodes recover to the operational-susceptible state (after churn has been taken into account) and λ_s^r is the effective rate at which nodes recover to the recharging susceptible state. Similarly, μ_s^o is the infection rate of susceptible operational nodes, while μ_s^r is the infection rate of susceptible recharging nodes. When combined, rate $\lambda_S = \lambda_s^o + \lambda_s^r$ is the effective recovery rate of nodes to the susceptible state and $\mu_S = \mu_s^o + \mu_s^r$ is the total infection rate of susceptible nodes (Fig. 4.22(b)). With respect to the lower tandem queues, λ_I is the effective infection rate (rate of infection of susceptible nodes after node churn

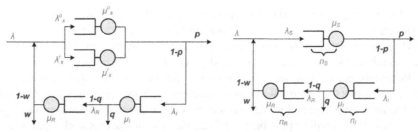

(a) Generic malware propagation queuing model.

(b) Norton equivalent queuing model.

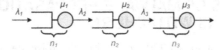

(c) Equivalent product-form serial queuing system.

FIGURE 4.22

Queuing models for malware spreading in networks with churn.

-From © 2015 IEEE. Reprinted, with permission, from Karyotis V, Papavassiliou S. Macroscopic malware propagation dynamics for complex networks with churn. IEEE Commun Lett 2015;19(4):577–80.

has been calculated), μ_R is the total rate of nodes out of the infected state, λ_R the effective rate under which nodes enter the recovery state (after node churn has been calculated), and μ_R the total rate out of the recovery state. In Fig. 4.22(b), n_S, n_I, n_R is the total number of susceptible, infected, and recovering nodes at the moment, respectively.

In order to analyze the generic network of Fig. 4.22(a), the Norton equivalent of the upper part with parallel queues may be employed, yielding eventually the queuing network of Fig. 4.22(b) [121, 122]. This does not harm the analysis because in malware dynamics and robustness analysis we are not particularly interested in which nodes are susceptible-operational and which are susceptible-recharging, focusing on the total number of susceptible users. We consider the number of susceptible nodes versus the number of infected and recovering, as in the queuing network of Fig. 4.22(b). Not shown in Fig. 4.22 is the parameter of the link infection rate of a susceptible (denoted by λ_e), which corresponds to the link infection rate λ_k for each link of a node in the malware to queuing mapping in Fig. 4.1. Also not shown, the service rates in the infected (μ_i) and recovering (μ_r) queues of each individual node. The rates defined previously and shown in Figs. 4.22(a) and 4.22(b) correspond to the cumulative queue service rates, which in turn depend on the partial rates of λ_e, μ_i, and μ_r of each node. Without loss of generality, these partial rates are considered the same for all users.

The model in Fig. 4.22(a) (and the Norton equivalent) can be analyzed using Jackson's theorem for product-form networks [25]. It should be also noted that all initial services are exponential with rates as shown in Fig. 4.22(a). The combined inputs are not Poisson, thus nor are the outputs. However, each queue behaves as an $M/M/r_i$ queue with specific input and output, as will be explained shortly, where r_i is the number of servers of each queuing stage ($r_1 = 2, r_2 = r_3 = 1$).

Regarding node churn, we study the system for a churn rate $\lambda/(p + q + w) \approx 1$, which means that the network will remain approximately the same, on average, avoiding degenerate topologies (too big or disconnected), while allowing numerous churn operations. Thus, it will be possible to evaluate various complex network types, where node churn emerges naturally, i.e. random, small-world, and scale-free networks.

4.4.3 ANALYSIS OF MALWARE PROPAGATION IN NETWORKS WITH CHURN

The queuing network of Fig. 4.22(b) has a product-form steady-state distribution and it is equivalent to a network of three cascade queues as shown in Fig. 4.22(c) [121, 122]. The service rates of the final cascade network are directly obtained from the Norton equivalent of Fig. 4.22(b), as $\mu_1 = \mu_S$, $\mu_2 = \mu_I$, $\mu_3 = \mu_R$. The arrival rates in the product-form network (Fig. 4.22(c)) in the general case can be obtained as

$$\lambda_i = \lambda_i' + \sum_{k=1}^{3} q_{ki} \lambda_k, \quad i, k = 1, 2, 3, \tag{4.24}$$

where λ_i' are potential external inputs to each stage (here only $\lambda_1' = \lambda$, corresponding to the input of new susceptible nodes and the rest external inputs are considered zero), and q_{ki} is the probability for a customer to move from stage k to stage i. From this, $1 - \prod_k (1 - \sum_i q_{ki})$ is the probability that the customer (network node in the malware context) stays in the system ($1 - \sum_i q_{ki}$ is the probability for the customer to leave the system at stage k). Solving the corresponding system of equations, we obtain

$$\lambda_1 = \frac{1}{1 - (1-p)(1-q)(1-w)} \lambda, \tag{4.25}$$

$$\lambda_2 = \frac{(1-p)}{1 - (1-p)(1-q)(1-w)} \lambda, \tag{4.26}$$

$$\lambda_3 = \frac{(1-p)(1-q)}{1 - (1-p)(1-q)(1-w)} \lambda. \tag{4.27}$$

The steady-state distribution of the cascade product-form network will be of the form

$$p(n_1, n_2, n_3) = p_1(n_1) p_2(n_2) p_3(n_3), \tag{4.28}$$

where n_1, n_2, n_3 are the number of users in the susceptible, infected, and recovering states, respectively, and at every time instant $n_1 + n_2 + n_3 = n(t)$. Jackson's theorem allows to treat each stage (queue in Fig. 4.22(c)) as an independent $M/M/r_i$ queue with input rate λ_i and service rate μ_i, $i = 1, 2, 3$. An input policy regulating the arrival of new susceptible nodes with respect to the death/removal rates should be employed to ensure $n(t) < \infty$, complying with realistic networks that typically have finite nodes. Distribution $p_i(n_i)$ provides the number of users in each queue, and in the general form where all service rates of a block (stage) with r_i parallel queues are the same, it is obtained as

$$
p_i(n_i) = \begin{cases} \frac{(\lambda_i/\mu_i)^{n_i}}{n_i!} p_{i,0}, \; n_i < r_i \\ \frac{r_i^{r_i} \rho_i^{n_i}}{r_i!} p_{i,0}, \; n_i \geq r_i \end{cases}, \tag{4.29}
$$

$$
p_{i,0} = \frac{1}{\sum\limits_{n=0}^{r_i-1} \frac{(\lambda_i/\mu_i)^n}{n!} + \frac{(\lambda_i/\mu_i)^{r_i}}{r_i!(1-\rho_i)}}, \tag{4.30}
$$

where $\rho_i = \lambda_i/r_i\mu_i$. In our case, $r_1 = 2, r_2 = r_3 = 1$ (Fig. 4.22(a)). However, the service rates of the two parallel queues in stage 1 are not the same. Thus, the above expressions may be used directly for obtaining the distributions of stages 2 and 3, while for the first stage, we consider the system as an $M/M/2$ queue with different service rates for the two servers and solve explicitly for its steady-state distribution. At this point, it should be noted that due to the simplification of the generic model in Fig. 4.22(a) into the Norton equivalent shown in Fig. 4.22(b), for the computation of the corresponding expression, arriving customers have no memory, i.e. they choose randomly which server to join if they find an empty system. This comes at the cost that with the equivalent model we cannot explicitly track which node recovers to the operational-susceptible state and which to the recharging susceptible, but as mentioned above, we sacrifice this analytic detail of the original model for the sake of obtaining explicit analytic results eventually. Consequently, the corresponding expression is obtained,

$$
p_1(n_1) = \begin{cases} \left[1 + \frac{C}{\rho_1(1-\rho_1)} \right]^{-1}, \; n_1 = 0 \\ \frac{C}{\rho_1 + \frac{C}{1-\rho_1}}, \; n_1 = 1, \\ \rho_1^{n-2} \frac{C}{1 + \frac{C}{\rho_1(1-\rho_1)}}, \; n_1 \geq 2 \end{cases} \tag{4.31}
$$

where $\rho_1 = \lambda_1/\mu_1$, C is a constant depending on λ_1, μ_1, given by

$$
C = \frac{\lambda_1^2}{2\mu_s^o \mu_s^r}, \tag{4.32}
$$

and for the second and third stages, we obtain the distributions directly as geometric expressions of $M/M/1$ queues, respectively,

$$
p_2(n_2) = \rho_2^{n_2}(1 - \rho_2), \tag{4.33}
$$

$$
p_3(n_3) = \rho_3^{n_3}(1 - \rho_3), \tag{4.34}
$$

where $\rho_2 = \lambda_2/\mu_2$, $\rho_3 = \lambda_3/\mu_3$, $\rho_1, \rho_2, \rho_3 < 1$, and $n_1, n_2, n_3 \geq 0$.

In Fig. 4.22(c), n_1 refers to susceptible, n_2 to infected, and n_3 to recovering nodes, respectively. The number of dead nodes is unimportant, since they do not participate in malware dynamics and in addition, it is assumed that new nodes are always available. The service rate of each queue in Fig. 4.22(c) is as shown in Fig. 4.22(b), which is the Norton equivalent service rate from Fig. 4.22(a). The service rates in the original queuing network depend on the infection model, malware potentials, and the topology of each network.

The average number of users in such system is given by

$$L = \sum_{i=1}^{m} L_i = L_1 + \sum_{i=2}^{3} p_{r_i} \frac{\rho_i}{(1 - \rho_i)^2},\tag{4.35}$$

where L_i is the average number of users in each queue and $p_{r_i} = \frac{(\lambda_i/\mu_i)^{r_i}}{r_i!} p_{i,0}$. In this case, the average number of susceptible (operational) and infected-recovering nodes is

$$L_1 = \frac{C(1 - \rho_1)}{\rho_1(1 - \rho_1) + C} \left[1 + \frac{\rho_1(2 - \rho_1)}{(1 - \rho_1)^2} \right],\tag{4.36}$$

$$L_2 = \frac{\rho_2}{1 - \rho_2},\tag{4.37}$$

$$L_3 = \frac{\rho_3}{1 - \rho_3}.\tag{4.38}$$

Using the customer distributions (4.31) and (4.34), other quantities of interest can be computed, e.g. the average throughput of stage 1, which provides the average total infection rate of the system. Similarly, the average throughput of stage 3 provides the average recovery rate of the system, while the average throughput of stage 2 the average healing rate of infected nodes. Summing up these throughput quantities, weighted by the corresponding loss probabilities p, q, w, the corresponding cumulative node loss churn rate can be obtained.

Until this point, the analysis is generic and applies to all types of networks with churn (centralized, multihop, small-world, scale-free, etc.). However, the model developed in Fig. 4.22 allows more specific results to be obtained on a per network case. For instance, apart from n_1, n_2, n_3, one may obtain analytical expressions of the average n_2 with respect to network parameters, such as the node transmission radius and node densities of a specific wireless network. Such task is network type/scenario specific depending on the topology and operation of each network. For instance, in order to obtain the above for an *ad hoc* network, the Norton equivalent of Fig. 4.22(b) will be required to be further substituted with another Norton equivalent closed queuing network were two adjacent queues are substituted with an equivalent queue, yielding eventually a two-queue network with feedback, which can be easily solved via standard Markov chain methods to obtain the steady-state distribution of the queue of interest with respect to all involved network parameters. Then, additional quantities/rates can yield other desired dynamics of the corresponding systems.

4.4.4 DEMONSTRATION OF QUEUING FRAMEWORK FOR MALWARE SPREADING IN COMPLEX AND WIRELESS NETWORKS

In this section, numerical and simulation results regarding the operation and behavior of complex and wireless distributed networks with churn when attacked by a single attacker are presented. Infected nodes are assumed to further propagate the infectious malware they received, while recovering nodes are prevented from doing so. This means that the spreading of malware is mainly due to the network, while the attacker has a smaller role in spreading dynamics, mostly needed to generate new infections in the event that a network manages to recover completely for an instance. Thus, the network spreading dynamics will be studied in the following.

The evaluation is based on a Matlab simulator, developed for studying the behavior of the attacked network [121, 122]. At each epoch (slot) of the simulator one event took place, according to the current state of the system $\{n_1, n_2, n_3\}$, the topology of the network and the corresponding infection (S→I transition), recovering (I→R transition), and full-recovery (R→S transition) rates. This was ensured by the nature of the system in Fig. 4.22.

For the multihop networks we focus on, the link infection rate λ_e of a susceptible node represents the probability that this user will become infected from a malicious neighbor. The multihop topology is considered for this case as a random geometric graph. Combining this with the link infection, a detailed analysis of the system in Fig. 4.22(a) yields the total infection rate, as the service rate of the single queue of susceptible nodes in the Norton equivalent (which is equal to μ_1 in the product-form equivalent). This infection rate will be $\sum_{m=1}^{n_1} k_m \lambda_e$, where k_m is the number of malicious neighbors of susceptible node m (counting both the attacker and infected legitimate nodes). The total recovering (corresponding to μ_2 in Fig. 4.22(c)) and full-recovery rates (corresponding to μ_3 in Fig. 4.22(c)) depend on n_2, n_3, and may be computed as $n_2\mu_i$, $n_3\mu_r$, respectively.

In this book, only some indicative results are provided, which however can be used for the assessment of network reliability and attack potentials. Regarding node churn, we study the behavior of the system for positive churn, i.e. for a $\lambda > (p+q+w)$, which means that the networks analyzed were considered to be growing on average. This is preferable for the study, as a decreasing network could sometimes lead to degenerate (disconnected) topologies, or even to the extinction of the network.

Evaluation of malware spreading in complex networks

In this subsection, we initially present some numerical and then simulation results exhibiting the operation of attacked complex networks with churn. For demonstration purposes, we consider some specific types of networks, but any other type of wireless or wired networks is applicable as well. We assume that the network churn parameters are fixed and equal to $\lambda = 0.25, p = w = 0.05, q = 0.1$ (for the rest of this chapter, all node rates are assumed to have units "node/time unit"). The loss of infected nodes

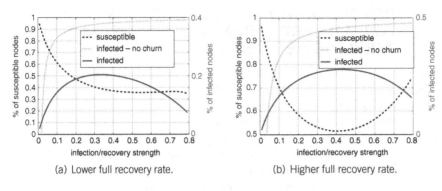

FIGURE 4.23

Percentage of susceptible and infected nodes versus network infection/recovery strength and comparison with networks with no churn for complex networks.

-From © 2015 IEEE. Reprinted, with permission, from Karyotis V, Papavassiliou S. Macroscopic malware propagation dynamics for complex networks with churn. IEEE Commun Lett 2015;19(4):577–80.

was considered double than the loss of the rest of node classes, due to two-fold death possibilities (i.e. malware and energy depletion).

Fig. 4.23 presents numerical results obtained from analysis (Eqs. (4.36) and (4.38)). Furthermore, it compares with the theoretical results for the same networks but with no churn, obtained previously in Section 4.3.2 and more specifically with numerical results already provided in Section 4.3.2 (also from [119] in the literature) for the case of 800 legitimate nodes. The corresponding curve is denoted in the comparison figure as "infected-no churn." The percentages of susceptible (L_1/L) and infected (L_2/L) nodes (obtained via Eqs. (4.35)–(4.37)) with respect to parameter λ_1/μ_1 denoted as infection-to-recovery strength (horizontal axis in the figure) is shown for two different combinations of the service rate of the infected and the recovering queues (stages 2 and 3). Notice the difference in the two vertical axes scales, while the third percentage of recovering nodes can be computed straightforwardly. In Fig. 4.23(b), the service rate of the recovery stage was double than that in Fig. 4.23(a), which implies a faster transition from recovering to susceptible state, i.e. less recovering but more susceptible nodes. Using results of these types, extensive analyses can be obtained with respect to the expected behavior of the system for various combinations of the parameters involved (churn rates, recovery/infection strengths, etc.), and thus the expected reliability of the network can be assessed. Fig. 4.23 characterizes network-related behavior with respect to attacks.

Fig. 4.24 presents similar results for three types of complex networks, a random [Erdos-Renyi (ER) model], a small-world [Watts-Strogatz (WS) model], and a scale-free [Barabasi-Albert (BA) model] network. However, now the infection-to-recovery strength is that of a single node, namely, the infection rate at a communication link to the recovery rate of a node λ_e/μ_r and not that of the network as before.

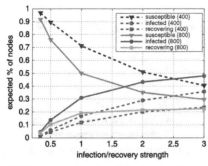

(a) Random network with edges scaling as $O(n \log n)$.

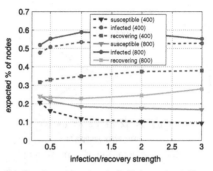

(b) Small-world network (initial regular lattice with 10 neighbors and $g_p = 0.1$ shortcut probability).

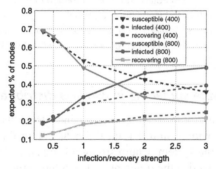

(c) Scale-free network (initially, all nodes connect to 10 other neighbors).

FIGURE 4.24

Expected percentage of susceptible, infected, and recovering nodes versus infection/recovery strength (simulation) for complex networks with 400 and 800 initial nodes.

-From © 2015 IEEE. Reprinted, with permission, from Karyotis V, Papavassiliou S. Macroscopic malware propagation dynamics for complex networks with churn. IEEE Commun Lett 2015;19(4):577–80.

Thus, Fig. 4.24 characterizes node-related behavior with respect to attacks. The total rates $\lambda_i, \mu_i, i = 1, 2, 3$ are functions of λ_e, μ_r. In general, it is observed that all these types of networks exhibit some form of robustness, maintaining sufficient percentages of susceptible nodes for infection/recovery strengths lower than unity. As the infection/recovery strength increases, the behavior of the three networks seems to converge. In addition, with respect to the percentage of susceptible nodes, the random network seems to be the more robust against generic attacks, followed by the scale-free and then the small-world, since the small-world, and the scale-free to a lesser degree, have structural vulnerabilities (shortcuts and node-hubs, respectively). The latter are in accordance with similar solutions reached in Chapter 5, following a completely different modeling approach.

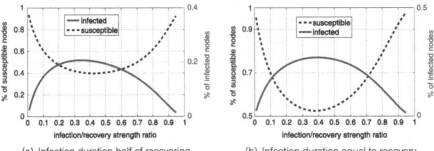

(a) Infection duration half of recovering. (b) Infection duration equal to recovery.

FIGURE 4.25

Percentages of susceptible and infected nodes as a function of infection to recovery strength (numerical) for wireless distributed (multihop) networks.

-Courtesy Karyotis V, Papavassiliou S. Evaluation of malware spreading in wireless multihop networks with churn. In: Proceedings of 6th international conference on ad hoc networks (EAI ADHOCNETS). Springer; 2014. With kind permission from Springer Science and Business Media.

Evaluation of malware spreading in wireless distributed networks

Fig. 4.25 presents some numerical results obtained from the analysis, valid for arbitrary networks, providing intuition on the behavior of the average number of nodes in the states of the system. Churn strength is equal to 1.67, which translates to a growing network. Notice the different scales in the vertical axes in both figures and for both y-axes of each figure, indicating how the expected number of susceptible and infected nodes varies with respect to the time each node is expected to spend in each of the three stages (service rates).

As expected, a decrease in susceptible nodes translates to an increase in the infected nodes. By comparing Fig. 4.25(a) and Fig. 4.25(b), it can be also observed that regarding the dependence on the infection/recovery strength, some symmetry (Fig. 4.25(b)) should be expected when the recovery (μ_I) and full-recovery (μ_R) rates are the same. These results, and many similar that can be obtained from the expressions provided before, can be used to assess the robustness of the network. Malware dynamics are represented via the infection and recovery rates, while the full-recovery rate represents the countermeasures' efficiency. Thus, given these parameters, the expected state of the system can be evaluated.

The following results have been obtained through simulations for a wireless distributed (random geometric) network, in which a square deployment region with size $L = 1000$ m was employed and all devices used a transmission radius $R_t = 150$ m. Fig. 4.26 presents the expected number of nodes in each state of the system as a function of network density (we fixed the deployment region and increased progressively network nodes). Fig. 4.26(a) regards a network with uneven recovery-full recovery rates, $\mu_I - \mu_R$, respectively. This means that the mean infection and recovery times will be uneven as well. Fig. 4.26(b) shows the corresponding results for even rates.

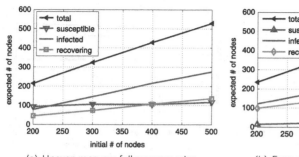
(a) Uneven recovery-full recovery rates.

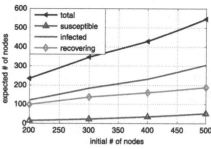
(b) Even recovery-full recovery rates.

FIGURE 4.26

Expected number of nodes in each state of a wireless distributed (multihop) network with respect to network density.

-Courtesy Karyotis V, Papavassiliou S. Evaluation of malware spreading in wireless multihop networks with churn. In: Proceedings of 6th international conference on ad hoc networks (EAI ADHOCNETS). Springer; 2014. With kind permission from Springer Science and Business Media.

It is observed that as the network density increases, so do the expected number of nodes in each state, and such increase is almost linear. However, the corresponding increase rates are different for uneven recovery-full recovery rates and similar for even rates. In both cases, the infection to full recovery strength is $\lambda_e/\mu_r = 2$. This scaling behavior explains the fact that the number of infected nodes is the smallest compared to infected and recovering nodes, revealing potential vulnerabilities for the network with respect to the specific malware dynamics and the network structure, as it was also possible to do with the numerical analysis we presented before.

However, different trends emerge regarding the expected number of nodes in each state with respect to the infection/recovery strength, as shown in Fig. 4.27. As before, the expected number of susceptible nodes has a complementary behavior to the expected number of infected and recovering nodes. The trend though is not linear. In fact, the number of recovering nodes, especially in Fig. 4.27(b), seems to saturate for increasing infection/recovery strength. Such results can be again used to evaluate the robustness of the network with respect to the expected behavior under various attack-countermeasure parameters.

Finally, Fig. 4.28 shows the average percentage difference of network size for even/uneven infection/recovery strengths, with respect to node density and the intensity of the infection/recovery strength. The first two bars in Fig. 4.28 are the average node increase as the density of the network increases, while the last two bars represent the average node count increase for increasing infection/recovery strength. As expected, average percentage difference is positive, even though small in some cases (since the churn strength was set slightly higher than 1 in all scenarios to ensure a proper topology). It can be observed that, in general, even infection-recovery strengths yield higher increase than uneven ones. More practically, equal

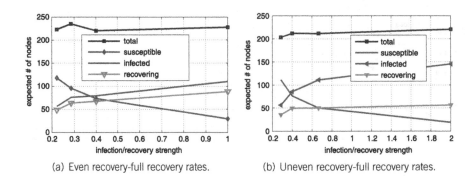

(a) Even recovery-full recovery rates. (b) Uneven recovery-full recovery rates.

FIGURE 4.27

Expected number of nodes in each state of a wireless distributed (multihop) network with respect to infection/recovery rates.

-Courtesy Karyotis V, Papavassiliou S. Evaluation of malware spreading in wireless multihop networks with churn. In: Proceedings of 6th international conference on ad hoc networks (EAI ADHOCNETS). Springer; 2014. With kind permission from Springer Science and Business Media.

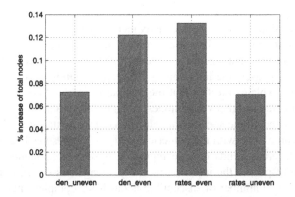

FIGURE 4.28

Expected percentage variation of the total number of nodes with respect to node density and infection/recovery strength for wireless distributed (multihop) networks.

-Courtesy Karyotis V, Papavassiliou S. Evaluation of malware spreading in wireless multihop networks with churn. In: Proceedings of 6th international conference on ad hoc networks (EAI ADHOCNETS). Springer; 2014. With kind permission from Springer Science and Business Media.

infection/recovery rates correspond to strategies providing countermeasures that match the effect of malware at the same time scales (fast-response), which in turn allow the network to maintain more nodes operational on average, by preventing some I→D transitions (infected nodes becoming dead) due to malware.

CHAPTER

Malware-propagative Markov random fields

5

5.1 INTRODUCTION

Chapter 4 presented a generic malware diffusion framework capable of describing the evolution of a malware attack in various and different types of network topologies. As shown, it can accommodate fixed and dynamic topology networks, propagative and nonpropagative (spreading) types of user devices (and corresponding networks). The queuing-based framework allows studying thoroughly the long-term robustness capabilities of networks, especially for wireless networks with topology control capabilities, where attackers can utilize their intelligence to generate smart topology control based attack strategies. This is shown in detail in Chapter 9 (Section 9.1), utilizing the results of Chapter 4. In addition, the queuing based malware diffusion framework enables dealing with such smart attack strategies by taking the network's perspective and formulating optimization problems for minimizing the anticipated damage with respect to those intelligent attack strategies (Chapter 9, Section 9.1.4).

Apart from the aforementioned features offered by the queuing based malware diffusion framework, specific complications may arise given special or ordinary situations. A change in the topology and mobility of legitimate users (even at slow velocity rates) can lead to nonergodic steady-states, which will make the explicit analytic treatment of the model tough or even intractable. Of course, the capability of the framework to tackle dynamic topologies with churn could allow the development of an implicit model to tackle various forms of dynamic networks. However, there exist focused approaches capable to tackle such issues seamlessly, in a more dedicated fashion.

In this chapter, we will focus on presenting such a holistic approach. More specifically, Chapter 5 will focus on introducing and analyzing a methodology, which exploits the concept of Markov Random Fields (MRFs) [136] and their applications in order to model in a generic manner malware diffusion in arbitrary complex networks. The corresponding approach bears all those characteristics mentioned above in terms of modeling malware diffusion. MRFs is a concept introduced for the first time in physics, and more specifically in statistical mechanics, in order to describe models relevant to the spin of particles and magnetic/glass materials. A more thorough introduction to the notion of MRFs and the properties exploited for modeling malware diffusion in complex communication networks is provided in Section 5.2. The MRFs are employed as a computational substrate, where the attacked network is represented via a spatial network graph (MRF topology) and the interactions between susceptible and infected neighbors of a node and the node itself affecting the node's state with respect to the diffusion are captured.

The proposed framework is also very generic and capable of capturing arbitrary types of network topologies and node interactions. Furthermore, it is easy to apply in practice, and thus it allows distributed and repetitive computation even when the network changes relatively often. In the following, we will first provide some background on MRFs and their properties and then explain the mapping of malware diffusion into a MRF framework. Subsequently, indicative applications and results on various types of complex networks will be provided. Several indications for extending this relatively newer malware modeling framework, compared to all other malware diffusion modeling frameworks, are postponed for Chapter 10.

5.2 MRFS BACKGROUND

In this section, we first introduce the concept of MRFs and the associated notation. Then, we present the Gibbs fields and their properties. Finally, we present the interplay between MRFs and Gibbs distributions, highlighting how they are used in practical applications.

5.2.1 MRFS

MRFs are a branch of probability theory with important theoretical properties and practical applications. Most prominent applications of MRFs lie in the areas of statistical physics [136], image processing [83, 228], and coordination of autonomous robots [206] until now. A MRF is a stochastic process that generalizes the Markov process in the sense that the time index of the original Markov process (characterized by the memoryless property) is substituted by a space index [136]. In the following, we introduce this concept more formally.

Assume a finite set S of cardinality n, with elements $s \in S$ referred to as sites (see Fig. 5.1 for an example of the notation introduced in the following paragraphs). Sites can be also referred to as nodes and typically correspond to nodes of a network for a MRF with $(n + 1)$ sites, as the term is self-explanatory. For instance, if the MRF is used to represent users of a network, sites may correspond to nodes/users of the network. If the MRF is used to represent services/applications of a network, sites correspond to the servers hosting or interconnecting such services/applications. Finally, when the MRF is used to represent images, sites correspond to pixels of an image [83].

Let Λ be the set of possible values of the state of each site $s \in S$, called the phase space (Fig. 5.1). The state of each site represents the condition in which a site may be at a time and it is typically quantified by the corresponding numerical value, i.e. the associated numerical value in the phase space. Thus, the correspondence is that each site may be in different states (each state representing different conditions) and each state is associated with a numerical value, which in turn comes from the phase

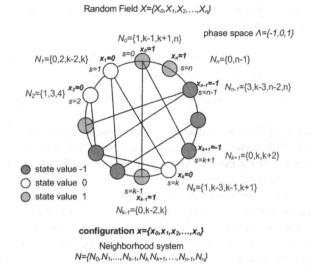

FIGURE 5.1

Random Field (RF) terminology over a random network of $n + 1$ sites and three phases.

space. In practical applications, the sites can be rather diverse, e.g. particles of a gas, pixels of an image, and state of a legitimate node of an attacked network.

A collection $X = \{X_s, s \in S\}$ of random variables with values in Λ is called a Random Field (RF) on S with phases in Λ (Fig. 5.1). The RF represents a whole system and the collection of the values (phases) of the sites of the RF corresponds to a state of the system at each time instant. A configuration $x = \{x_s, s \in S\}$, where $x_s \in \Lambda$, corresponds to one of the all possible states of the system (Fig. 5.1). It should be noted that $X_s = x_s, \forall s \in S$ is a realization of each random variable in X. Thus, successive configurations of a RF give the evolution of the system. The approach employed in using RFs, in general, is to find a representation of the system and then study successive configurations as the RF evolves, thus obtaining implicitly an evolution of the system. The product space Λ^n is called the configuration space and provides the complete span of cumulative states that a system may be at any time instant. Cumulative system states are the states defined by the partial state of each site, as a vector with n components, element of the configuration space, with components the state of each site.

A neighborhood system on S is defined as a family $\mathcal{N} = \{\mathcal{N}_s\}_{s \in S}$ of subsets $\mathcal{N}_s \subset S$, such that for every $s \in S$, $s \neq \mathcal{N}_s$ (the site is excluded from its neighborhood) and $r \in \mathcal{N}_s$ if and only if $s \in \mathcal{N}_r$. $\mathcal{N}_s$ is called the neighborhood of site (node) s (Fig. 5.1). The neighborhood system is the substrate defining the relations (direct links) between pairs of sites. Consequently, the definition of the employed neighborhood system is crucial for determining the neighborhood of each site, and thus identifying the relation between sites that determine the evolution of a RF and the system it is associated to.

Definition 5.1 *(MRF). The RF X is called a MRF with respect to the neighborhood system $\mathcal{N}$, if for every site $s \in S$,*

$$\mathbb{P}(X_s = x_s \mid X_r = x_r, r \neq s) = \mathbb{P}(X_s = x_s \mid X_r = x_r, r \in \mathcal{N}_s). \qquad (5.1)$$

From the above definition, a number of observations can be made. First of all, the aforementioned MRF framework abstracts connectivity and node (site) interactions through the neighborhood system $\mathcal{N}$, as it is the case with more general RFs. A second and most important observation is that Eq. (5.1) essentially expresses a spatial form of the Markov property (memoryless property in stochastic process that is defined with respect to time). In processes defined with respect to time, e.g. Poisson process, which have the "memoryless" property, the notion is that only the most recent (proximate) event counts, i.e. the next event depends only on the previous one and not the whole history of the process. In spatial processes [14, 92, 93, 140, 193, 205], the memoryless property would be interpreted in a sense that only the local neighborhood counts in decision making. Entities farther away from the immediate neighborhood do not have a direct effect on decisions made. Thus, MRFs may be considered an analog of the Markov process in space, namely, a spatial process that describes not only actually spatial systems but also systems where the relations between entities involve generic definitions of metric "distances."

The spatial Markov property expressed by a MRF is very fitting for the basic fact that malware infections propagate in a localized fashion that depends on 1-hop communications. The last observation has been realized and studied in many works in the literature [117, 118, 134]. This approach explicitly excludes the cases of email contamination, which allows for a form of "multihop" user infection. However, even in such cases, one can adapt the neighborhood definition and implicitly define a proper neighborhood system between the malware source and infected host, suppressing the intermediate noninfected nodes.

5.2.2 GIBBS DISTRIBUTION AND RELATION TO MRFS

Apart from MRFs, there are other useful types of RFs. One of the most prominent ones, especially in the image and video processing communities are the Gibbs Random Fields (GRFs). More specifically, the following definition is applicable for a GRF.

Definition 5.2. *A RF X is called a GRF [136, 228] if it satisfies*

$$\mathbb{P}(X = x) = \frac{1}{Z} e^{-\frac{U(x)}{T}}, \qquad (5.2)$$

where $Z := \sum_{x \in \Lambda^n} e^{-\frac{U(x)}{T}}$ is a normalizing constant called the partition function of the system and $T = T(t)$, with t the time index, is called the temperature of the system.

In the above definition, $U(x)$ is called the *potential function* and represents an "energy" metric of configuration x. The temperature $T(t)$ is a measure of the decrease of the "energy" of the system as the latter evolves, which means that as time passes by, the initial temperature will decrease allowing the system to remain close to the desired equilibrium distribution. Supposing that the MRF is used in order to search for optimal solutions in a search space of possible system states, a greater value of the temperature in the beginning will allow sampling a greater search space for the energy-minimizing distribution. The search is then essentially fine-tuned via a lower temperature parameter.

In general, the potential function is not unique and different types of functions can be employed. In fact, the selection of the most suitable potential function can be critical for obtaining a GRF that is easy to implement and compute. Given that, a very useful class of potential functions is one in which $U(x)$ is decomposed into a sum of clique potentials

$$U(x) = \sum_{c \in C_s} \Phi_c(x), \tag{5.3}$$

where each clique potential depends only on the states of the cliques formed in the underlying graph of the system. As explained in Section 1.3.3, a clique is a set of nodes in a graph where all are connected among them. C_s denotes the set of cliques formed.

As will be shown in the following, the clique-based representation of the potential function is very useful in malware propagation, when infections are transmitted through one-hop neighbor interactions.

A very useful result established in the literature is the eventual equivalence of GRFs with MRFs [136]. More specifically, the Hammersley-Clifford theorem ensures this equivalence as follows:

Theorem 5.1 *(Hammersley–Clifford). A GRF with distribution* $\mathbb{P}(X = x) = \frac{1}{Z}e^{-\frac{U(x)}{T}}$ *and potential function expressed in terms of clique potentials* $U(x) = \sum_{c \in C_s} \Phi_c(x)$ *leads to a MRF with conditional probabilities* $\mathbb{P}(X_s = x_s \mid X_r = x_r, r \neq s) = \mathbb{P}(X_s = x_s \mid X_r = x_r, r \in \mathcal{N}_s)$ *and vice versa.*

Thus, even though GRFs were introduced as different type of RFs, they are not essentially different to MRFs. However, as it will become more evident in the sequel, GRFs allow more implementation and computational convenience than MRFs in the applications, thus most of the times, modeling starts with a mapping to a MRF model and then computation proceeds with a suitably selected GRF, equivalent to the initial MRF.

5.2.3 GIBBS SAMPLING AND SIMULATED ANNEALING

Gibbs sampling

Gibbs sampling was introduced in statistics and in statistical physics. In general, *Gibbs sampling* or a Gibbs sampler is a Markov chain Monte Carlo (MCMC) algorithm for obtaining a sequence of observations, which are approximated from a specified multivariate probability distribution (i.e. from the joint probability distribution of two or more random variables), when direct sampling is difficult. This sequence can be used to approximate the joint distribution, to approximate the marginal distribution of one of the variables, or some subset of the variables (for example, the unknown parameters or latent variables), or to compute an integral (such as the expected value of one of the variables). Typically, some of the variables correspond to observations whose values are known, and hence do not need to be sampled.

The Gibbs sampling setting is similar to the one described above with the distribution of a MRF, and Gibbs sampling can be used to obtain approximate solutions of the corresponding MRF distribution. Gibbs sampling is commonly used as a means of statistical inference, especially Bayesian inference. It is a randomized algorithm (i.e. an algorithm that makes use of random numbers, and hence may produce different results each time it is run), and it is an alternative to deterministic algorithms for statistical inference such as variational Bayes or the expectation-maximization (EM) algorithm [102].

The concept behind Gibbs sampling is that it generates a Markov chain of samples, each of which is correlated with nearby samples. As a result, care must be taken if independent samples are desired (typically by thinning the resulting chain of samples by only taking every nth value, e.g. every 100th value). In addition samples from the beginning of the chain (the burn-in period) may not accurately represent the desired distribution.

In the basic flavor of Gibbs sampling, it is a special case of the Metropolis-Hastings algorithm [186]. It capitalizes on the fact that given a multivariate distribution it is simpler to sample from a conditional distribution than to marginalize by integrating over a joint distribution. Suppose one wants to obtain k samples of $\mathbf{X} = (x_1,\ldots,x_n)$ from a joint distribution $p(x_1,\ldots,x_n)$. Denote the ith sample by $\mathbf{X}^{(i)} = (x_1^{(i)},\ldots,x_n^{(i)})$. The algorithm proceeds as follows:

- Begin with some initial value $\mathbf{X}^{(0)}$.
- To get the next sample (call it the $(i+1)$th sample for generality) sample each component variable $x_j^{(i+1)}$ from the distribution of that variable conditioned on all other variables, making use of the most recent values and updating the variable with its new value as soon as it has been sampled. This requires updating each of the component variables in turn. Up to the jth component, it is updated according to the distribution specified by $p(x_j|x_1^{(i+1)},\ldots,x_{j-1}^{(i+1)},x_{j+1}^{(i)},\ldots,x_n^{(i)})$. It should be noted that for the $(j+1)$th component, the value used is the one it had in the ith sample not the $(i+1)$th.
- Repeat the above step k times.

If such sampling is performed, three key facts hold:

1. The samples approximate the joint distribution of all variables.
2. The marginal distribution of any subset of variables can be approximated by simply considering the samples for that subset of variables, ignoring the rest.

3. The expected value of any variable can be approximated by averaging over all the samples.

Consequently, given the aforementioned facts, when performing the sampling, the following hold:

- The initial values of the variables can be determined randomly or by some other algorithm such as EM.
- It is not actually necessary to determine an initial value for the first variable sampled.
- It is common to ignore some number of samples at the beginning (the so-called burn-in period), and then consider only every nth sample when averaging values to compute an expectation. For example, the first 1000 samples might be ignored, and then average every 100th sample, throwing away all the rest. The reason for this is that successive samples are not independent of each other but form a Markov chain with some amount of correlation and the stationary distribution of the Markov chain is the desired joint distribution over the variables, but it may take a while for that stationary distribution to be reached. Sometimes, algorithms can be used to determine the amount of autocorrelation between samples and the value of n (the period between samples that are actually used) computed from this, but in practice there is a fair amount of rule of thumb experience involved.
- The process of simulated annealing (SA) is often used to reduce the "random walk" behavior in the early part of the sampling process (i.e. the tendency to move slowly around the sample space, with a high amount of autocorrelation between samples, rather than moving around quickly, as is desired). Other techniques that may reduce autocorrelation are collapsed Gibbs sampling, blocked Gibbs sampling, and ordered overrelaxation; see below.

Numerous variations of the basic Gibbs sampler exist. The more interested reader is referred to an extensive literature, e.g. [83, 186] and references therein for more details.

Simulated annealing

Simulated Annealing (SA) [139] is a generic probabilistic meta-heuristic for the global optimization problem of locating a good approximation to the global optimum of a given function in a large search space. It is often used when the search space is discrete, as it is the case with MRF configuration in malware diffusion. For certain problems, SA may be more efficient than exhaustive enumeration — provided that the goal is merely to find an acceptably good solution in a fixed amount of time, rather than the best possible solution.

The name and inspiration come from annealing in metallurgy, a technique involving heating and controlled cooling of a material to increase the size of its crystals and reduce their defects. This notion of "slow cooling" is implemented in the SA algorithm as a slow decrease in the probability of accepting worse solutions as it explores the solution space. The temperature parameter defined for Gibbs distributions and MRFs previously in Section 5.2.1 is closely connected to this concept. Accepting worse solutions is a fundamental property of meta-heuristics because it allows for a more extensive search for the optimal solution. The method is an adaptation of the Metropolis-Hastings algorithm, a Monte Carlo method to generate sample states of a thermodynamic system [158].

At each step, the SA heuristic considers some neighboring state s' of the current state s, and probabilistically decides between moving the system to state s' or staying in state s. These probabilities ultimately lead the system to move to states of lower energy. Typically this step is repeated until the system reaches a state that is good enough for the application, or until a given computation budget has been exhausted. The neighboring states are new states of the problem that are produced after altering a given state in some well-defined way. The well-defined way in which the states are altered in order to find neighboring states is called a "move" and different moves give different sets of neighboring states. These moves usually result in minimal alterations of the last state, as the previous example depicts, in order to help the algorithm keep the better parts of the solution and change only the worse parts.

Searching for neighboring states is fundamental to optimization because the final solution will come after a sequence of successive neighbors. Simple heuristics move by finding best neighbor after best neighbor and stop when they have reached a solution which has no neighbors that are better solutions. The best solution found by such algorithms is called a local optimum in contrast with the actual best solution which is called a global optimum. Meta-heuristics use the neighbors of a solution as a way to explore the solutions space and although they prefer better neighbors they also accept worse neighbors in order to avoid getting trapped in local optima. As a result, if the algorithm is run for an infinite amount of time, in principle, the global optimum will be found. In practice, the desired solution, or a very close approximation sufficient for the purposes of each problem, is found rather quickly.

The probability of making the transition from the current state s to a candidate new state s' is specified by an acceptance probability function $P(e, e', T)$ that depends on the energies $e = E(s)$ and $e' = E(s')$ of the two states, and on a global time-varying parameter T called the temperature. The MRF temperature has a similar role to T for SA. States with a smaller energy are better than those with a greater energy. The probability function P must be positive even when e' is greater than e. This feature prevents the method from becoming trapped at a local minimum that is worse than the global one. When T tends to zero, the probability $P(e, e', T)$ must tend to zero if $e' > e$ and to a positive value otherwise. For sufficiently small values of T, the system will then increasingly favor moves that go "downhill" to lower energy values. With $T = 0$, the procedure reduces to the greedy algorithm, which makes only the downhill transitions. The temperature T plays a crucial role in controlling the evolution of the state s of the system with regard to its sensitivity to the variations of system energies. To be precise, for a large T, the evolution of s is sensitive to coarser energy variations, while it is sensitive to finer energy variations when T is small.

The name and inspiration of the algorithm demand an interesting feature related to the temperature variation to be embedded in the operational characteristics of the algorithm. This necessitates a gradual reduction of the temperature as the simulation proceeds. The algorithm starts initially with T set to a high value (or infinity), and then it is decreased at each step following some annealing schedule that may be specified by the user, but must end with $T = 0$ toward the end of the allotted time budget. In this way, the system is expected to wander initially toward a broad region of the search space containing good solutions, ignoring small features of the energy function; then drift toward low-energy regions that become narrower and narrower; and finally move downhill according to the steepest descent heuristic. For any given finite problem, the probability that the SA algorithm terminates with a global optimal solution approaches 1 as the annealing schedule is extended [28]. However, it should be noted this is a result of theoretical value, since the time required to ensure a significant probability of success will usually exceed the time required for a complete search of the solution space and in practice, acceptable solutions may be found with good accuracy much faster.

In the sequel, we first introduce the use of MRFs and Gibbs distributions/sampling for modeling malware diffusion in arbitrary types of complex networks and then provide some applications of the framework for specific types of such wireless or wired complex communications networks, such as regular, random, SW, SF, and random geometric, examples of which are shown in Fig. 5.2. Fig. 5.2 demonstrates a potential configuration of susceptible-infected nodes, where the shaded nodes correspond to infected nodes and the rest to susceptible.

5.3 MALWARE DIFFUSION MODELING BASED ON MRFS

One of the key observations regarding the dynamics of malware diffusion in various types of networks, and especially highlighted in wireless *ad hoc* networks as shown in the previous chapter, is dictated by interactions of adjacent nodes (corresponding to users directly connected with each other). Such local interactions can be interpreted as local dependencies of the corresponding users/nodes and this triggers employing MRFs for modeling such interactions/dependencies. Since MRFs generalize the notion of the spatial Markov property [136], it will become possible to capture the aforementioned dependencies in a holistic manner, uniformly for all types of complex communications networks. In this section, we present such techniques utilizing MRFs, which in addition, characterize malware behavior in the long-run irrespectively of the specific type(s) of threat(s), similarly to the macroscopic analysis of the previous chapter.

Most of the previously presented models (especially epidemics) are devoted to the development of malware-specific models [232, 235]. The MRF-based framework presented in this chapter is closer to models such as [82] and [80] sharing the same objectives in terms of modeling goals and essentially analyzing the "macroscopic" behavior of the attacked system, i.e. the cumulative states of the studied systems, not the exact states of all nodes in the legitimate network. However, the MRF-based

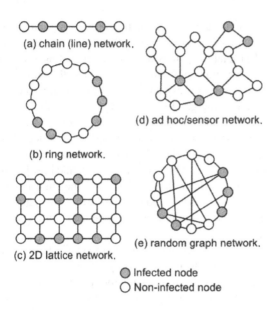

(a) chain (line) network.

(b) ring network.

(c) 2D lattice network.

(d) ad hoc/sensor network.

(e) random graph network.

● Infected node
○ Non-infected node

FIGURE 5.2

Examples of complex network topologies of interest.

framework differs, in that it is able to treat more general topologies and application scenarios (i.e. types of attack) in a uniform approach, as opposed to the previous ones [80, 82]. Compared to [82] and [80], the MRF framework focuses on local interactions and system transitions, the aggregated system state, and its steady state, while topological information is embedded in the MRF framework by design. The MRF-based framework allows to identify the local topological subtleties that affect malware diffusion better than previous approaches, and most importantly, it allows controlling those factors in a simple and efficient manner in terms of implementation complexity.

The MRF formulation is especially suitable for all types of complex networks. The first step is to properly define a suitable neighborhood system irrespective of the type of network and application objectives. In the following, we analyze the details of the framework and briefly show how it can be applied for various types of complex wireless communications networks.

Initially, an arbitrary malware-propagative network with n nodes in total is considered. Thus, the infected legitimate nodes will be capable of further spreading the malware they received to other legitimate nodes in the susceptible state. Given that, the network is assumed to follow the SIS infection paradigm [177], in which legitimate nodes oscillate between the susceptible (noninfected) and infected state.

The mapping of malware diffusion in arbitrary types of complex networks is straightforward. Each of the legitimates nodes of the analyzed network, corresponds to a MRF site. Thus, the sites $s \in S$ of the MRF will correspond to the nodes of the network, and the collection $X = \{X_s, s \in S\}$ of random variables to the states of the sites/nodes (network users). The values of the phase space Λ will be as many as the states of the infection model. Thus, each phase will denote a different state, and the random variable associated with each site of the MRF will denote the current state of the corresponding legitimate node. This means that a configuration $x = \{x_s, s \in S\}$, where $x_s \in \Lambda$ of the MRF, will correspond to one of all the possible states of the legitimate network, i.e. a cumulative state that denotes in which state of the infection model each legitimate user lies.

In this book, we will present and explain the simple case, where the assumed infection model is SIS; thus, only two states are possible for legitimate nodes. Consequently, the phase space will have cardinality two and the infected legitimate nodes (sites) will have the same phase as the sites corresponding to the malicious (attack) nodes. It should be also noted that attackers can be treated as normal sites, with the only difference that their state does not change according to the infection model, but rather it is fixed to the phase value denoting the infected state.

For the malware diffusion MRF model, a neighborhood system on S is defined as a family $\mathcal{N} = \{\mathcal{N}_s\}_{s \in S}$ of subsets $\mathcal{N}_s \subset S$, dictated by the underlying physical topology of the complex legitimate network. It is noted that as it was mentioned in the first part of the book, this framework of malware modeling approaches is focused on describing malware diffusion through direct (physical) connections as well, leaving the indirect cases for future research endeavors that can be based on the direct models or not. In the above sense, the neighborhood system of the malware diffusion modeling MRF is determined by the network graph of the communication network, and the neighborhood of each site s coincides with the neighborhood of each node in the network graph. Some indicative examples with respect to the topologies presented in Fig. 5.2 are shown in Fig. 5.3, where the neighborhood of a randomly selected site s (darkly blue shaded) is shown in each case. The neighborhood of each site is denoted as $\mathcal{N}_s$, as noted above.

Assuming all the above regarding topology, neighborhood systems, and sites, it remains to specify a suitable potential function for the considered MRF. A suitable potential function can be of the form $U(x) = \sum_k \Phi_k(x)$, $x = \{x_k, 1 \le k \le n\}$, where $\Phi_k(x)$ depends only on x_k (state of the considered site) and the state random variables in $\mathcal{N}_k$, i.e. states of site k neighbors. The specification of each $\mathcal{N}_k$ varies for each network type considered. For an arbitrary network, the clique potential functions can be obtained as

$$\Phi_k(x) = \hat{\Phi}_k(x_k, \{x_{k'} : k' \in \mathcal{N}_k\}), \tag{5.4}$$

where $\mathcal{N}_k$ is defined by each network topology, and $\{x_{k'} : k' \in \mathcal{N}_k\}$ may be considered as a state subvector corresponding to the neighborhood of a site k. In that sense, the above expression of potential function depends only on the state of the site k and the state subvector of its neighborhood.

Furthermore, depending on the neighborhood system formed in each network, the clique potential may contain singleton, pairwise, and other higher-order clique terms, corresponding to the cardinality of local cliques forming in a network. Thus,

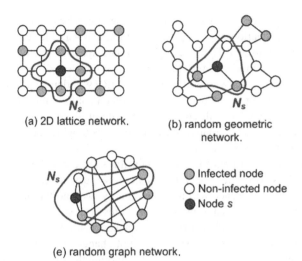

(a) 2D lattice network.

(b) random geometric network.

(e) random graph network.

- ○ Infected node
- ○ Non-infected node
- ● Node s

FIGURE 5.3

Examples of neighborhood for the darkly blue shaded (black in print versions) node (site) s in topologies of interest.

if a networks forms cliques up to size three (single nodes, pairs of nodes, triangles of nodes), the clique potential can have up to three terms. The choice of clique terms to include depends on the specific application context and the operation of the actual nodes modeled by the sites of the MRF.

In the following sections, according to the specific topology of each type of considered network, a specific expression for clique potential (5.4) is obtained and then used to derive the cumulative potential function that will be used for performing Gibbs sampling and obtain solutions for the steady-state behavior of the systems. With respect to the order of cliques considered, the employed clique potentials include only pairwise terms, since singleton terms are typically used to represent external factors affecting each node individually, and higher-order terms (three and above) do not have a straightforward interpretation in malware terms (machines typically interact on a point-to-point basis for malware diffusion not in interacting groups).

5.4 REGULAR NETWORKS

In this section, we consider regular networks, i.e. networks where the node degree distribution is either a constant value for all nodes, or a very restricted subset of specific constant values (e.g. three distinct values as in finite grids studied in

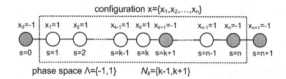

FIGURE 5.4

SIS malware-propagative chain network and MRF notation.

-From © 2010 IEEE. Reprinted, with permission, from Karyotis V. Markov random fields for malware propagation: the case of chain networks. IEEE Commun Lett 2010;14(9): 875–7.

Section 5.4.2) [23]. In other words, all nodes have the same degree, or there exist a few subsets of nodes, where in each subset, the nodes have the same node degree. Regular networks can be of many types, i.e. chains, rings, grids, and in general lattices in higher-dimension spaces, some examples of which are shown in Fig. 5.2. In the following, we present an exhaustive analysis of the chain network case, and then extend it to regular lattices. Extensions for higher dimensional lattices and other regular networks can be performed in a straightforward manner.

5.4.1 CHAIN NETWORKS

Chain networks, which are similar to paths, arise often in wired networks [18], especially at the transport layer under the TCP, and similarly in various types of multihop networks, such as sensor and *ad hoc*. A protocol generic example of a chain network is depicted in Fig. 5.4 [116], consisting of nodes that connect only to two neighbors, except for two nodes, denoted as endpoints, which connect only to one neighbor. In the legitimate network of Fig. 5.4, nodes-sites 1 and n are the endpoints of the chain, and nodes 0, $n + 1$ correspond to attackers that attach to the chain in order to diffuse malware to legitimate users sites.

In order to use the previously presented MRF model for malware diffusion, every node of the chain corresponds to a site of the malware diffusion modeling MRF. For each site $k = 1, 2, ..., n$, its neighborhood includes $N_k = \{k - 1, k + 1\}$, i.e. two neighbors, since even the endpoints connect to the attack nodes, in addition to their one legitimate neighbor. For the phase space, and given the SIS assumed paradigm, it needs only have two possible values, $\Lambda = \{-1, 1\}$, where "-1" corresponds to an infected state and "1" corresponds to the noninfected (susceptible) state. Alternatively, $\Lambda = \{0, 1\}$ with "0" corresponding to the infected state, as will be employed in the case of stochastic topologies in the next section. Such assumption for the phase values does not harm the generality of the formulation and it is a matter of implementation convenience.

Given the above neighborhood system, admittedly a very simple one, the potential function considered has the form $U(x) = \sum_k \Phi_k(x)$, $x = \{x_k, 1 \le k \le n\}$, where

$\Phi_k(x)$ depends only on x_k and the state random variables in $\mathcal{N}_k$, i.e. x_{k-1}, x_{k+1} (i.e. it is a form of clique potential mentioned above). Potential function $U(x)$ is decomposed into clique potentials, in order to facilitate simple computation and highlight the locality of interaction in malware diffusion. Thus, for chain networks, the clique potentials can be obtained as

$$\Phi_k(x) = \hat{\Phi}_k\left(x_k, \{x_{k'} : k' \in \mathcal{N}_k\}\right) = \hat{\Phi}_k\left(x_k, \{x_{k-1}, x_{k+1}\}\right). \tag{5.5}$$

The MRF model will be focused on accurately modeling the interactions between neighboring sites-nodes at different states, i.e. S(usceptible)-I(nfected) and I-S. Those pairs of sites, i.e. S-I and I-S, essentially drive the evolution of the system with respect to malware propagation. Assuming that $\sigma_i(x) = \sigma_i(x_i)$ is a function of the current state of site i, and that D is a proper scaling factor employed to collect all emerging scaling issues in the computations, one may select the clique potential as $\Phi_k(x) = \Phi_k(x_k) = \sigma_k \sigma_{k-1} + \sigma_k \sigma_{k+1}$ for each site (node). Then, the overall system potential function may be obtained in the more general form: $U(x) = -D \sum_{(i,j)} \sigma_i \sigma_j$ for all neighboring pairs (i,j) in the chain network. For simplicity and without loss of generality, we assume $\sigma_i(x_i) = x_i$, i.e. linear dependence on the system state, and in this case D accounts for the product of linear scaling factors in interactions of the type $\sigma_i \sigma_j \forall i, j \in \{1, 2, ..., n\}$ applicable. However, any bijective function of x_i would work similarly under proper scaling, yielding analytic results. Again D would collect all scaling factors involved. Consequently, in this case the form of the potential function of the chain network will be

$$U(x) = -J \sum_{k=0}^{n} x_k x_{k+1}, \tag{5.6}$$

where J is a proper scaling factor that now collects all the involved linear factors emerging in the summation of partial clique potentials.

As already explained in Sections 5.2.1 and 5.2.2, computing directly the steady-state distribution of the designed MRF is tough, mainly due to the exponential increase of the state space of the system, especially as the network becomes large-scale, which complicates the computation of the partition function of the specific MRF. Alternatively, SA can be combined with the Gibbs sampler in order to analyze the evolution of the system state, as it was suggested in [83] for the case of image processing.

In order to compute the steady-state distribution of the MRF, the sites of the MRF (nodes of the network) are visited once and sequentially in a random fashion (such a visit is denoted as *sweep*) and the state of each site is updated only according to the state of its neighboring sites (network neighbors) and parameters associated with the infection and recovery rates. The process is repeated for a number of sweeps, until the system converges to its steady state, or a fairly large number of sweeps performed that yield satisfactory performance (the latter typically discovered by simple trial-and-error). This is an example of a sequential implementation, where all updates in sweep k are performed according to the states of sweep $k - 1$. On the contrary, in a parallel implementation of the sweep process, updates in sweep k take into account states at sweep $k - 1$ for states not updated currently, and states at sweep k for those sites that the state has been updated already. This process resembles that of distributed consensus among distributed agents [17, 106, 167]. Of course, hybrid intermediate implementations are also applicable combining benefits and drawbacks of the sequential and parallel approaches. In this book, we only consider sequential implementations.

In practice, convergence is achieved quickly for sequential sweep paradigms, a lot faster than the maximum number of sweeps (a practical setting for the maximum sweep number is in the order of thousands or tens of thousand in demanding scenarios). Prior to starting the "sweeping" of the system, a temperature $T(\cdot)$ and the total number of sweeps I are selected. For each sweep, a randomized visiting scheme is determined. For each node in the random sequence obtained, a decision is made on whether its state should remain the same for the next sweep or change. For demonstration purposes, we focus on binary decisions, but more states could be defined reflecting different infection or noninfection states, i.e. recovering, removed, or other states of the node infection model. For instance, if one is interested on how many infections a node currently suffers, $L + 1$ states are required, state 0 corresponding to no infection and the rest $1, 2, \cdots, L$, denoting the number of different infections a node suffers (L being the maximum number of infections propagating in the network). For $\Lambda = \{-1, 1\}$ (two states as explained above) and $\ell \in \Lambda$ we have $\Phi_k(x_k^{(\ell)}) \doteq \hat{\Phi}_k(x_k = \ell, \{x_{k-1}, x_{k+1}\})$. Then, from Gibbs distribution (5.2), the probability that the state of node k will become $x_k = \ell$, may be obtained as

$$\mathbb{P}(x_k = \ell) = \frac{e^{-\frac{\Phi_k(x_k^{(\ell)})}{T(t)}}}{\sum_{\ell' \in L_k} e^{-\frac{\Phi_k(x_k^{(\ell')})}{T(t)}}}, \tag{5.7}$$

where L_k refers to the possible states of the nodes $k' \in \mathcal{N}_k$. Furthermore, given the potential $\Phi_k(x_k)$, we have

$$\begin{aligned}
\Phi_k\left(x_k^{(-1)}\right) &= \hat{\Phi}_k(x_k = -1, x_{k-1}, x_{k+1}) \\
&= \sigma_k(x)\sigma_{k-1}(x) + \sigma_k(x)\sigma_{k+1}(x) \\
&= x_k x_{k-1} + x_k x_{k+1} = -(x_{k-1} + x_{k+1})
\end{aligned} \tag{5.8}$$

and

$$\Phi_k\left(x_k^{(1)}\right) = \hat{\Phi}_k(x_k = 1, x_{k-1}, x_{k+1}) = x_{k-1} + x_{k+1}. \tag{5.9}$$

Consequently, the probability that the next state of node k is $x_k = 1$ is obtained as

$$\mathbb{P}(x_k = 1) = \frac{1}{1 + e^{-\frac{\left(\Phi_k(x_k^{(-1)}) - \Phi_k(x_k^{(1)})\right)}{T(t)}}} = \frac{1}{1 + e^{\frac{2(x_{k-1} + x_{k+1})}{T(t)}}}. \tag{5.10}$$

Similarly, the probability that the next state of node k is $x_k = -1$ is obtained as

$$\mathbb{P}(x_k = -1) = \frac{1}{1 + e^{-\frac{\left(\Phi_k(x_k^{(1)}) - \Phi_k(x_k^{(-1)})\right)}{T(t)}}} = \frac{1}{1 + e^{-\frac{2(x_{k-1} + x_{k+1})}{T(t)}}}, \tag{5.11}$$

where it is straightforward to verify that $\mathbb{P}(x_k = 1) = 1 - \mathbb{P}(x_k = -1)$. Consequently, in each sweep of the chain, the state of each site is updated according to $\mathbb{P}(x_k = 1)$ or $\mathbb{P}(x_k = -1)$. If one employed the mapping

$$\sigma_k(x) = \sigma(x_k) = \begin{cases} 1, & \text{if } x_k = 1 \\ 0, & \text{if } x_k = -1 \end{cases}, \tag{5.12}$$

the expressions obtained for $\mathbb{P}(x_k = 1)$, $\mathbb{P}(x_k = -1)$ would be similar to those of (5.10) and (5.11), respectively, without the factor 2 in the exponential term of the denominator.

The effect of SA is more evident when the whole system is considered. Defining configuration $\bar{x}$ for a site k according to configuration x, where in $\bar{x}$ the state of site k has been switched, then expression

$$\frac{e^{-bU(\bar{x})}}{e^{-bU(\bar{x})} + e^{-bU(x)}} = \frac{1}{1 + e^{b(U(\bar{x}) - U(x))}} \tag{5.13}$$

may be used to determine the next state of k as well, where $U(x)$ is given by Eq. (5.6). Parameter b is a proper constant, so that $b = J/T$ eventually acts as an inverse temperature factor, with the corresponding effects as that of the T anticipated. It should be noted that here a constant value of the temperature T is used, as opposed to the case of complex networks in the next section, where sweep-varying temperatures will be employed.

The case of a ring network is identical to the chain, where now the original chain network folds around and the two attackers can collapse to just one malware source to propagate the malware toward both directions of the chain. Thus, if nodes $s = 0, n+1$ are considered as one in Fig. 5.4, the above analysis holds and the results obtained apply straightforwardly with minor adaptations on the neighborhoods of legitimate edge nodes.

In the following, we provide some indicative results regarding MRFs modeling malware diffusion in chain networks. Without loss of generality random sweeps over the chain network are performed in a topology similar to that presented in Fig. 5.4, i.e. nodes are visited randomly and only once in each sweep. A total of $S = 10,000$ sweeps are performed in each scenario and the results are averaged over 50 different scenarios.

Fig. 5.5 presents the distribution of the ergodic probabilities of the system, i.e. probabilities $\pi(i)$ that the system is at state i. The state of the system is defined as the number of infected nodes. The main observation is that as the size of the chain increases, it becomes tougher to increase the number of infected nodes (nonvanishing

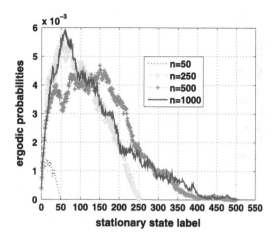

FIGURE 5.5

Steady-state system distributions for $T/J = -0.2$.

-From © 2010 IEEE. Reprinted, with permission, from Karyotis V. Markov random fields for malware propagation: the case of chain networks. IEEE Commun Lett 2010;14(9): 875–7.

$\pi(i)$'s remain concentrated at lower states). This is because in a lengthy chain, the state of a node in a specific location has little effect on the state of another node far away. Thus, it is tough for sources at the two ends to eventually propagate malware to nodes located around the middle of the chain. In relatively small chains, such as the one with $n = 50$ nodes, propagation becomes easier with nonvanishing values for all states.

Fig. 5.6 shows the average number of infected nodes for various chain sizes and different temperatures T/J, thus revealing the impact of network size and the T/J parameter (directly affecting sampling and representing the intensity of malware in the potential function) on the behavior of the system and the results obtained, respectively. The greater n is, the greater the expected number of infected nodes in the network, as expected, signifying the validity of the model. However, this does not mean necessarily a greater percentage of infected nodes. In fact, there seems to exist a critical n value, for which system behavior changes drastically in the range $-0.5 < T/J < 0$. Above the critical n threshold the percentage of infected nodes drops and increases sharply, whereas below the threshold, only a small drop is observed as $T/J \to 0^-$. Furthermore, since T/J may be considered as the analog of the infection over recovery rate (indicative of infection-recovery capabilities), it is shown that for the range $-0.5 < T/J < 0$ attack potentials are practically limited, irrespective of network size. This is an important factor that should be taken into account in designing efficient countermeasures against generic attack types. In addition, such outcome should be taken into account in practical applications, calling

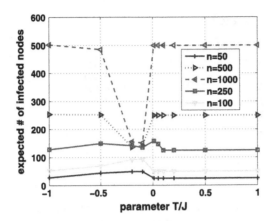

FIGURE 5.6

Expected number of infected nodes.

-From © 2010 IEEE. Reprinted, with permission, from Karyotis V. Markov random fields for malware propagation: the case of chain networks. IEEE Commun Lett 2010;14(9): 875–7.

for accurate mapping of the infection/recovery processes of the analyzed malware to the parameters of the model and especially T/J, in order to capture accurately the exact rates.

5.4.2 REGULAR LATTICES: FINITE AND INFINITE GRIDS

In two-dimensional regular networks [23], i.e. lattices as the one shown in Fig. 5.7, it is straightforward to extend the formulation developed for the chain network. In the two-dimensional lattice, each node is indexed by i, j, denoting the row and column the node belongs to, respectively. Thus, each site has the form $s_{i,j} = (i, j)$. The state of site $s_{i,j}$ is denoted by $x_{i,j}$. The neighborhood of node $s_{i,j}$ is $\mathcal{N}_{i,j} = \{(i - 1, j), (i + 1, j), (i, j - 1), (i, j + 1)\}$. Similarly to the chain case, the potential function expressed as $\Phi_{i,j}(x) = \hat{\Phi}_{i,j}\left(x_{i,j}, \{x_{(i,j)'} : (i, j)' \in \mathcal{N}_{i,j}\}\right)$ can be obtained in the form

$$\Phi_{i,j}(x_{i,j}^{(\ell)}) = \hat{\Phi}_{i,j}\left(x_{i,j} = \ell, \{x_{i-1,j}, x_{i+1,j}, x_{i,j-1}, x_{i,j+1}\}\right)$$
$$= \ell(x_{i-1,j} + x_{i+1,j} + x_{i,j-1} + x_{i,j+1}). \tag{5.14}$$

Consequently, the probability that the next state of site $s_{i,j}$ is $x_{i,j} = \ell$ is given by

$$\mathbb{P}(x_{i,j} = \ell) = \frac{1}{1 + e^{-\frac{\left(\Phi_{i,j}(x_{i,j}^{(\ell')}) - \Phi_{i,j}(x_{i,j}^{(\ell)})\right)}{T(t)}}}, \tag{5.15}$$

which for $\ell = 1$ equals

$$\mathbb{P}(x_{i,j} = 1) = \cfrac{1}{1 + e^{\frac{2(x_{i-1,j}+x_{i+1,j}+x_{i,j-1}+x_{i,j+1})}{T(t)}}}, \tag{5.16}$$

while for $\ell = -1$ equals

$$\mathbb{P}(x_{i,j} = -1) = \cfrac{1}{1 + e^{\frac{-2(x_{i-1,j}+x_{i+1,j}+x_{i,j-1}+x_{i,j+1})}{T(t)}}}. \tag{5.17}$$

In practical scenarios, finite lattices arise in applications, as the one shown in Fig. 5.7, where the attacker(s) may roam in the network under the infection models described in the previous chapters (e.g. Chapter 2). We consider such a $I \times J$ lattice, where each node is indexed as in the infinite lattice case by $1 \leq i \leq I$, $1 \leq j \leq J$. In a finite lattice network, nodes belong to one of the three groups as shown in Fig. 5.7. Group I includes the four corner nodes, each of which has only two neighbors. Group II can be separated in four subgroups, each in one of the four boundaries of the finite lattice, excluding the nodes of Group I, where each node has exactly three neighbors. Finally, nodes in Group III include all nodes with $2 \leq i \leq I - 1$, $2 \leq j \leq J - 1$, each of which has exactly four neighbors and behaves as if in an infinite lattice. The neighborhoods of nodes in Group I are

$$\mathcal{N}_{1,1}^I = \{x_{1,2}, x_{2,1}\}, \qquad \mathcal{N}_{1,J}^I = \{x_{1,J-1}, x_{2,J}\}, \tag{5.18}$$

$$\mathcal{N}_{I,1}^I = \{x_{I,2}, x_{I-1,1}\}, \qquad \mathcal{N}_{I,J}^I = \{x_{I-1,J}, x_{I,J-1}\}. \tag{5.19}$$

The neighborhoods of nodes in Group II are

$$\mathcal{N}_a^{II} = \mathcal{N}_{i,1}^{II} = \{x_{i-1,1}, x_{i+1,1}, x_{i,2}\}, \qquad \mathcal{N}_b^{II} = \mathcal{N}_{1,j}^{II} = \{x_{1,j-1}, x_{1,j+1}, x_{2,j}\}, \tag{5.20}$$

$$\mathcal{N}_c^{II} = \mathcal{N}_{I,j}^{II} = \{x_{I,j-1}, x_{I,j+1}, x_{I-1,j}\}, \qquad \mathcal{N}_d^{II} = \mathcal{N}_{i,J}^{II} = \{x_{i-1,J}, x_{i+1,J}, x_{i,J-1}\}, \tag{5.21}$$

and finally, the neighborhood of nodes in Group III is

$$\mathcal{N}_{i,j}^{III} = \{x_{i-1,j}, x_{i+1,j}, x_{i,j-1}, x_{i,j+1}\}. \tag{5.22}$$

Following a similar approach as the one for the chain network, it is possible to construct a potential function based on the above neighborhood systems and then obtain an update scheme for a proper Gibbs sampling procedure.

In Group I, each node has essentially a different potential function, however of similar type, and the same holds for the state transition probabilities. Thus, we have the following potential function expressions:

$$\Phi_{1,1}(x^{(\ell)}) = \ell(x_{1,2} + x_{2,1}), \qquad \Phi_{1,J}(x^{(\ell)}) = \ell(x_{1,J-1} + x_{2,J}),$$

$$\Phi_{I,1}(x^{(\ell)}) = \ell(x_{I-1,1} + x_{I,2}), \qquad \Phi_{I,J}(x^{(\ell)}) = \ell(x_{I,J-1} + x_{I-1,J}),$$

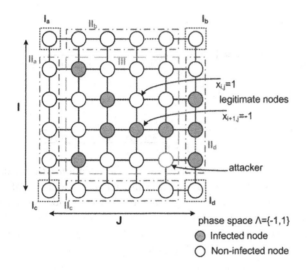

FIGURE 5.7

Lattice network and MRF malware diffusion model notation.

and corresponding state transition probabilities

$$\mathbb{P}(x_{1,1} = 1) = \frac{1}{1 + e^{\frac{2(x_{1,2}+x_{2,1})}{T(t)}}} = 1 - \mathbb{P}(x_{1,1} = -1),$$

$$\mathbb{P}(x_{1,J} = 1) = \frac{1}{1 + e^{\frac{2(x_{1,J-1}+x_{2,J})}{T(t)}}} = 1 - \mathbb{P}(x_{1,J} = -1),$$

$$\mathbb{P}(x_{I,1} = 1) = \frac{1}{1 + e^{\frac{2(x_{I-1,1}+x_{I,2})}{T(t)}}} = 1 - \mathbb{P}(x_{I,1} = -1),$$

$$\mathbb{P}(x_{I,J} = 1) = \frac{1}{1 + e^{\frac{2(x_{I,J-1}+x_{I-1,J})}{T(t)}}} = 1 - \mathbb{P}(x_{I,J} = -1).$$

Similarly, for nodes in Group II, we obtain

$$\Phi_{i,1}(x^{(\ell)}) = \ell(x_{i-1,1} + x_{i+1,1} + x_{i,2}) \text{ (II.a)},$$

$$\Phi_{1,j}(x^{(\ell)}) = \ell(x_{1,j-1} + x_{1,j+1} + x_{2,j}) \text{ (II.b)},$$

$$\Phi_{I,j}(x^{(\ell)}) = \ell(x_{I,j-1} + x_{I,j+1} + x_{I-1,j}) \text{ (II.c)},$$

$$\Phi_{i,J}(x^{(\ell)}) = \ell(x_{i-1,J} + x_{i+1,J} + x_{i,J-1}) \text{ (II.d)},$$

and corresponding state transition probabilities

$$\mathbb{P}(x_{i,1} = 1) = \frac{1}{1 + e^{\frac{2(x_{i-1,1}+x_{i+1,1}+x_{i,2})}{T(t)}}} = 1 - \mathbb{P}(x_{i,1} = -1) \text{ (II.a)},$$

$$\mathbb{P}(x_{1,j} = 1) = \frac{1}{1 + e^{\frac{2(x_{1,j-1}+x_{1,j+1}+x_{2,j})}{T(t)}}} = 1 - \mathbb{P}(x_{1,j} = -1) \text{ (II.b)},$$

$$\mathbb{P}(x_{I,j} = 1) = \frac{1}{1 + e^{\frac{2(x_{I,j-1}+x_{I,j+1}+x_{I-1,j})}{T(t)}}} = 1 - \mathbb{P}(x_{I,j} = -1) \text{ (II.c)},$$

$$\mathbb{P}(x_{i,J} = 1) = \frac{1}{1 + e^{\frac{2(x_{i-1,J}+x_{i+1,J}+x_{i,J-1})}{T(t)}}} = 1 - \mathbb{P}(x_{i,J} = -1) \text{ (II.d)}.$$

Finally, for nodes in Group III, the potential function is

$$\Phi_{i,j}(x_{i,j}^{(\ell)}) = \ell(x_{i-1,j} + x_{i+1,j} + x_{i,j-1} + x_{i,j+1}) \tag{5.23}$$

for $\ell = 1$ or $\ell = -1$. Such potential yields

$$\mathbb{P}(x_{i,j} = 1) = \frac{1}{1 + e^{\frac{2(x_{i-1,j}+x_{i+1,j}+x_{i,j-1}+x_{i,j+1})}{T(t)}}} \tag{5.24}$$

and $\mathbb{P}(x_{i,j} = -1) = 1 - \mathbb{P}(x_{i,j} = 1)$. This expression is identical to the infinite lattice case, as expected.

Numerical and qualitative results for lattice networks are similar to those provided for chain networks. However, the second dimension and the relative size of the lattice do have a role in the speed-up/slow-down of malware diffusion. The second dimension further aids in the diffusion of malware up to a certain degree.

In general, the larger the lattice is, the more similar the behavior of the finite lattice to that of the infinite, as the corner effects of the finite grid become negligible. Furthermore, for long chains, their behavior in terms of emerging trend resembles that of a single row of a very large finite lattice (or equivalently the row of an infinite lattice). There is only a constant malware diffusion acceleration factor observed on the long-term behavior of lattice row case, due to the effect of the neighboring rows.

5.5 COMPLEX NETWORKS WITH STOCHASTIC TOPOLOGIES

In this section, we extend the previous MRF framework for the case of complex networks characterized by stochastic topologies. We specifically focus on purely random networks, small-world (SW) topologies, scale-free (SF) networks, and random geometric graphs (RGGs). Each of these topologies corresponds to a different type of wireless network. Random networks describe successfully *ad hoc* or sensor networks where all nodes can potentially connect to each other, i.e. they are within transmission range of each other. SW wireless networks correspond to distributed networks where nodes connect to a few neighbors close to them; however, some nodes can develop some longer term connections between wireless terminals, e.g. the majority of nodes can be smartphones, while a few of them laptops with 4G/5G capabilities [3, 98, 108, 111, 184, 196, 217]. SF networks represent wireless networks where a few nodes connect to a significant number of nodes, while most of them connect to only a few nodes, e.g. sensor networks with sink nodes. Finally, RGGs represent the typical wireless distributed (multihop) paradigm, where connections are spatially defined. Of course, the random, SW, and SF topologies can be encountered in wired networks, and thus, the methodology presented in this section can be straightforwardly adapted and used to obtain results for arbitrary random, SW, and SF networks of interest.

In the following, we provide the general expressions for the malware diffusion modeling MRF framework for a network of arbitrary type, and then present more specific results for each different topology. For each network, we consider a graph $G(V,E)$ representing the topology with V the set of nodes and E the set of links between them. This defines a neighborhood system N of subsets of V and each node k is associated with its neighborhood N_k, which is the set of one-hop neighbors of node k. Each node corresponds to a site of the malware diffusion modeling MRF.

The states of each site can be susceptible, denoted by zero, or infected, denoted by one. Thus, the phase space can be binary $\Lambda = \{0,1\}$ as for regular networks and the state of each site x_k takes values in Λ, i.e. $x_k \in \{0,1\}$. This phase space is slightly different than the one employed in lattice networks, but this does not change modeling or analysis at all. It is a matter of implementation convenience, as long as the same number of phase values is retained. The potential function can be again decomposed in clique potentials as in expression (5.4), where x denotes a configuration of the state of the whole system. In order to study the impact of each modeling parameter, we assume that each clique function is determined only by the pairwise interaction of sites. This means that the corresponding clique potentials will be pairwise functions of the states of sites, and as with the chain network, we consider a linear relationship as with regular networks. The corresponding pairwise potentials will be given by expression $U(x) = -J \sum_{m \in N_k} x_k x_m$, where $m \in N_k$ and J is a proper scaling constant. Thus, the clique potentials can be calculated by

$$\Phi_k(x^{(\ell)}) = \hat{\Phi}_k\left(x_k = \ell, \{x_m : m \in N_k\}\right). \tag{5.25}$$

Since the state space is binary, the expressions for the potential for each value $\ell \in \{0,1\}$ are obtained as

$$\Phi_k(x^{(0)}) = -J \sum_{m \in N_k} \sigma_k(x_k = 0)\sigma_m(x_m) = -J \sum_{m \in N_k} x_k x_m = 0, \tag{5.26}$$

$$\Phi_k(x^{(1)}) = -J \sum_{m \in N_k} \sigma_k(x_k = 1)\sigma_m(x_m) = -J \sum_{m \in N_k} x_k x_m = -J \sum_{m \in N_k} x_m. \tag{5.27}$$

We note again that in arbitrary complex networks, as in regular networks, we consider only the pairwise potentials for the computation of clique potentials, before we compute the cumulative potential function. In this process, we omit singleton potentials, i.e. contributions to the potential function by the site itself. Such contributions are typically used to represent external forces to the sites of MRFs, e.g. an external magnetic force in MRFs developed for studying the spin of particles in physics [136]. In the malware diffusion setting we study, such force would correspond to processes that make a legitimate susceptible node infected by itself, without the need of attackers. We do not consider such cases, even though their analysis would be straightforward, but rather focus only the pairwise interactions between attackers/infected nodes and susceptible ones. Similarly, higher-order cliques than node pairs, i.e. triangles, are also not considered in this study, due to the focus on malware types that diffuse through pairwise interactions.

In each sweep, each site is visited sequentially and its state is updated with appropriate probabilities obtained as

$$Pr(x_k = 1) = \frac{e^{-\frac{\Phi_k(x^{(1)})}{T(t)}}}{\sum\limits_{\ell' \in \{0,1\}} e^{-\frac{\Phi_k(x^{(\ell')})}{T(t)}}} = \frac{1}{1 + e^{-\frac{\Phi_k(x^{(1)})}{T(t)}} e^{-\frac{\Phi_k(x^{(0)})}{T(t)}}} = \frac{1}{1 + e^{-\frac{\Phi_k(x^{(0)}) - \Phi_k(x^{(1)})}{T(t)}}},$$

and by taking into account Eqs. (5.26) and (5.27), the above expression becomes

$$Pr(x_k = 1) = \frac{1}{1 + e^{-\frac{\Phi_k(x^{(0)}) - \Phi_k(x^{(1)})}{T(t)}}} = \frac{1}{1 + e^{\frac{-J \sum\limits_{m \in N_k} x_m}{T(t)}}} = \frac{1}{1 + e^{-\frac{J}{T(t)} \sum\limits_{m \in N_k} x_m}}. \quad (5.28)$$

The corresponding expression for the state $x_k = 0$ is obtained as

$$Pr(x_k = 0) = 1 - Pr(x_k = 1) = \frac{1}{1 + e^{\frac{J}{T(t)} \sum\limits_{m \in N_k} x_m}}. \quad (5.29)$$

Consequently, in each sweep, the state of each site is updated, becoming infected (equal to one) with probability given by (5.28) and susceptible (equal to zero) with probability given by (5.29). The annealing rule we use is $T(t) = \frac{c_0}{\log(1+t)}$, where $c_0 = 1000$ is a constant and t is the number of sweeps performed. Parameter c_0 is usually determined empirically [136, 228], and the value $c_0 = 1000$ was employed for all the results presented in the following subsections. From the above expressions, it is evident that the parameter J/c_0 has an impact on the modeling behavior and we will initially present some results signifying the sensitivity of the system, as well as the impact of J/c_0 of the behavior modeled. We will study each of the aforementioned topologies with respect to parameter J/c_0. Also, for each type of network, we examine the behavior of malware diffusion with respect to the specific parameters of the corresponding topology that determine connectivity and network density. For all scenarios, $t = 2000$ sweeps have been performed and all results have been averaged for 50 different topologies.

5.5.1 RANDOM NETWORKS

We consider random graph topologies following the E-R model and more specifically the $\mathcal{G}(N, M)$ model, where a graph is chosen uniformly at random from the collection of all graphs which have N nodes and $M = |E|$ edges [67, 125, 164]. We choose M

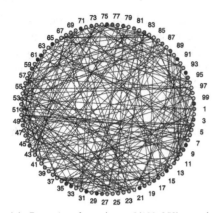

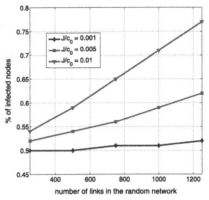

(a) Example of random $\mathcal{G}(100, 250)$ graph topology.

(b) Percentage of infected nodes for increasing network density.

FIGURE 5.8

ER random networks and malware modeling MRFs.

in the range of $\frac{N \log(N)}{2} \leq M \leq \frac{N(N-1)}{2}$. The lower bound $\frac{N \log(N)}{2}$ ensures that the studied topology is fully connected, while the upper $\frac{N(N-1)}{2}$ signifies the case where the graph is massively dense, in this case a complete graph [161]. Thus, the range of M employed covers a complete range of random graphs of interest, starting at the nontrivially sparse regime where the network is sparse but still connected and, moving to more dense topologies, up to the complete graph one.

We consider a network of $N = 100$ nodes (including one attacker), with a topology similar to the one shown in Fig. 5.8(a). The numerical labels indicate the nodes (sites) of the network, and by inspection of the figure it can be seen that connections between nodes seem completely random without a special emerging trend.

Fig. 5.8(b) shows the percentage of infected nodes in the steady-state (y-axis) for different values of links in the random graph (x-axis). Thus in the horizontal axis, the increasing number of links corresponds to denser networks as well. Also, the depicted different curves correspond to a different value of parameter J/c_0. Such parameter determines the combined Gibbs sampling with SA for computing the steady-state solution of the MRF modeling malware diffusion. From Fig. 5.8(b), higher values of parameter J/c_0 correspond to less robust networks with higher percentages of infected nodes.

With respect to network parameters, it can be observed that as the number of links in the random graph increases, namely, the density of the network increases as well, the percentage of infected nodes increases in the long-term. This means that the robustness of the network decreases, as the network becomes more dense. In fact, for higher values of J/c_0, i.e. for a less robust network by nature, the percentage increase in the number of infected nodes is higher for more dense networks, signifying quite a lesser performance in terms of robustness.

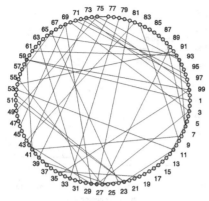
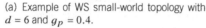

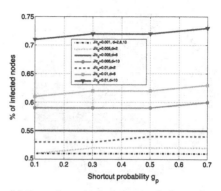

(a) Example of WS small-world topology with $d = 6$ and $g_p = 0.4$.

(b) Percentage of infected nodes for increasing network density.

FIGURE 5.9

MRF malware diffusion modeling for WS SW networks.

5.5.2 SMALL-WORLD NETWORKS

We consider a specific category of SW networks that were identified as a class of random graphs by Watts and Strogatz in 1998 [225], and denoted as WS SW networks. WS noted that graphs could be classified according to two independent structural features, namely, the clustering coefficient, and average node-to-node distance (also known as average shortest path length). Purely random graphs, built according to the ER model, exhibit a small average shortest path length (varying typically as the logarithm of the number of nodes) along with a small clustering coefficient. WS then proposed a novel graph model, the WS model, with a small average shortest path length, and a large clustering coefficient. An example of such a network we considered for $N = 100$ nodes in total (including one attacker) is shown in Fig. 5.9(a), where the labels denote the nodes and the links between them signify the locality of connections, with some "shortcut" links between nonlocal nodes that are responsible for the low average node distance cross the network.

The considered WS SW networks are characterized by two parameters, the shortcut probability g_p which determines the probability of a link acting as a shortcut between distinct nodes, as explained before, and the number of nearest neighbors each node is connected d. Initially, the higher both the d and g_p are, the denser the final network is, but for very high g_p values, the network starts losing its SW properties, since a high number of shortcuts the locality of connections is fading.

Fig. 5.9(b) presents the percentage of infected nodes obtained for a WS SW network in the steady-state with respect to g_p (x-axis) and d and J/c_0 (different curves provided). As with the random networks before, again higher J/c_0 values signify a less robust network that yields more infected nodes in the long-run, and a denser network (higher g_p, d) is also more prone to diffuse malware.

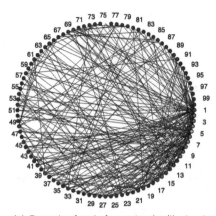

(a) Example of scale-free network with $d = 6$.

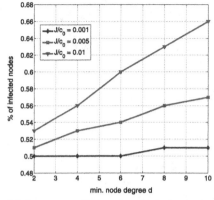

(b) Percentage of infected nodes for increasing network density.

FIGURE 5.10

MRF malware diffusion modeling for SF networks.

5.5.3 SCALE-FREE NETWORKS

In our study, we consider the Barabasi-Albert (BA) model [1, 15], which is an algorithm for generating random SF networks using a preferential attachment mechanism. Preferential attachment means that the more connected a node is, the more likely it is to receive new links. Nodes with higher degree have stronger ability to attract links added to the network. If a network begins with an initial connected network of m_0 nodes and new nodes are added to the network one at a time, each new node will connect to $m \leq m_0$ existing nodes with a probability that is proportional to the number of links that the existing nodes already have. Such probability p_i that the new node is connected to node i is $p_i = \frac{k_i}{\sum_j k_j}$, where k_i is the degree of node i and the sum is made over all pre-existing nodes j (i.e. the denominator results in twice the current number of edges in the network). Heavily linked nodes (hubs) tend to quickly accumulate even more links, while nodes with only a few links are unlikely to be chosen as the destination for a new link. We consider such a network, as the one shown for $N = 100$ nodes in Fig. 5.10(a), where a node hub with label "1" is evident.

The considered BA SF networks are characterized by the minimum number of nearest neighbors each node is connected to initially d. The higher d is, the denser the network becomes.

Fig. 5.10(b) presents the percentage of infected nodes obtained for a BA SF network in the steady-state with respect to d (x-axis) and J/c_0 (different curves provided). As with the random and WS networks before, again higher J/c_0 values signify a less capable network yielding more infected nodes in the long-run, and a denser network (higher d) is also more prone to diffuse malware more. The less

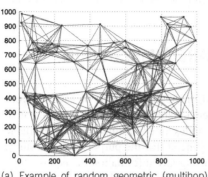

(a) Example of random geometric (multihop) topology for $L = 1000$ m, $R = 250$ m.

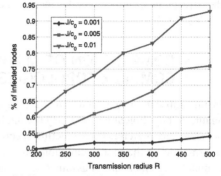

(b) Percentage of infected nodes for increasing network density.

FIGURE 5.11

MRF malware diffusion modeling for random geometric (multihop) networks.

robust to infection attacks the network is, i.e. lower J/c_0, the higher the percentage of infected nodes, as expected.

5.5.4 RANDOM GEOMETRIC NETWORKS

A random geometric graph (RGG) is the simplest spatial network, namely, an undirected graph constructed by randomly placing N nodes in some topological space (according to a specified probability distribution) and connecting two nodes by a link if their distance (according to some metric) is in a given range, e.g. smaller than a certain neighborhood radius, R [180]. The simplest choice for the node distribution is to sprinkle them uniformly and independently in the embedding space. Such an example of a considered network for $N = 100$ nodes, distributed over a square region of side $L = 1000$ m and a neighborhood (transmission) radius $R = 250$ m, is given in Fig. 5.11(a).

The considered RGG networks are characterized by their density which is defined by the number of nodes N over the area of the network L^2. Furthermore, the higher the connectivity radius R, the more neighbors each node will have and thus, the more clustered the network will be. Since in this initial study we consider fixed density (i.e. constant N, L) we vary the clustering parameter of the network, i.e. R. Of course a more clustered network also leads to denser topology for fixed N, L.

Fig. 5.11(b) presents the percentage of infected nodes obtained for a RGG network in the steady-state with respect to R (x-axis) and J/c_0 (different curves provided). As with all the previous networks, higher J/c_0 values signify a less capable network yielding more infected nodes in the long-run. A more clustered network, i.e. better connected nodes, yields also more infected nodes, since it allows easier/faster diffusion of malware. As can be observed in Fig. 5.11(b), for higher J/c_0, the percentage of infected nodes increases even more as network nodes cluster more (higher R values).

5.5.5 COMPARISON OF MALWARE DIFFUSION IN COMPLEX TOPOLOGIES

In this subsection, we focus on the comparison of the behavior of each network against nonintelligent malware. By nonintelligent malware, we denote attacks that do not show any preference on infecting specific nodes, or exploiting specially available information on the more vulnerable ones. In general, for the purposes of this study, nonintelligent malware is assumed to avoid using any information or methodology that would aid in the diffusion of malware. The attacker and infected nodes can only have point-to-point contacts with their direct neighbors. Susceptible nodes are homogeneously infected, i.e. they become infected with the same mechanism and have the same infection probability when in contact with infected/attack nodes. Consequently, such nonintelligent attacks may be considered as "blind" (random) attacks against susceptible direct neighbors.

We model the behavior of random ER (RG), multihop-random geometric (RGG), BA SF, and WS SW networks via the MRF framework under the attacks described above and compare their long-term performance. We consider $c_0 = 1000$ and average all results for 50 different topologies of each network type, randomly generated. For each type, we consider three broad topology regimes. The first is the sparse network regime with results shown in Fig. 5.12. The second is the moderate-density regime with results shown in Fig. 5.13, and the third is the dense network regime with results shown in Fig. 5.14.

For each type of network, the three regimes are defined differently. For random networks (RG), the sparse considers $E = 250$ links, very close to the connectivity threshold of the topology. The moderate regime considers $E = 500$ links and the dense $E = 750$. Similarly, for random geometric topologies (RGG), the sparse regime considers transmission radius $R = 200$ m, the moderate $R = 300$ m, and the dense $R = 400$ m for a square deployment region with side $L = 1000$ m. For BA SF networks, the sparse, moderate, and dense regimes are defined for $d = 2, 6, 10$ values of the initial node degree, respectively. Finally, for WS SW networks employed, the sparse regime is defined by $d = 2$, the moderate by $d = 6$, and the dense by $d = 10$ values of the minimum initial degree of each node, all for shortcut probability $g_p = 0.3$. Thus, SF and SW topologies are directly comparable with respect to density and clustering of nodes.

Fig. 5.12 provides results of the percentage of infected nodes for each topology for $J/c_0 = 0.005$ (Fig. 5.12(a)) and $J/c_0 = 0.01$ (Fig. 5.12(b)). As mentioned in the previous subsection, $J/c_0 = 0.005$ corresponds to a more capable network in terms of restricting malware. However, as it can be observed by comparing the two figures, in the sparse regime the results are very similar and with small differences between the scenarios with $J/c_0 = 0.005$ and those with $J/c_0 = 0.01$. The worse performance is exhibited by RGG topologies, which as the network size and density

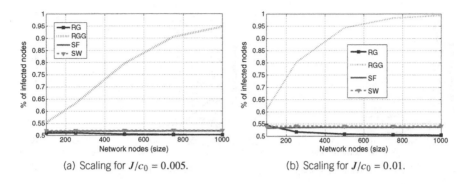

(a) Scaling for $J/c_0 = 0.005$.   (b) Scaling for $J/c_0 = 0.01$.

FIGURE 5.12

Scaling of percentage of infected nodes with respect to network density: the sparse network regime.

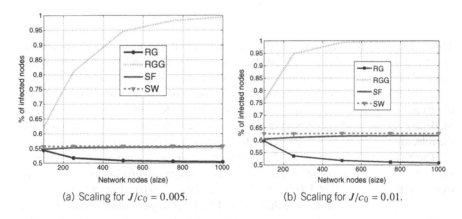

(a) Scaling for $J/c_0 = 0.005$.   (b) Scaling for $J/c_0 = 0.01$.

FIGURE 5.13

Scaling of percentage of infected nodes with respect to network density: the moderate-density regime.

increase, become dominated by malware diffusion. The contrary happens for RG networks, while SF and SW have an intermediate and in fact, stable behavior. SF and SW are relatively insensitive to the increase of network density and only slight increases in the number of infected nodes in the steady-state can be observed as the density increases.

Fig. 5.13 provides results of the percentage of infected nodes for each topology for $J/c_0 = 0.005$ (Fig. 5.13(a)) and $J/c_0 = 0.01$ (Fig. 5.13(b)). In the moderate-density regime, performance is different for $J/c_0 = 0.005$ and $J/c_0 = 0.01$, where all networks exhibit better performance for $J/c_0 = 0.005$ cumulatively. Again, the worse performance is exhibited by RGG topologies, which as the network size and

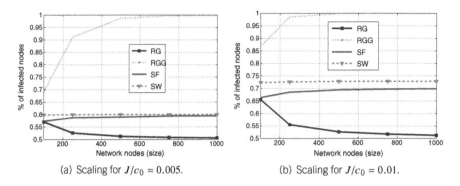

(a) Scaling for $J/c_0 = 0.005$.

(b) Scaling for $J/c_0 = 0.01$.

FIGURE 5.14

Scaling of percentage of infected nodes with respect to network density: the dense network regime.

density increase, become dominated by malware diffusion. RG networks retain the best performance, while SF and SW have an intermediate and stable behavior. SF and SW are again relatively insensitive to the increase of network density, but now some notable increase in the number of infected nodes in the steady-state can be observed for $J/c_0 = 0.01$. This signifies that for moderately dense networks, the modeling parameter J/c_0 can have an impact and it can be safely used to characterize the robustness capabilities of the network against simple (random) attacks.

The latter can be verified via the results in the dense regime as well, shown in Fig. 5.14. Fig. 5.14 provides results of the percentage of infected nodes for each topology for $J/c_0 = 0.005$ (Fig. 5.14(a)) and $J/c_0 = 0.01$ (Fig. 5.14(b)). Once more, the worse performance is exhibited by RGG topologies, which as the network size and density increase, become completely dominated by malware diffusion. The contrary happens for RG networks, while SF and SW have again an intermediate, more stable behavior. In the case of $J/c_0 = 0.01$, there is an evident deterioration for the performance of SF and SW networks. Nevertheless, their scaling trend for increasing network densities remains similar to the previous cases. In all three regimes, SF has a slightly better performance than SW topologies.

Considering all the aforementioned results cumulatively, we have the following list of important remarks regarding the modeling of malware propagation in wireless complex networks and their behavior under random attack.

- The density (and clustering degree) of a network topology is very important for the long-term outcome of attacks and network behavior under the SIS model. It could determine strongly the final tally sustained by each network in its steady-state.
- For random networks of the ER type, the robustness of the network actually increases as the network scales become more dense.

- SF (BA model) and SW (WS model) networks are fairly stable with respect to network density, with only minor increase in the percentage of infected nodes, when network density and clustering increase. Thus, such networks are fairly robust against random attacks in terms of network density.
- RGG (multihop) network suffers dearly from random attacks when their density increases. The percentage of infected nodes increases rapidly, even for small increases in the clustering degree of the network.
- Comparing cumulatively the performance of the studied complex networks and their scaling performance with respect to network density/clustering degree, ER random networks seem to have the best performance (lower percentage of infected nodes in their steady-state), followed by BA SF networks. WS SW follows SF exhibiting very slight deterioration of performance, and finally, random geometric (multihop) topologies are the most prone to suffer from random attacks as their topologies become dense and more clustered.

All the aforementioned observations and associated conclusions can be taken into account for driving future research in securing better these complex networks and ensuring efficient and simple countermeasures.

Optimal control based techniques

6.1 INTRODUCTION

Mathematical and numerical models for capturing the dynamics of a system are useful in understanding the forces at work and the efficacy of different environmental/contextual and behavioral parameters on the dynamics. A next step, often the ultimate goal of such models, is to use them in the design, assessment, and implementation of intervention mechanisms to influence the dynamics in one's favor. This can be done by any entity that can influence the environmental parameters. In the context of malware epidemics, this can, for instance, be done through changing the communication powers (and hence the communication ranges) of the nodes by the system manager, or behavioral parameters of the malware, e.g. the media scanning rate of the malware, implemented in its code by its creator. What differentiates such scenarios from simple *optimization* or *control* problems, specially in the context of SIR models and its variations, is the fact that the characteristic nature of the SIR-type epidemic spread is that it is a phenomenon of *transient* and *evanescent* dynamics. The *process* of SIR epidemic is not a stationary one: the outbreak of a given worm starts with only a handful of infected nodes, which then over time grows in number to a sizable fraction of the whole network, before eventually dying out through the removal process such as patching, or replacement of the infected nodes. The "steady state" or measures of "long-term averages" of such a system are not of much interest: Given a malware, the steady state of a SIR epidemic is zero infection (absence of the malware), because any given malware will be eventually weeded out of the network. Indeed, the transient phase of a SIR malware epidemic determines its aggregate "success" and, therefore, the overall damage to the network.

A static optimization framework deals with problems where the domain of optimization (space of the *inputs*) is a subset of the Euclidean space $\mathbb{R}^n$ for a given n, where n is the number of distinct control parameters. The optimization problem is specified by a number of *constraints*, and an *objective function*—also known as the *utility*, *cost*, or *fitness* function—which is a mapping from $\mathbb{R}^n$ to the real numbers. The goal is to find *feasible* solutions that yield the smallest (or the largest) values of the objective function. However, as we discussed, a relevant objective function dealing with a SIR-like epidemic process should entail its transient and evanescent evolution.[1] This implies that, here, the objective function, which should now more accurately be referred to as the objective *functional*, is rather a mapping from the epidemic state, as a function of time, to the real numbers. Solving optimization problems with an objective functional, where a change in an input parameter can influence the "path" of the evolution and thus the value of optimization, is the subject of the field of *calculus of variations*. Moreover, it is often the case that the input controls themselves can also be dynamically varied over time, perhaps subject to some path constraints, in anticipation of the temporal evolution of the state of the process. *Optimal control* theory is the branch of mathematics that extends calculus of variations to analyze such optimization problems, whose objective functionals are mappings from states and inputs (controls) as functions of time, to the real numbers, along with a set of initial, final, and/or pathwise constraints on input parameters or the state of the system. In this context, the input parameters are often referred to as *controllers*, or simply, *controls*.

The field of optimal control theory can be seen as an extension of the classical control theory as well, which also deals with influencing dynamic systems (the "plant") to achieve a measure of desirability. The major difference is in the type of desirability measures that are the focuses of these two branches: in control theory, the objective is primarily a measure of qualitative *long-term behavior* of the systems and no strict measure of "optimality" is enforced. The task there is often designing "controllers" as feedback systems that modify the input in order to force the output to "eventually" follow a trajectory, i.e. the "reference." That is, typically, the main goal in control theory is "stability" of the dynamics with a guarantee of convergence to a desired stationary state using concepts such as "controllability" and "observability" and tools such as Lyapunov stability theorem and frequency domain analysis. The desirability of the transient behavior is typically treated as a secondary objective with looser measures such as absence of oscillations, quicker rise time, smaller peak "overshoot" and "ringing," and "robustness" to external random disturbance or errors in parameter estimations (system identification). In contrast, in optimal control theory, the measure of optimality of an input is a "true optimality" measured by its detailed effect on potentially the whole path over even a finite window of interest.

The main result in static optimization is the duality theorem and the *Karush-Kuhn-Tucker (KKT)* conditions [189].[2] These conditions provide a set of straightforward

---

[1]Note that static optimization problems, with powerful tools from convex, linear and mixed-integer programming, are sometimes used in relation to "dynamic" systems too. A notable example is the area of Network Utility Maximization (NUM) with applications in wired and wireless network scheduling (see e.g. [128]). The dynamic nature of the system is related to the fact that the arrival of flows, packets, channel fading, channel shadowing, mobility, etc. are realized over time, and measures of optimality such as average end-to-end delay or energy consumption rate make sense in the context of time. However, the utility functions in such problems depend on measures of long-term averages of the dynamics, and the underlying stochastic processes such as flow/packet arrival, channel availability, fading, etc., are stationary–ergodic processes. None of these assumptions are well-justified in the context of controlling a transient evanescent outbreak of a malware.

[2]Along with the following two classes of convergence theorems: (1) ergodicity theorems that provide conditions to relate the long-term "time-average" performance measures of a dynamic stochastic system involving stationary–ergodic processes to its statistical measures (such as first and second moments); and

necessary (and sometimes even sufficient) first-order conditions for optimality of a solution given some mild "regularity" conditions. Search for candidate solutions can hence be restricted to those satisfying the KKT conditions, or iteratively updated toward satisfying them. The equivalent key result in optimal control theory is the *Pontryagin's maximum principle* (PMP).[3] Much in the spirit of the KKT conditions, the maximum principle (informally) states that any optimal solution, referred to as an *optimal controller* (or simply as an *optimal control*), must pointwise maximize the *Hamiltonian along an optimal path*, i.e. if all the state and "co-state" variables are according to their optimal values. The Hamiltonian of an optimization over a dynamic system captures both the instantaneous effect a control has on the utility rate and its indirect future effects on the utility.

Another well-celebrated result in optimal control is the *Hamilton-Jacobi-Bellman (HJB)* equation, which provides necessary and sufficient conditions for an optimal control through first obtaining the optimal "cost to go" of a dynamic system, which then in turn yields the optimal controls. Both methods require solving ODEs with mixed boundaries (initial and final conditions). However, working with the PMP, although only providing necessary conditions, is usually simpler, as the optimization of the Hamiltonian is pointwise and along the optimal path, as opposed to the HJB optimization, which needs to be carried out over the whole state space.

Optimal control theory, using either PMP or HJB, has had many applications in engineering and economics. Indeed, closed-form solutions are developed for a frequently used class of optimal control problems referred to as *linear quadratic (LQ)*. Namely, LQ problems arise where the dynamics of the system is governed (or approximated) by linear differential equations (linear in the vector of states and controls at any given instance), and the utility rate (the integrand in the objective functional) is quadratic with respect to the state and controls.[4] The optimal solution of a LQ problem as an optimal state-feedback controller, known as the *linear quadratic regulator (LQR)*, can be derived readily from the maximum principle. Namely, for the case of optimal control over a finite horizon, finding the LQR reduces to solving a *continuous time Riccati equation,* and in infinite-horizon settings, even simpler, it reduces to solving an *algebraic Riccati equation.*

However, a point of complication in the problem of optimally manipulating SIR-like epidemics is that, even in their simplest manifestation as deterministic

---

(2) algorithmic convergence theorems that establish that classes of distributed iterative update rules can lead to a global optimal point. The first class of results enable us to find a solution of a carefully constructed deterministic static optimization and establish its optimality in the underlying dynamic and stochastic system. On the other hand, the second class of results enable distributed implementation of the solutions using methods such as gradient or sub-gradient descent numerical algorithms. This is attractive in practical situations where aggregation of the required information and carrying out a centralized computation and relaying back the new solutions is less desirable than a distributed and local update (usually with a small tradeoff in the nonasymptotic performance) – see e.g. [50, 151].

[3] The prominent result in calculus of variations is the *Euler-Lagrange* equation developed by the two name-bearing mathematicians in the 1750s. We will only focus on the PMP from optimal control theory (developed in 1950s by Lev Pontryagin) that entails the Euler-Lagrange as a special case.

[4] A quadratic function can for instance represent the magnitude of the distance of the state from a desired path.

ODEs, they are not linear. In fact, nonlinearity of the state dynamics in epidemics, specially SIR-like scenarios, is a defining feature of such systems, and attempts in approximating the system dynamics through linearization would destroy the characteristic behavior of the spread.[5] Therefore, analysis of epidemic spreads, exactly due to their nonlinearity, does not follow from *á la carte* application of classic results. However, even in the absence of closed-form solutions, only by investigating the necessary conditions that PMP provides, we can extract substantial deductions about the structure of the optimal solutions. We will showcase this approach through a rather detailed treatment of a stylized example from the viewpoint of an attacker.

Before we delve into the chapter, note that in situations where multiple decision-makers with misaligned incentives are at play in influencing the dynamics of the spread, the "optimal" control of each player has to heed the (current and future) decisions of the other players as well. This demands a new notion of optimality in the face of the optimal decisions of the other interacting entities, where decisions of each player are functions of time, and are hence infinite-dimensional. Analysis of such scenarios in the framework of *differential game theory* is the subject of our next chapter.

6.2 EXAMPLE—AN OPTIMAL DYNAMIC ATTACK: *SEEK AND DESTROY*

Assume that the goal of an attacker writing the code for a worm is to infect as many nodes as possible, in order to disrupt the functionality of the hosts as well as of the network. Consider the following stylized setting: The malware can use an infective node to find new susceptible nodes to spread to, while performing malicious activities, until either the battery of the host is completely drained or the presence of the malware is detected and removed from the host by the network administrator (the defender). Moreover, we assume that the malware can also *kill* an infective host and make it completely dysfunctional. This killing process can be performed by executing a specific code payload which inflicts irretrievable hardware or software damage on the node. The worm can determine the time to kill, or equivalently the rate of killing the hosts, by regulating the rate at which it triggers such codes.

A notable proof-of-concept example of such malware, albeit spreading only among MS Windows 9x based PCs, was the CIH (also known as the Chernobyl or Spacefiller) virus. It first appeared in Taiwan in 1998 and reached worldwide

---

[5]To see this, note that the SIR epidemics starts with a very slow growth since the fraction of the nodes that are infected in the beginning is small and the initial spreading occasions, whose rate is proportional to the "product" of the number of infective and susceptible nodes, are rare. However, once the fraction of infected nodes reaches a critical mass, there will be a period of avalanche of spread, but eventually, the number of spreading occasions decreases once again since there will be few susceptible nodes left. This leads to an almost S-shape evolution, which is not present in linear systems, and is exactly due to the presence of nonlinear terms such as the product of susceptible and infective fractions in the state dynamics.

spread with an estimated damage of 1 billion USD.[6] The CIH virus was able to corrupt the host's boot drive and overwrite the Flash BIOS chip of certain machines. Both of these payloads served to render the host computer inoperable, essentially destroying the PC for a typical user. Moreover, the CIH virus code had activation dates for triggering its destructive phase, which in its first generation was set to exactly one year after its inception. This in turn shows that malware writers can indeed incorporate dynamic decision-making, casting the problem as an optimal control problem.

In the generic scenario considered, the worm starts with an initial number of infected nodes. Then, at each moment of time, the worm at each infected node faces the following tradeoff decision: whether to kill the host and inflict a guaranteed large cost on the network (reward to the worm) at the expense of losing the opportunity to spread and infect new nodes, or deferring the killing in the anticipation of infecting more susceptible nodes at later times but at the risk of being detected and removed through patching by the network administrator. Specifically, at time t, the worm at an infective node can kill the host by invoking its detrimental payload with rate $v(t)$, where $0 \leq v(t) \leq v_{max}$.

The dependence of the control on time is critical. Indeed, if the worm restricts itself to assume a static killing rate, then the above tradeoff suggests an intermediate value to be optimal. As we will see, when the control is dynamic, the optimal solution behaves substantially differently. In fact, intermediate levels of killing rate are *never* optimal.

6.2.1 DYNAMICS OF STATE EVOLUTION

Let the instantaneous *fractions* of susceptible, infective, recovered, and dead nodes at time t be $S(t)$, $I(t)$, $R(t)$, and $D(t)$, respectively. Then, for all t, we have $S(t) + I(t) + R(t) + D(t) = 1$. We assume that at the onset of the epidemic outbreak, that is at time zero, some but not all nodes are infected: $0 < I(0) = I_0 < 1$. For simplicity, we assume $R(0) = D(0) = 0$, which implies $S(0) = 1 - I_0$.

Infective nodes spread the malware during communication with susceptible nodes. Each pair of infective-susceptible nodes initiates communication at rate $\hat{\beta}$. The security patches are installed at an infective (susceptible, respectively) and transform their state into recovered. The rates of patch installation at any given time t are $Q(S(t), I(t))$ and $B(S(t), I(t))$ for susceptible and infective nodes, respectively, where to preserve a level of generality, we allow them to depend on the fraction of susceptible and infective nodes as well.[7] We will use the notations: $Q_S(S,I) := \partial Q(S,I)/\partial S$ and $B_I(S,I) := \partial B(S,I)/\partial I$. Also, whenever not ambiguous, we will drop the arguments S, I, and simply use Q and B. The functions Q and B only need

---

[6]The following anecdote shows the extent of the reach of the virus: During March 1999, IBM discovered that it had accidentally shipped a batch of new Aptiva PCs in the number of several thousands that were infected with the CIH virus. The virus was set to be activated in a month!

[7]For instance, elevated network traffic due to the media scanning activities of the infected nodes may reduce the rate of patching as the network administrator has to compete for the same bandwidth resources and shared media as the worm.

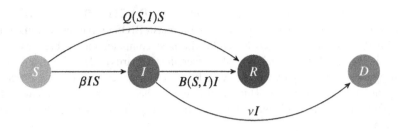

FIGURE 6.1

Transitions: S, I, R, D, respectively, represent fraction of the susceptible, infective, recovered, and dead. $v(t)$ is the dynamic control parameter of the malware.

to satisfy the following mild assumptions: (a) on the region of $0 \leq S, I$ & $S + I \leq 1$, both Q and B are finite, nonnegative, and differentiable; (b) for $0 < S, I$ & $S + I < 1$ we have $Q + B > 0$; and (c) for $0 < S, I$ & $S + I < 1$, we have $Q + Q_S S > 0$. These rates are associated with delays in detection of infection and fetching the appropriate security patch, etc. Note that the functions B and Q are likely to be constants (e.g. $B(S, I) = B_0, Q(S, I) = Q_0$ for all S, I) in practice,[8] and any constant function satisfies all of the above properties. Nevertheless, we allow more general functions so as to accommodate more general scenarios. Let $\beta = \lim_{N \to \infty} N\hat{\beta}$ where N is the total number of nodes in the network. The system of differential equations associated with the dynamics of the state is thus the following:

$$\dot{S}(t) = -\beta I(t)S(t) - Q\left(S(t), I(t)\right)S(t), \qquad S(0) = 1 - I_0, \qquad (6.1a)$$
$$\dot{I}(t) = \beta I(t)S(t) - B\left(S(t), I(t)\right)I(t) - v(t)I(t), \qquad I(0) = I_0, \qquad (6.1b)$$
$$\dot{D}(t) = v(t)I(t), \qquad D(0) = 0, \qquad (6.1c)$$

and also satisfy the following constraints at all t:

$$0 \leq S(t), I(t), D(t), \qquad (6.2a)$$
$$S(t) + I(t) + D(t) \leq 1. \qquad (6.2b)$$

Henceforth, wherever not ambiguous, we drop the dependence on t and make it implicit. Fig. 6.1 illustrates the transitions between different type fractions.

---

[8]This is because the users are likely to receive the security patches either from software stores or from servers distributed in the area. In the first case, the rates are naturally constants. In the latter case, the dissemination rates of the patches depend on the host's reception gains, servers' transmission gains, etc.; and none of the above depend on the infective and susceptible fractions, as long as the bandwidths are not critically used.

6.2.2 OBJECTIVE FUNCTIONAL

The attacker seeks to inflict the maximum possible damage in a time window $[0,T]$ of its choice. An attack can benefit over time from the infected hosts, by using the worms to (i) eavesdrop and analyze traffic that is generated or relayed by the infected hosts, or the traffic that traverses in the hosts' vicinity and (ii) alter or destroy the traffic that is generated or relayed by the infected hosts. An attacker also benefits by inflicting a large death-toll by the end of the desired time window. These motivate the following damage function:

$$J = \kappa D(T) + \int_0^T f\left(I(t)\right) dt, \tag{6.3}$$

where κ is an arbitrary nonnegative coefficient, and $f(\cdot)$ is an arbitrary nondecreasing, convex function such that $f(0) = 0$. Note that the assumptions on κ and $f(\cdot)$ are mild and natural, and a large class of functions, e.g. $f(I) = KI^\alpha$ for $\alpha \geq 1$ and $K \geq 0$, $f(I) = K(e^{\alpha I} - 1)$ for $\alpha, K \geq 0$ satisfy them. Finally, an attacker that simply seeks to maximize the final tally of the dead without any other agenda is readily represented by taking $f \equiv 0$.

The attacker seeks to maximize the aggregate damage by appropriately regulating its killing rate, $v(t)$, subject to $v \in \Omega$ for all $t \in [0,T]$, where Ω represent the feasible region of controllers, which is simply the set of piecewise continuous functions $v(\cdot)$ such that at almost all times satisfy $v(t) \in [0, v_{\max}]$. For simplicity of exposition, we consider $v_{\max} = 1$, with the understanding that all of our results hold for any maximum rate of killing $v_{\max} > 0$. We first show that for any $v \in \Omega$, the state constraints in (6.2) are automatically satisfied throughout $[0 \ldots T]$. Thus, we ignore (6.2) henceforth.

Lemma 6.1. *For any $v \in \Omega$, the state functions $(S, I, D) : [0,T] \to \mathbb{R}^3$ that satisfy the state equations and initial states in (6.1), also satisfy the state constraints in (6.2). Moreover, $S(t) \geq (1 - I_0)e^{-K_1 t} > 0, I(t) \geq I_0 e^{-K_2 t} > 0$ for $t \in [0,T]$ and some finite K_1, K_2.*

The proof, coming next, reveals that

$$K_1 = \beta + \max_{x,y \geq 0 \ \& \ x+y \leq 1} Q(x,y), \qquad K_2 = \max_{x,y \geq 0 \ \& \ x+y \leq 1} B(x,y) + 1.$$

Proof. Since $0 < I_0 < 1$, the initial conditions in (6.1) ensure that state constraints (6.2) are strictly met at $t = 0$. Note that for any choice of v, all S, I, and D, resulting from (6.1) (and hence any continuous functions of them) are continuous functions of time. The continuity of S and I functions ensures that there exists an interval of nonzero length starting at $t = 0$ on which both S and I are strictly positive. Thus, from (6.1c) and since $v(t) \geq 0$, $\dot{D} \geq 0$ in the above interval. Thus, since $D(0) = 0$, we have $D \geq 0$ in this interval as well. Since $\frac{d}{dt}(S + I + D)|_{t=0} = -Q(S_0, I_0) - B(S_0, I_0) < 0$ and $S(0) + I(0) + D(0) = 1$, there exists an interval after $t = 0$ over which the constraint in (6.2b) is strictly met.

Now, suppose the first statement of the lemma did not hold, and consequently, let $t_0 \leq T$ be the first time after $t = 0$ at which, at least one of the constraints of $S, I \geq 0$ and $S + I + D \leq 1$ becomes active or $D \geq 0$ is violated right after it. That is, at t_0, we have (1) $S = 0$ OR (2) $I = 0$ OR (3) $S + I + D = 1$ OR (4) there exists an $\epsilon > 0$ such that $D < 0$ on $(t_0 \ldots t_0 + \epsilon)$; AND throughout $(0, t_0)$, we have $0 < S, I$ and $S + I + D < 1$ and $D \geq 0$. Therefore, for $0 \leq t < t_0$, from $0 < S, I$ and the positivity and finiteness of Q, by referring to (6.1a), we have $\dot{S} \geq -\beta S - Q(S, I) \geq -K_1 S$ where $K_1 = \beta + \max_{x, y \geq 0 \; \& \; x+y \leq 1} Q(x, y)$. Hence, $S(t) \geq S(0)e^{-K_1 t} \geq S(0)e^{-K_1 t_0}$ for all $0 \leq t < t_0$. Since S is continuous in time, this implies $S(t_0) \geq S(0)e^{-K_1 t_0}$ as well. Similarly, since $0 < S, I$ over $(0, t_0)$ and following (6.1a), we have $\dot{I} \geq -B(S, I) - vI \geq -K_2 I$ where $K_2 = \max_{x, y \geq 0 \; \& \; x+y \leq 1} B(x, y) + 1$, and hence $I(t) \geq I(0)e^{-K_2 t}$ for any $t \in [0, t_0)$, and by continuity of I in time, also at t_0. Therefore, in short, we have $S(t_0) \geq S_0 e^{-K_1 t_0} > 0$ and $I(t_0) \geq I_0 e^{-K_2 t_0}$. Thus, since $S_0, I_0 > 0$, neither (1) nor (2) could have happened. Also, $\frac{d}{dt}(S + I + D) = -Q(S, I) - B(S, I) \leq 0$ throughout $[0 \ldots t_0]$. This, along with the facts that $S(0) + I(0) + D(0) = 1$ and $-Q(S_0, I_0) - B(S_0, I_0) < 0$, establishes that $(S + I + D)|_{t=t_0} < 1$. Hence, case (3) could not have occurred either. Moreover, since $I(t_0) > 0$ and I is continuous, from (6.1c) we can deduce that there exists an $\epsilon' > 0$ such that $\dot{D} \geq 0$ over $(t_0 \ldots t_0 + \epsilon')$, and hence $\dot{D} \geq 0$ over $(0, t_0 + \epsilon')$. Since $D(0) \geq 0$, this in turn implies $0 \leq D$ over $(t_0 \ldots t_0 + \epsilon')$, dismissing the possibility of (4). This negates the existence of t_0. Thus, the first statement of the lemma holds by contradiction. Moreover, the second statement of the lemma now follows from (6.1) and the fact that $S, I > 0$.   □

Once the control $v(\cdot)$ is selected, the state vector (S, I, D) is specified at all t as a solution to (6.1) and hence the value of the damage function J is determined as well. Thus, the control v is considered only as a function of time rather than that of the states, and since the value of J is determined only by the selection of v, we will henceforth denote J as $J(v)$ instead.

The state and control functions pair $((S, I, D), v)$ is called an *admissible pair* if (i) $v \in \Omega$, (ii) v is continuous except for possibly finite number of time epochs such that the left and right hand limits exist at the points of discontinuity, and (iii) equations in (6.1) hold. The function v is then called an admissible control. Let $((S, I, D), v)$ be an admissible pair. If

$$J(v) \geq J(\underline{v}) \quad \text{for any admissible control } (\underline{v})$$

then $((S, I, D), v)$ is called an *optimal solution* and v, an *optimal control* of the problem.

6.3 WORM'S OPTIMAL CONTROL

Let $((S, I, D), v)$ be an optimal solution. Let the *Hamiltonian* $\mathcal{H}$, and *costate* or *adjoint* functions $\lambda_S(t)$, $\lambda_I(t)$, and $\lambda_D(t)$, and a scalar $\lambda_0 \geq 0$ be defined as the following:

$$\mathcal{H} := \lambda_0 f(I) + (\lambda_I - \lambda_S)\beta IS - \lambda_S Q(S, I)S - \lambda_I B(S, I)I + (\lambda_D - \lambda_I)vI, \quad (6.4)$$

$$\dot{\lambda}_S = -\frac{\partial \mathcal{H}}{\partial S} = -(\lambda_I - \lambda_S)\beta I + \lambda_S Q(S,I) + \lambda_S Q_S(S,I)S,$$

$$\dot{\lambda}_I = -\frac{\partial \mathcal{H}}{\partial I} = -\lambda_0 f' - (\lambda_I - \lambda_S)\beta S + \lambda_I B(S,I) + \lambda_I B_I(S,I)I - (\lambda_D - \lambda_I)v,$$

$$\dot{\lambda}_D = -\frac{\partial \mathcal{H}}{\partial D} = 0$$

$$(6.5)$$

along with the *transversality* (final) conditions,

$$\lambda_S(T) = 0, \qquad \lambda_I(T) = 0, \qquad \lambda_D(T) = \lambda_0 \kappa. \qquad (6.6a)$$

Then according to Pontryagin's maximum principle with terminal constraints—[70, P.111 theorem 3.14]—there exist continuous and piecewise continuously differentiable costate functions λ_S, λ_I, and λ_D, and a constant $\lambda_0 \geq 0$ that at every point $t \in [0 \dots T]$ where $v(\cdot)$ is continuous, satisfy (6.5) and transversality conditions (6.6), and we have

$$\vec{\lambda} \not\equiv \vec{0}, \qquad (6.7a)$$

$$v \in \arg\max_{(\underline{v}) \in \Omega} \mathcal{H}(\vec{\lambda},(S,I,D),\underline{v}). \qquad (6.7b)$$

First, we argue that λ_0 has to be strictly positive. This is because if $\lambda_0 = 0$, then the system of ODE in (6.5) becomes a homogeneous ODE with the final conditions of $\lambda_S(T) = \lambda_I(T) = \lambda_D(T) = 0$. However, this ODE has the unique solution of $\vec{\lambda} \equiv \vec{0}$, which contradicts (6.7a). Hence, combined with the property that $\lambda_0 \geq 0$, we have $\lambda_0 > 0$.

Define the switching function φ as follows:

$$\varphi := (\lambda_D - \lambda_I)I \qquad (6.8)$$

which is a continuous and piecewise continuously differential function of time and referring to (6.6) has the following final value:

$$\varphi(T) = \lambda_0 \kappa I(T) > 0. \qquad (6.9)$$

The positivity comes from the facts $\lambda_0 > 0$, $\kappa > 0$, and $I > 0$ according to Lemma 6.1. Introduction of φ allows us to rewrite the Hamiltonian in (6.4) as follows:

$$\mathcal{H} = \lambda_0 f(I) + (\lambda_I - \lambda_S)\beta IS - \lambda_S Q(S,I)S - \lambda_I B(S,I)I + \varphi v. \qquad (6.10)$$

According to PMP in (6.7b), we have

$$\mathcal{H}(S,I,D,v,\lambda_S,\lambda_I,\lambda_D) \geq \mathcal{H}(S,I,D,\underline{v},\lambda_S,\lambda_I,\lambda_D) \quad \forall \underline{v} \in [0,1]. \qquad (6.11)$$

Hence, the optimal v satisfies $\varphi v \geq \varphi \underline{v}$, for all $\underline{v} \in [0,1]$. Thus, to find the optimal controller, one needs to maximize the linear function φv over the admissible set

$v \in [0,1]$, which yields

$$v = \begin{cases} 0, & \varphi < 0, \\ \\ 1, & \varphi > 0, \end{cases} \tag{6.12}$$

hence the name switching function. An immediate observation of the above relation is the following important property:

$$\varphi v \geq 0. \tag{6.13}$$

Also note that according to (6.9), $\varphi(T) > 0$ and thus by continuity of φ and following (6.12), $v = 1$ over an interval of nonzero length toward the end of $(0 \ldots T)$ interval which extends until time T.

6.3.1 STRUCTURE OF THE MAXIMUM DAMAGE ATTACK

Whether in practice the worm can indeed inflict the maximum damages developed in this section depends on implementability of the optimal strategies. Specifically, if the optimal policies that inflict the maximum damage are complex to execute, then the worm may not be able to perform them since they are limited by the capabilities of their resource constrained hosts as well. Inauspiciously though, we show that optimal attack strategies follow simple structures (Theorem 6.1) which make them conducive to implementation. Fig. 6.2 provides visualization of the theorem. Note that in the left figure, the patching only immunizes the susceptible nodes ($Q(S,I) \equiv 0.2$ but $B(S,I) \equiv 0$), while in the right figure, patching can equally immunize the susceptible and clean and immunize the infective nodes ($Q(S,I) = B(S,I) \equiv 0.2$).

Recall that one of the basic tradeoff/s decisions that the attacker was dynamically faced with was the best timing to kill an infective node. Specifically, should an attacker kill a node as soon as it is infected so as to have claimed a casualty and secured a large damage on the network but losing the chance to further the spread? Or should it wait in anticipation of contacting new susceptible nodes and extend the contagion but at the risk of being detected and removed by the defender before it gets to destroy the host. Is the tradeoff broken by choosing an intermediate probability of killing the host? Theorem 6.1 states not.

Theorem 6.1. *Consider an optimal solution v that maximizes the worm's damage function in (6.3) subject to the control constraint of $v(t) \in [0,1]$ for all $t \in [0,T]$. Then $v(t)$ has the following structure: $\exists t_1 \in [0 \ldots T)$ such that $v(t) = 0$ for $0 < t < t_1$ and $v(t) = 1$ for $t_1 < t < T$.*

In words, an optimal $v(\cdot)$ is of *bang-bang* form, that is, it possesses only two possible values 1 and 0, and switches abruptly between them. It has at most one such jump, which necessarily culminates at 1. Thus, the theorem says that although killing a node early on would ensure a partial damage, the overall damage is more if this decision is deferred until toward the end of the attacking period despite the

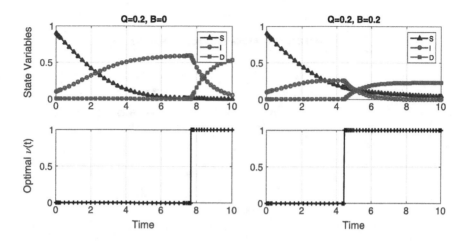

FIGURE 6.2

Evaluation of the optimal controller and the corresponding states as functions of time. The parameters are time horizon: $T = 10$, initial infection fraction: $I_0 = 0.1$, contact rate: $\beta = 0.9$, instantaneous reward rate of infection for the malware: $f(I) = 0.1I$, reward per each killed node: $\kappa = 1$. Also, we have taken $Q(S, I) \equiv 0.2$, and $B(S, I) \equiv 0$ in the left and $B(S, I) \equiv 0.2$ in the right figures. That is, in the left figure, patches can only immunize the susceptible nodes but in the right figure, the same patch can successfully remove the infection, if any, and immunize the node against future infection. We can see that when patching can recover the infective nodes too (right figure), then the malware starts the killing phase earlier. This makes sense as deferring the killing in the hope of finding a new susceptible is now much riskier.

risk of recovery of the infective nodes by the system. Specifically, at the start of the outbreak, the number of susceptible nodes is high and infective nodes can be used to further propagate the infection. As time passes by, the level of susceptible nodes drops due to both spread of infection and immunization effort by the system. At a certain threshold, the risk of recovery of the infective nodes in the remaining time outweighs the potential benefit by spreading the infection. At this point, whose exact value depends on the parameters of the case, the malware starts killing the nodes with maximum possible rate. This will ensure that infective nodes are maximally used for spread of the infection and for attacker's malicious activities.

In summary, Theorem 6.1 provides the optimal attack as follows: initially, the effort of the malware is focused on spreading the worm and amassing infective nodes without killing any. Subsequently, the reverse course of action is taken: at a threshold time, the amassed infected nodes are slaughtered at the highest rate which lasts till the end of the interval.

Note that the optimal killing policy (ν) will be completely specified by the (only possible) jump points (trigger epoch). Given the flexibility provided by software-driven devices, the infective nodes can subsequently execute these strategies without

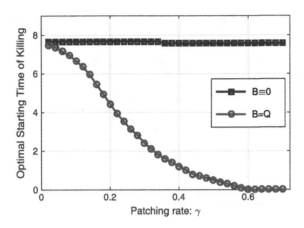

FIGURE 6.3

The jump (up) point of optimal v, i.e. the starting time of the slaughter period, for different values of the patching and rates. For both curves, we have taken the recovery rate of the susceptible nodes, i.e. $Q(S,I)$ as γ, and the recovery rate of the infective nodes, i.e. $B(S,I)$, once as zero and once as the same as $Q(S,I)$ where γ is varied from 0.02 to 0.7 with steps of 0.02. The rest of the parameters are $f(I) = 0.1I$, $\kappa = 1$, $T = 10$, $\beta = 0.9$, and $I_0 = 0.1$. Note that when $B(S,I) \equiv \gamma$, then for $\gamma \geq 0.6$, the malware starts killing the infective nodes from time zero.

coordinating any further among themselves or with any central entity. The transition time can be determined by solving a system of differential equations, as described in the previous sections. Such systems can be solved very fast due to the existence of efficient numerical algorithms for solving differential equations, and the computation time is constant in that it does not depend on the number of nodes N. Note also that our algorithms do not require any local or global information as time progresses and only the initial information is sufficient to determine the decision of infective nodes for the entire interval.

Fig. 6.3 shows the effect of increasing the patching rates Q and B on the trigger epoch. We observe that increasing the patching rate generally decreases the jump time. Intuitively, in a system with a large recovery rate, both the susceptible and infective nodes recover rapidly. Hence, the worm should start killing them earlier in order to not lose too many nodes in the competition with the network administrator to the pool of recovered. Note also that the starting time of the killing is sensitive to the value of recovery rate when the patching can impact both infective as well as susceptible nodes ($B = Q$). This is because when $B \equiv 0$, once a node is infected, then it will not be recovered by the system and is safely in the tally of the worm, but when $B > 0$, the worm is in competition with the system and excessively deferring the killing is ever more risky if the speed of recovery of the victims is increased.

In the next section, we provide the proof of the theorem.

6.3.2 PROOF OF THEOREM 6.1

We first obtain some useful properties of the Hamiltonian and system states.

Lemma 6.2. $\mathcal{H} = constant > 0$.

Proof. First, the system is *autonomous*, i.e. the Hamiltonian and the control region do not have any explicit dependence on the independent variable t. Hence ([138, P.236]),

$$\mathcal{H}(S(t), I(t), D(t), \nu(t), \lambda_S(t), \lambda_I(t), \lambda_D(t)) \equiv constant. \tag{6.14}$$

Therefore, from (6.10),

$$\mathcal{H} = \mathcal{H}(T) = \lambda_0 f(I(T)) + \lambda_0 \kappa \nu(T) I(T). \tag{6.15}$$

We showed (after (6.7)) that $\lambda_0 > 0$, and following Lemma 6.1, $I(T) > 0$; also $\nu(T) = 1 > 0$, as we argued after (6.12). Hence, $\mathcal{H}(T) > 0$. $\square$

The second observation is that I satisfies the following condition:

Lemma 6.3. $(f'(I)I - f(I)) \geq 0$ *for all* $t \in [0 \ldots T]$.

Proof. By Lemma 6.1, I and S are nonnegative. Define $\xi(I) = f'(I)I - f(I)$. Since $f(0) = 0$, we have $\xi(0) = 0$. Also,

$$\frac{d}{dI}\xi(I) = \xi' = f''(I)I + f'(I) - f'(I) = f''(I)I.$$

Following Lemma 6.1 and properties of f, we observe that $\xi' \geq 0$ for all $t \in [0 \ldots T]$. Thus, since $\xi(0) = 0$, $\xi(I) = f'(I) - f(I)I \geq 0$ for all $t \in [0 \ldots T]$. $\square$

We will also use the following key lemma in the sequel.

Lemma 6.4. *For all* $t \in (0 \ldots T)$, *we have* $\lambda_S \geq 0$ *and* $(\lambda_I - \lambda_S) > 0$.

Proof. **Step-1.** Following (6.6), $\lambda_I(T) = (\lambda_I(T) - \lambda_S(T)) = 0$ and from (6.5) and (6.6), $(\dot{\lambda}_I(T) - \dot{\lambda}_S(T)) = -\lambda_0 f'(I(T)) - \kappa \nu(T)$, which is strictly negative. Thus, there exists an $\epsilon_1 > 0$ such that on the interval of $(T - \epsilon_1 \ldots T)$, we have $(\lambda_I - \lambda_S) > 0$. Also recall from (6.6) that $\lambda_S(T) = 0$.
Step-2. Proof by contradiction. Let t^* be defined as follows:

$$t^* := \inf_{0 \leq t \leq T} \{t | \lambda_S(t) \geq 0 \text{ and } (\lambda_I(t) - \lambda_S(t)) > 0 \text{ on the interval } (t \ldots T)\}.$$

If $t^* = 0$ then we are done. Suppose $t^* > 0$. According to the continuity of λ_S and λ_I, and following step-1, we must have

$$\lambda_I(t^*) - \lambda_S(t^*) = 0 \quad \text{OR} \quad \lambda_S(t^*) = 0.$$

- Case 1: $\lambda_I(t^*) - \lambda_S(t^*) = 0$. From (6.5) and continuity of λ_S, $\lambda_S(t^*) \geq 0$. We have

$$\left[\frac{d}{dt}(\lambda_I - \lambda_S)\right](t^{*+}) = \left[\frac{d}{dt}(\lambda_I - \lambda_S)\right](t^{*-})$$

$$= -\lambda_0 f' + \lambda_I(B + B_I I) - \frac{\varphi}{I}v - \lambda_S(Q + Q_S S) \quad [\because (6.5)]$$

$$= -\lambda_0 f' + \lambda_I(B + B_I I) - \frac{\varphi}{I}v - \lambda_S(Q + Q_S S)$$

$$- \frac{\mathcal{H}}{I} + \lambda_0 \frac{f}{I} - \frac{\lambda_S Q S}{I} - \lambda_I B + \frac{\varphi}{I}v \quad [\because (6.10)]$$

$$= \frac{\lambda_0}{I}[f - f'I] + \lambda_I B_I I - \lambda_S(Q + Q_S S) - \frac{\lambda_S Q S}{I} - \frac{\mathcal{H}}{I}.$$
$$(6.16)$$

From Lemma 6.3, $[f - f'I] \leq 0$, and following the properties of B and Q, we have $B_I \leq 0$ and $Q, (Q + Q_S S) \geq 0$. Also in this case, $\lambda_I(t^*) = \lambda_S(t^*)$ and $\lambda_S(t^*) \geq 0$ (by assumption of the case). Now following Lemmas 6.1 and 6.2, and Eq. (6.16), we observe that $[\frac{d}{dt}(\lambda_I - \lambda_S)](t^{*+}) = [\frac{d}{dt}(\lambda_I - \lambda_S)](t^{*-}) < 0$. According to Property 6.1, this is a contradiction. Thus, case 1 could not occur.

- Case 2: $\lambda_I(t^*) - \lambda_S(t^*) > 0$, and $\lambda_S(t^*) = 0$, and $\forall \delta > 0$, there exists $t_1 \in (t^* - \delta \ldots t^*)$ such that $\lambda_S(t_1) < 0$. From continuity of λ_S and λ_I, $\exists \epsilon > 0$ such that on $(t^* - \epsilon \ldots t^*)$, $\lambda_I - \lambda_S > 0$, and hence according to (6.5) and Lemma 6.1, wherever v is continuous, $\dot{\lambda}_S \leq \lambda_S(Q + Q_S S)$. Now consider a $\delta < \epsilon$, and define $\hat{t}$ to be the point which has the lowest value of λ_S on the interval of $[t^* - \delta \ldots t^*]$. According to the assumption of case 2, $\lambda_S(\hat{t})$ is strictly negative. Thus, $\dot{\lambda}_S(\hat{t}^+) \leq [\lambda_S(Q + Q_S S)]|_{t=\hat{t}} < 0$. This, along with continuity of λ_S, imply that in the right neighborhood of $\hat{t}$, λ_S has lower values than $\lambda_S(\hat{t})$. This contradicts the definition of $\hat{t}$.

Therefore, none of the two cases could occur, which is a contradiction with the existence of t^*. The lemma hence follows by contradiction. □

We are now ready to proceed to the proof of the theorem.

6.3.3 PROOF OF THEOREM 6.1: OPTIMAL RATE OF KILLING

Proof. To establish the statement of the theorem, we will show that the switching function φ is equal to zero on at most one time epoch. The theorem subsequently follows from the relation between φ and v given by (6.12).

Let us begin by studying two simple real analysis properties.

Property 6.1. Let $f(t)$ be a continuous and piecewise continuously differentiable function of t. Assume $f(t_0) > L$. Now if $f(t_1) = L$ for the first time before t_0, i.e. $f(t_1) = L$ and $f(t) > L$ for all $t \in (t_1 \ldots t_0]$, then $\dot{f}(t_1^+) \geq 0$.[9]

Property 6.2. Let $f(t)$ be a continuous and piecewise continuously differentiable function of t. Assume t_1 and t_2 to be its two consecutive L-crossing points, that is, $f(t_1) = f(t_2) = L$ and $f(t) \neq L$ for all $t_1 < t < t_2$. Now if $\dot{f}(t_1^+) \neq 0$ and $\dot{f}(t_2^-) \neq 0$, then $\dot{f}(t_1^+)$ and $\dot{f}(t_2^-)$ must have opposite signs.

Now, we proceed with the proof of the theorem. Let us calculate the time derivative of the φ function wherever v is continuous,

$$\dot{\varphi} = (\dot{\lambda}_D - \dot{\lambda}_I)I + i\frac{\varphi}{I} \qquad [\because(6.8)]$$

$$= \left(\lambda_0 f' + (\lambda_I - \lambda_S)\beta S - \lambda_I(B + B_I I) + (\lambda_D - \lambda_I)v\right)I + i\frac{\varphi}{I} \quad [\because(6.5)]$$

$$= \lambda_0 f' I + (\lambda_I - \lambda_S)\beta IS - \lambda_I(B + B_I I)I + \varphi v + i\frac{\varphi}{I}$$

$$\quad + (\mathcal{H} - \lambda_0 f - (\lambda_I - \lambda_S)\beta IS + \lambda_S QS + \lambda_I BI - \varphi v) \qquad [\because(6.10)]$$

$$= \mathcal{H} + \lambda_S QS + \lambda_0(f'I - f) - \lambda_I B_I I^2 + i\frac{\varphi}{I}. \qquad (6.17)$$

Let a time at which $\varphi = 0$ be denoted by τ. From (6.17), we obtain

$$\dot{\varphi}(\tau^+) = \dot{\varphi}(\tau^-) = \mathcal{H} + \lambda_S QS + \lambda_0(f'I - f) - \lambda_I B_I I^2. \qquad (6.18)$$

Eq. (6.18), positivity of λ_0, and Lemmas 6.1–6.4 show that $\dot{\varphi}(\tau^-)$, $\dot{\varphi}(\tau^+) > 0$ wherever v is continuous. First, this shows that φ cannot be equal to zero over an interval of nonzero length. To see this, note that otherwise, due to piecewise continuity of v, there exists a subinterval inside the interval of $\varphi = 0$ over which, v is continuous. Thus, φ is differentiable over this subinterval, and necessarily $\dot{\varphi} = 0$ for any time inside that subinterval, which is not possible. Thus, referring to (6.12), v is bang-bang, i.e. $v \in \{0, v_{\max}\}$.

Second, referring to Property 6.2, we conclude that $\varphi = 0$ at at most one point inside $(0 \ldots T)$ interval. Since (from (6.9)) $\varphi(T) > 0$ and because φ is a continuous function of time, $\varphi(t) > 0$ for an interval of nonzero length toward the end of $(0 \ldots T)$. If $\varphi(t) > 0$ for all $0 \leq t \leq T$, then $v = 0$ throughout the interval. Otherwise, there exists a $t_0 \in [0 \ldots T]$ such that $\varphi(t) < 0$ for $t_1 < t \leq T$ and $\varphi(t) > 0$ for $0 \leq t < t_1$. Theorem 6.1 now follows from the relation between optimal v and φ in (6.12).   □

---

[9]For a general function $f(x)$, the notations $f(x_0^+)$ and $f(x_0^-)$ are defined as $\lim_{x \downarrow x_0} f(x)$ and $\lim_{x \uparrow x_0} f(x)$, respectively. We denote $\dot{f}(t^+)$ as the right-side time derivative of f at time t.

SUMMARY

In this chapter, we discussed the idea that once a reasonable model of spread of worms and malware in a network is developed, how it can help the actors to utilize its predictive and analytic power to manipulate the dynamics in one's favor. In particular, we introduced the tools from optimal control theory, specifically, the PMP. We showed through a stylized example how PMP can be applied in a SIRD worm epidemic in which the worm can dynamically control when to destroy an infective node, facing an evolving tradeoff between infecting more nodes and being recovered by the system manager. We showed how, even in the absence of closed-form solutions that plagues nonlinear systems, a careful analysis of the necessary conditions that optimal solutions must satisfy can reveal significant structural properties of the solution. In particular, in our stylized example, we observed that the optimal killing strategy is to refrain completely to maximally exploit them for spreading until a certain threshold after which killing them with maximum intensity.

The system manager can similarly use optimal control tools to develop dynamic countermeasures that take into account the evolving state of a malware epidemic. These dynamic countermeasures include, e.g. reducing the communication rates of the nodes as a means of containing the spread of the worm but at the expense of introducing additional delays in the normal communications, rate of disseminating the security patches at the expense of taxing both the shared media of the network as well as the energy resources of the mobile nodes, etc. However, the assumption in optimal control problems is that there is only one entity that can manipulate the system in its own favor. In the next chapter, we will discuss the case where multiple decision-makers with distinct, and potentially antagonistic, incentives are at play. Specifically, when both the network defender and worm can dynamically and strategically manipulate the dynamics of a malware epidemic.

Game-theoretic techniques

7.1 INTRODUCTION

Given the flexibility that software-based operation provides, it can be expected that future malware will demonstrate a dynamic behavior over time in response to the dynamics of the network, in order to maximize its success chance. In return, the network can also dynamically change its countermeasure policies (such as rate reduction and patching rates) to more effectively oppose the spread of the infection. Game theory is the branch of mathematics that provides tools to analyze such confrontations between independent decision-makers, to specifically, tackle the strategic uncertainty of each player about the actions of the other.

As we argued in the previous chapter, due to the dynamic, transient and evanescence nature of SIRD epidemic outbreaks, a static approach that concerns only the steady-state of the epidemic is inadequate. Moreover, the utility of both players often depends on the dynamics of evolution (as opposed to just the initial or the final state). Therefore, the most relevant class of game theory in this context is "dynamic game theory." Specifically, since the evolution of the epidemic dynamics in the most basic form can be captured by differential equations, we will focus on *differential games*. Differential games can be thought of as the immediate extension of optimal control theory to the realm of game theory.

A widely accepted notion of a solution, specially in cases where cooperation is not relevant or feasible, is that of Nash equilibrium (NE), where each player is assumed to play its best response while fixing the strategy of other players. In other words, NE characterizes equilibria that are resistant against "unilateral" deviations of any of the players. Here, the actions of each player are itself a function of time. In fact, a higher level of complication is present: even when the evolution of the system is deterministic, depending on whether or not the players can dynamically "observe" the state of the dynamic, the conditions for equilibrium vary. Specifically, when the players can observe the state, a setting known as "closed-loop," the opportunities of update (or deviation) of a strategy of a player in response to the strategy of other player are "along the whole path" of evolution, and the equilibrium profile should leave no incentive for deviation along the whole path, knowing the instantaneous state. In contrast, in an "open-loop" scenario, the players choose their whole strategies at the onset of the player and have no further information to *a posteriori* deviate from it in response to the evolution of the dynamics. In this section, we present in detail an example of a differential game between a defender (network administrator) and an attacker (a malware) assuming open-loop information using differential game theory.

The worm may disrupt the normal functionalities of the hosts, steal their private information, and use them to eavesdrop on other nodes. A worm can also render the host dysfunctional by deliberately draining its battery, or by executing a pernicious code that incurs irretrievable critical hardware or software damage. We specifically consider the latter to be the decision action of the malware over time.

Upon the outbreak of a new worm, anomaly detection techniques can be used to identify the presence of malicious activities and generate security patches that can then be distributed among the nodes. Such patches either *immunize* susceptible nodes against future attacks, by rectifying their underlying vulnerability, or *heal* the infectives of the infection and render them robust against future attacks. Meanwhile, reducing the communication rates in the network can quarantine the worm by slowing down its spread. Specifically, hosts can simply drop packets sent to them before processing them, or even refuse some connection requests, or reduce the reception gain of their antennas.

Since the media in the wireless network are common and the channels are unreliable, the bandwidth consumed for distribution of the security patches can itself disrupt the normal functionality of the network. Hence, excessive quarantining through reception rate reduction also deteriorates the QoS by introducing delays for the data traffic. Such quarantining cannot usually discriminate based on the identity of the transmitters, since the hosts applying the reception rate control, in general, do not know which other nodes are infected; the reception rate itself may however be judiciously selected. The network's challenge now is to achieve a guaranteed performance by selecting the instantaneous (a) rate of patching and (b) reception rate that jointly minimize the overall damage due to (i) the subversive activities of the malware that is capable of annihilating infectives and (ii) the additional resource consumption and deterioration of QoS owing to the application of the countermeasures. The design must adapt over time, remaining cognizant of the malware's ability to dynamically optimize its spread in response to the network's dynamic strategy.

First, we construct a mathematical framework which models the strategic confrontations between the malware and the network as a zero-sum dynamic game (Section 7.3.1), drawing from (i) a deterministic epidemic model for worm propagation in a wireless networks (Section 7.2) and (ii) damage functions that we introduce to investigate the tradeoffs resulting from different decisions of the entities concerned (Section 7.2). We then prove the existence of the *robust* (i.e. *saddle-point*) strategies of the network and the malware (Section 7.2) and compute them (Section 7.3.2). We prove that the robust defense strategy has a simple two-phased structure (Section 7.3.3): (i) patch at the maximum possible rate until a threshold time, and then stop patching (ii) choose the minimum possible reception rate (i.e. the maximum packet drop rate at the receivers) until a threshold time and subsequently revert to the normal reception rate. The initial aggressive defense limits the spread of infection and thereby the pool of nodes that can potentially be compromised or killed; this guarantees an upper bound on the damage inflicted irrespective of the malware's choice of annihilation strategy. From a game-theoretical point of view, the structural results are somewhat surprising given the nonlinear dynamics of the state evolution and the nonmonotonicity of the state functions.

7.2 SYSTEM MODEL

Dynamics of state evolution

A susceptible accepts a communication request with a probability $u^{N_r}(t)$, where the subscript r represents reception, and the superscript N designates control functions of the network. At any given time t, there are $n_S(t)n_I(t)$ infective-susceptible pairs. Susceptible nodes are hence transformed to infectives at rate $\hat{\beta}u^{N_r}(t)n_S(t)n_I(t)$, where $\hat{\beta}$ is the rate at which a particular pair of nodes "meet," which is assumed to be the same for all pairs (i.e. homogeneous mixing assumption). Propagation of the worm, therefore, can be contained through appropriate regulation of $u^{N_r}(t)$ subject to $0 < u^{N_r}_{\min} \leq u^{N_r}(t) \leq u^{N_r}_{\text{norm}}$ at each t. The lower bound $u^{N_r}_{\min}$ arises due to the minimum QoS requirements for data traffic, since the acceptance probability has to be the same irrespective of whether the request arrives from another infective, susceptible, or recovered node. The latter is due to the fact that a recipient node cannot distinguish the type of a transmitter in advance and has no choice but to treat all requests the same — since otherwise all infective nodes can be trivially blacklisted. The upper bound $u^{N_r}_{\text{norm}}$ (which can be normalized to 1) provides the reception rate that nodes use for providing the desired QoS in the absence of security considerations, i.e. during the "normal" operation of the network.

We now consider the dissemination of security patches in the network. A predetermined set of nodes, referred to as dispatchers (e.g. BS for cellular and exit-points for delay-tolerant networks), are preloaded with the patches. We assume that the dispatchers cannot be infected, and that there are NR_0 dispatchers where N is as usual the total number of nodes in the network and parameter R_0 is between 0 and 1. Each node communicates with the dispatchers, and thereby fetches security patches, at the overall rate of $\tilde{\beta}NR_0u^{N_i}(t)$ at time t. The parameter $\tilde{\beta}$ depends on node density, mobility parameters, allowable transmission rates, etc. The control function $u^{N_i}(t)$, with subscript i denoting immunization, can be used to regulate the bandwidth consumed in propagation of patches: the higher the value of $u^{N_i}(t)$, the higher is the recovery rate but so is the rate of resource consumption in patch transmissions. Clearly, if the node that receives the patch is a susceptible node, it installs the patch and its state changes to recovered. If an infective receives the patch, the patch may fail to heal it, or, the worm may prevent its installation. We capture the above possibility, by introducing a coefficient $0 \leq \pi \leq 1$: $\pi = 0$ occurs when the patch is completely unable to remove the worm from infectives and only immunizes the susceptibles, whereas $\pi = 1$ represents the other extreme scenario where a patch can equally well immunize and heal susceptibles and infective nodes.[1] Now, if the patch heals an infective, its state changes to recovered, else it continues to remain an infective.

The worm at an infective host "kills" it with rate proportional to $u^M(t)$ at a given time t, where superscript M designates this is a control function of the malware; this is accomplished by executing specific codes with a probability of choice. The worm regulates the death process by appropriately choosing $u^M(t)$ at each t, subject to $0 \leq u^M(t) \leq u^M_{\max}$ at each t. The upper bound arises due to processor constraints and the resulting limitations on the maximum rate of execution of such codes. Let $\beta_0 := N\hat{\beta}$, $\beta_1 := N\tilde{\beta}$. Our discussions lead to the following system of differential equations representing the dynamics of the system:

$$\dot{S}(t) = -\beta_0 u^{N_r} I(t) S(t) - \beta_1 u^{N_i}(t) R_0 S(t), \quad I(0) = \lim_{N \to \infty} n_I(0)/N = I_0 > 0, \quad (7.1\text{a})$$

$$\dot{I}(t) = \beta_0 u^{N_r} I(t) S(t) - \pi \beta_1 u^{N_i}(t) R_0 I(t),$$
$$- u^M(t) I(t) \qquad\qquad S(0) = 1 - I_0, \qquad\qquad (7.1\text{b})$$

$$\dot{D}(t) = u^M(t) I(t), \qquad\qquad D(0) = 0, \qquad\qquad (7.1\text{c})$$

and also satisfy the following constraints at all t:

$$0 \leq S(t), I(t), D(t), \qquad S(t) + I(t) + D(t) \leq 1. \qquad\qquad (7.2)$$

Thus, $(S(\cdot), I(\cdot), D(\cdot))$ constitute the system state functions, $u^N(\cdot) = (u^{N_r}(\cdot), u^{N_i}(\cdot))$ constitutes the network control functions and $u^M(\cdot)$ constitutes the malware's control function. Note that nodes use *identical* reception, patching, and killing rate functions irrespective of the states in their neighborhoods since they do not know these states. Nevertheless, since these rates are allowed to vary with time, they can be chosen in accordance with how the *overall* network states are expected to evolve.

Henceforth, wherever not ambiguous, we drop the dependence on t and make it implicit. Fig. 7.1 illustrates the transitions between different states of nodes and the notations used.

Defense and attack objectives

The total damage inflicted by the malware during the network operation interval $[0, T]$ is due to the presence of infectives, the death of nodes, the resources consumed for spreading the security patches, and the QoS deterioration due to the reduction of reception rate. Infectives can perform harmful activities over time. Dead nodes are inoperative and thus inflict a time-accumulative cost on the network. The bandwidth overhead at time t due to the media scanning and transmission of the security packets by the dispatchers is $R_0 u^{N_i}(t)$. Due to the reception rate control, the susceptibles lose a $1 - u^{N_r}(t)/u^{N_r}_{\text{norm}}$ fraction of packets transmitted by all nodes which degrades the overall QoS. We therefore consider the aggregate network damage at time t as a combination of $I(t), D(t), u^{N_i}(t), u^{N_r}(t)$. We adopt a linear cost function in this

---

[1] In order to avoid immediate detection and blacklisting, the infective nodes may choose not to refuse all connection requests from the dispatchers.

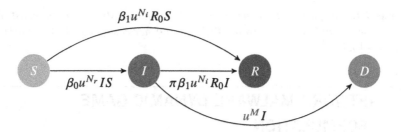

FIGURE 7.1

State transitions. $u^{N_i}(t)$ and $u^{N_r}(t)$ are the control parameters of the network while $u^M(t)$ is the control parameter of the malware.

chapter for analytical tractability. Note that the damage function can be scaled so that one of the coefficients may be chosen as unity: we choose the one associated with the instantaneous bandwidth overhead. Thus, the damage over the interval $[0,T]$ is[2]

$$J(u^N(t), u^M(t)) = \int_0^T [\kappa_I I(t) + \kappa_D D(t) + R_0 u^{N_i}(t) - \kappa_r u^{N_r}(t)] \, dt + K_D D(T),$$

$$(7.3)$$

where $K_D D(T)$ relates to the final tally of the dead nodes. The coefficients are all non-negative and represent the relative importance of each corresponding term in the overall damage, e.g. if the worm gains the most by killing, and thereby completely disabling nodes, $\kappa_D \gg \kappa_I$. Let $\kappa_I > 0, \kappa_r > 0$.

The network seeks to choose its control vector $u^N(\cdot)$ so as to minimize the above while the malware seeks to choose its control $u^M(\cdot)$ so as to maximize the above, subject to satisfying state constraints (7.2) and ensuring that

$$u_{\min}^{N_r} \le u^{N_r}(t) \le u_{\text{norm}}^{N_r}, \quad 0 \le u^{N_i}(t) \le 1, \qquad (7.4\text{a})$$

$$0 \le u^M(t) \le u_{\max}^M. \qquad (7.4\text{b})$$

In Section 7.3, we model their interactions resulting from opposing objectives as a dynamic game. The formulation relies on the following result that allows us to ignore the state constraints without any loss of generality. The proof is similar to that of Lemma 6.1 in the previous chapter and is hence omitted for brevity.

---

[2]Note that $(1 - u^{N_r}/u_{\text{norm}}^{N_r})$ inside the integral is replaced by $-u^{N_r}$, since $\int_0^T dt = T$, the time horizon of optimization, does not depend on the states or the controls, but is a fixed exogenous parameter to the optimization (and κ_r can be scaled).

Lemma 7.1. *Any pair of strategies* $(u^N(\cdot), u^M(\cdot))$ *that satisfy the control constraints* (7.4a), (7.4b), *satisfy state constraints* (7.2) *and further ensure that* $I(t) > 0, S(t) > 0$ *for all* $t \in [0,T]$.

7.3 NETWORK-MALWARE DYNAMIC GAME

7.3.1 FORMULATION

Following the setup described in Section 7.2, consider a system with two players N (network) and M (malware), specified by a system of n differential equations [146, P.83]: $\dot{x}(t) = f(t, x(t), u^N(t), u^M(t))$ $t \in [t_0, T]$, where $u^N(t) \in U^N \subset \mathcal{R}^m$, and $u^M(t) \in U^M \subset \mathcal{R}^s$, and the initial condition $x(t_0) = x_0$, and a damage function (the functional) $J(u^N, u^M) = g(x(T)) + \int_{t_0}^T h(x, u^N, u^M, t) \, dt$, where $x(t)$ is the n-dimensional state vector. *Player N* seeks to minimize J by choosing the m dimensional *control function $u^N(\cdot)$*, and player M seeks to maximize J by choosing the s-dimensional control function $u^M(\cdot)$. The game is therefore referred to as a dynamic two-player zero-sum game. The players' payoffs and the set of strategies available to them are called *rules of the game*. Both players know the rules of the game and each player knows that its opponent knows the rule and ad infinitum.[3] As we mentioned in the Introduction to this chapter, we consider open-loop strategies, that is, the controls only depend on time (and not on the state, or the previous history). This is appropriate in the context of the security in networks, as the instantaneous state of the network (exact fraction of the nodes of each type) is impossible (or very costly) to follow, for both of the players.

In our context, (7.1) provides the $f(\cdot)$ functions, the initial conditions are provided by (7.1), (7.3) provides the $g(\cdot), h(\cdot)$ functions, (7.4a), (7.4b) provide U^N, U^M. Also, we have $n = 3, m = 2, s = 1$. Note that the $f(\cdot), h(\cdot)$ functions in our context depend on time t only implicitly, that is through the state and control functions. Also, the formulation does not capture any other constraints on the state functions, and in our context it does not need to either, owing to Lemma 7.1.

We now consider the values of the game. The *lower value* denoted by V_* is the overall damage when the minimizing player (N) is given the upper-hand, i.e. selects its strategy after learning its opponent's strategy. Mathematically, $V_* = \max_{u^M} \min_{u^N} J(u^N, u^M)$. Conversely, the *upper value* of the game V^* is defined as $V^* = \min_{u^N} \max_{u^M} J(u^N, u^M)$ Thus, V_* (V^*, respectively) is the maximum (minimum, respectively) damage that the malware (network, respectively) can inflict (incur, respectively) if the other player has the upper-hand. Hence, $V_* \leq V^*$. A pair of strategies (u^{N*}, u^{M*}) is called a *saddle-point* if $J(u^{N*}, u^M) \leq J(u^{N*}, u^{M*}) = V \leq J(u^N, u^{M*})$ for any strategy u^N of the network and u^M of the malware, and then V is the value of the game, and $V = V_* = V^*$.

Thus, if the network selects its saddle-point strategy u^{N*}, irrespective of the strategy of the malware, the damage it incurs is at most V, which is also the minimum

---

[3]Each player knows that the opponent knows, *etc.*

damage that the malware can inflict if it has the upper-hand. Thus, the network's saddle-point strategy is also its *robust* strategy, in the sense, that it minimizes the maximum possible damage it can incur. Conversely, the malware's saddle-point strategy is also its robust strategy, since it maximizes the minimum possible damage it can inflict. Also, the network's and the malware's saddle-point strategies are their respective best responses to the other's robust strategy. We prove the existence of saddle-point pair in the next theorem.

Theorem 7.1. *The dynamic game defined above has a saddle-point pair of strategies.*

Proof. This theorem directly follows from Theorem 2 in page 91 of [146]. The necessary conditions of the theorem are readily satisfied in our game. Namely,

(i)   the system function $f(t,x,u^N,u^M)$ in this game is continuous in states and controls and is moreover bounded. Note that this is sufficient for the condition in page 83 of [146] to hold;

(ii)   the instantaneous pay-off function $h(t,x,u^N,u^M)$ and the terminal pay-off function $g(x(T))$ are continuous in the states and controls (page 84 of [146]);

(iii)   the system function $f(t,x,u^N,u^M)$ is linear in controls. The set defining controls are convex sets, and the instantaneous pay-off function $h(t,x,u^N,u^M)$ is linear in controls (page 91 of [146]).   □

7.3.2 A FRAMEWORK FOR COMPUTATION OF THE SADDLE-POINT STRATEGIES

Since the set of deterministic strategies of each player is uncountably infinite, the saddle-point strategies and the value of the game cannot be computed using convex or linear programming. We now present a framework for numerical computation of the saddle-point strategies.

   Define the *Hamiltonian* for a given policy pair (u^N,u^M) in an arbitrary two-person dynamic game as $\mathcal{H}(u^N,u^M) = \langle \lambda, f(x,u^N,u^M,t) \rangle + h(x,u^N,u^M,t)$, where the state functions $x(\cdot)$ are those that correspond to the strategy pair (u^N,u^M), and λ, the *costate* (or *adjoint*) functions, are continuous and piecewise differentiable functions of time that satisfy the following system of differential equations wherever the controls (u^N,u^M) are continuous: $\dot{\lambda} = -\frac{\partial}{\partial x}\mathcal{H}(x,\lambda,t)$, and the final value (transversality) condition $\lambda(T) = \frac{\partial(g(x))}{\partial(x)}|_{x=x(T)}$. In our context,

$$\mathcal{H}(u^M,u^N) = \kappa_I I + \kappa_D D + u^{N_i} R_0 - \kappa_r u^{N_r} + (\lambda_I - \lambda_S)\beta_0 u^{N_r} IS$$
$$- \lambda_S \beta_1 R_0 u^{N_i} S - \lambda_I \beta_2 R_0 u^{N_i} I + (\lambda_D - \lambda_I)u^M I,$$

where again the state functions $(S(\cdot),I(\cdot),D(\cdot))$ are obtained from (7.1) with $(u^N(\cdot),u^M(\cdot))$ as the control functions, and the costate functions $(\lambda_S(\cdot),\lambda_I(\cdot),\lambda_D(\cdot))$ are obtained from the following system of differential equations (with $u^N(\cdot),u^M(\cdot)$ as the control functions) with the final conditions,

$$\dot{\lambda}_S = -\frac{\partial \mathcal{H}}{\partial S} = -(\lambda_I - \lambda_S)\beta_0 u^{N_r} I + \lambda_S \beta_1 R_0 u^{N_i}, \qquad \lambda_S(T) = 0, \qquad (7.5a)$$

$$\dot{\lambda}_I = -\frac{\partial \mathcal{H}}{\partial I} = -\kappa_I - (\lambda_I - \lambda_S)\beta_0 u^{N_r} S$$
$$+ \lambda_I \beta_2 u^{N_i} R_0 - (\lambda_D - \lambda_I)u^M, \qquad \lambda_I(T) = 0, \qquad (7.5b)$$

$$\dot{\lambda}_D = -\frac{\partial \mathcal{H}}{\partial D} = -\kappa_D, \qquad \lambda_D(T) = K_D. \qquad (7.5c)$$

Then, following [146, P.31], a necessary condition for the pair (u^N, u^M) to be a saddle-point strategy pair is that for all $t \in [0, T]$,

$$(u^N, u^M) \in \arg\min_{\tilde{u}^N} \max_{\tilde{u}^M} \mathcal{H}(\tilde{u}^N, \tilde{u}^M) \text{ and} \qquad (7.6a)$$

$$(u^N, u^M) \in \arg\max_{\tilde{u}^M} \min_{\tilde{u}^N} \mathcal{H}(\tilde{u}^N, \tilde{u}^M). \qquad (7.6b)$$

Henceforth, we denote the saddle-point strategy pair as $(u^N(\cdot), u^M(\cdot))$, and $(S(\cdot), I(\cdot), D(\cdot))$, $(\lambda_S(\cdot), \lambda_I(\cdot), \lambda_D(\cdot))$ as the corresponding state and costate functions and $\mathcal{H}$ as the corresponding Hamiltonian. We now express $(u^N(\cdot), u^M(\cdot))$ in terms of $(S(\cdot), I(\cdot), D(\cdot)), (\lambda_S(\cdot), \lambda_I(\cdot), \lambda_D(\cdot))$ using necessary conditions (7.6). Define $\psi^{N_r} := (\lambda_I - \lambda_S)\beta_0 I S - \kappa_r$, and $\psi^{N_i} := R_0 - \lambda_S \beta_1 R_0 S - \lambda_I \beta_2 R_0 I$, and $\psi^M := (\lambda_D - \lambda_I)I$. Now, the Hamiltonian can be written as

$$\mathcal{H} = \kappa_I I + \kappa_D D + \psi^{N_r} u^{N_r} + \psi^{N_i} u^{N_i} + \psi^M u^M. \qquad (7.7)$$

Thus, the Hamiltonian is a separable function of different components of the defense controls $(u^{N_r}(\cdot), u^{N_i}(\cdot))$ and the attack control $u^M(\cdot)$, that is, each of these appears in different terms in the right-hand side of the above characterization. Now, from the necessary conditions in (7.6) subject to the control constraints in (7.4), the saddle-point strategies are derived as

$$u^{N_r} = u^{N_r}_{\min} \text{ if } \psi^{N_r} > 0 \quad \text{and} \qquad u^{N_r} = u^{N_r}_{\text{norm}} \text{ if } \psi^{N_r} < 0, \qquad (7.8)$$

$$u^{N_i} = 0 \text{ if } \psi^{N_i} > 0 \quad \text{and} \qquad u^{N_i} = 1 \text{ if } \psi^{N_i} < 0, \qquad (7.9)$$

$$u^M = u^M_{\max} \text{ if } \psi^M > 0 \quad \text{and} \qquad u^M = 0 \text{ if } \psi^M < 0. \qquad (7.10)$$

Since $\psi^{N_r}, \psi^{N_i}, \psi^M$ are uniquely specified once the state and the costate functions are known, the above relations express the saddle-point strategies in terms of the state and costate functions. The strategies $u^{N_r}(\cdot), u^{N_i}(\cdot), u^M(\cdot)$ can be substituted by the above characterizations in (7.1) and (7.5), resulting in a system of *six* differential equations involving only the state and the costate functions. Using standard numerical methods for solving differential equations, this system can be solved (very fast) using initial and final conditions (7.1) and (7.5). The state and costate functions obtained as solutions will now provide the $\psi^{N_r}, \psi^{N_i}, \psi^M$ functions, and thereby the saddle-point strategies via (7.8)–(7.10). The resulting set of differential equations is nonlinear and a closed-form solution is unknown. However, as we will show in the next section, using novel techniques, even without access to the closed-form solution, we can establish the type of behavior that the saddle-point strategies exhibit.

7.3.3 STRUCTURAL PROPERTIES OF SADDLE-POINT DEFENSE STRATEGY

We establish that the saddle-point defense strategy has a simple threshold-based structure that ought to facilitate its implementation in a localized manner in resource constrained wireless devices. Specifically, we prove the following.

Theorem 7.2. *For the saddle-point defense strategy $u^N(\cdot) = (u^{N_r}(\cdot), u^{N_i}(\cdot))$, there exists times $t_1, t_2, 0 \leq t_1 < T, 0 \leq t_2 < T$ such that*

- $u^{N_r}(t) = u^{N_r}_{\min}$ *for* $0 < t < t_1$, *and* $u^{N_r}(t) = u^{N_r}_{\text{norm}}$ *for* $t_1 < t < T$;
- $u^{N_i}(t) = 1$ *for* $0 < t < t_2$, *and* $u^{N_i}(t) = 0$ *for* $t_2 < t < T$.

The overall strategy therefore has the following three phases: In the initial *aggressive defense* phase, i.e. during $(0, \min(t_1, t_2))$, the susceptibles select the minimum possible reception rate, and the dispatchers transmit the patches whenever they are in contact with any other node. Thus, the quarantining is the most stringent, and the recovery most rapid during this phase. Then, in the interim *watchful* phase, i.e. during $(\min(t_1, t_2), \max(t_1, t_2))$, one of the defense controls subsides while the other continues as before. Finally, in the terminal *relaxed* phase, i.e. in $(\max(t_1, t_2), T)$, both defense controls subside, that is, the susceptibles select their normal reception rate and the dispatchers do not transmit the patches. Thus, the QoS in data traffic is back to its normal value and the resource consumption overhead due to patch transmission ends.

Note that the defense strategy always chooses either the maximum or the minimum values of the parameters except possibly at t_1, t_2. Such strategies are referred to as *bang-bang* in the control literature. The durations of the phases (i.e. the values of the threshold times t_1, t_2) and which defense subsides in the interim watchful period depend on the damage coefficients $\kappa_I, \kappa_D, K_D, \kappa_r, \kappa_i$. We will shed more light on the latter in Theorem 7.3 later. But first, we conclude this subsection by proving Theorem 7.2.

Proof. The continuity and piecewise differentiability of $\psi^{N_r}(\cdot), \psi^{N_i}(\cdot)$ follow from those of the costate functions. From the final conditions on the costate functions, i.e. (7.5), $\psi^{N_r}(T) = -\kappa_r < 0$, $\psi^{N_i}(T) = R_0 > 0$. We show that $\psi^{N_r}(\cdot)$ ($\psi^{N_i}(\cdot)$, respectively) is a strictly decreasing (increasing, respectively) function of time. Thus, each has at most one zero-crossing point in $(0, T)$; denote these as t_1, t_2. If ψ^{N_r} (ψ^{N_i}, respectively) has no zero-crossing point in $(0, T)$, $t_1 = 0$ ($t_2 = 0$, respectively). Thus, from the continuity of the $\psi(\cdot)$ functions, and from their terminal values, (i) $\psi^{N_r}(\cdot)$ is negative in (t_1,T) and positive in $(0,t_1)$ and (ii) $\psi^{N_i}(\cdot)$ is positive in (t_2,T) and negative in $(0,t_2)$. The theorem follows from (7.8) and (7.9). We prove the strict monotonicity of $\psi^{N_r}(\cdot), \psi^{N_i}(\cdot)$, using the following.

Lemma 7.2. $\lambda_S > 0$ and $\lambda_I > \lambda_S, \lambda_D \geq 0 \; \forall \; t, 0 < t < T$.

The lemma is intuitive since the shadow prices (i.e. costate variables) associated with the susceptibles and dead nodes ought to be positive, and also the shadow price associated with the infectives ought to be at least as high as that associated with susceptibles. However, as we will see next, the proof requires detailed analysis of the state and costate differential equations (7.1) and (7.5), respectively, and is less direct.

Proof. Since $\lambda_D(T) = K_D \geq 0$ (from (7.5)), and $\frac{d}{dt}\lambda_D \leq 0$, $\lambda_D \geq 0$. Now, for the rest, we argue in two steps.

Step 1: $\lambda_S(T) = 0$ and $\lambda_I(T) = K_I = 0$, also $\dot{\lambda}_I(T) - \dot{\lambda}_S(T) = \dot{\lambda}_I(T) = -\kappa_I - K_D u^M(T) < 0$. Therefore, $\exists \epsilon > 0$ such that on $(T - \epsilon \ldots T)$ we have $\lambda_S > 0$ and $(\lambda_I - \lambda_S) > 0$.

Step 2: Proof by contradiction. Let τ be such that $\lambda_S > 0, (\lambda_I - \lambda_S) > 0$ on $(\tau \ldots T)$ and $\lambda_S(\tau) = 0$ OR $\lambda_I(\tau) = \lambda_S(\tau)$. From the continuity of the costate functions, $(\lambda_I(\tau) - \lambda_S(\tau)) \geq 0$ and $\lambda_S(\tau) \geq 0$.

We first prove that $(\lambda_I(\tau) - \lambda_S(\tau)) > 0$. Suppose not. Then, $\lambda_I(\tau) = \lambda_S(\tau)$. Thus, $\dot{\lambda}_I(\tau) - \dot{\lambda}_S(\tau) = -\kappa_I + \lambda_I \beta_2 u^{N_i} - (\lambda_D - \lambda_I)u^M - \lambda_S \beta_1 u^{N_i} = -\kappa_I - \lambda_S u^{N_i}(\beta_1 - \beta_2) - (\lambda_D - \lambda_I)u^M$. Here, (i) the first term is strictly negative,[4] (ii) the second term is negative because $\lambda_S(\tau) \geq 0$ and $\beta_2 \leq \beta_1$, and (iii) the third term is negative because of (7.10). Thus, $\dot{\lambda}_I(\tau) - \dot{\lambda}_S(\tau) > 0$. But, then both $\lambda_I(\tau) = \lambda_S(\tau)$, and $(\lambda_I - \lambda_S) > 0$ on $(\tau \ldots T)$ cannot happen. Thus, $(\lambda_I(\tau) - \lambda_S(\tau)) > 0$.

Now, suppose $\lambda_S(\tau) = 0$. $\dot{\lambda}_S(\tau) = -(\lambda_I - \lambda_S)\beta_0 u^{N_r} I|_\tau < 0$. The last inequality follows since $(\lambda_I(\tau) - \lambda_S(\tau)) > 0$, $\beta_0 > 0$, $u^{N_r} \geq u^{N_r}_{min} > 0$ and $I(\tau) > 0$ (Lemma 7.1). This again contradicts the assumptions that $\lambda_S(\tau) = 0$ and $\lambda_S > 0$ on $(\tau \ldots T)$. Thus, $\lambda_S(\tau) \neq 0$, and hence $\lambda_S(\tau) > 0$. $\qquad \square$

Strict monotonicity of $\psi^{N_r}(\cdot)$

We show that $\dot{\psi}^{N_r}(t)$ is strictly *negative* at all $t \in (0,T)$[5]

$$\dot{\psi}^{N_r} = \frac{\partial}{\partial t}\psi^{N_r} = (\dot{\lambda}_I - \dot{\lambda}_S)\beta_0 IS + (\lambda_I - \lambda_S)\beta_0 \dot{I}S + (\lambda_I - \lambda_S)\beta_0 I\dot{S}$$

---

[4]Negative in this proof is distinguished from strictly negative.
[5]Partial derivative w.r.t time, only because of the dependence also on the initial values for the states. Otherwise, t is the only independent variable.

which after replacement and simplification yields

$$\frac{\dot{\psi}^{Nr}}{\beta_0 IS} = -\kappa_I - (\lambda_D - \lambda_S)u^M - \beta_1\lambda_I u^{N_i} + \beta_2\lambda_S u^{N_i}$$

$$= -\kappa_I - (\lambda_D - \lambda_I)u^M - (\lambda_I - \lambda_S)u^M - (\beta_1 - \beta_2)\lambda_I u^{N_i} - (\lambda_I - \lambda_S)\beta_2 u^{N_i}.$$

From (7.10), Lemma 7.2 and since $\kappa_I > 0$, $\beta_1 \geq \beta_2$, $u^M(t) \geq 0$, $u^{N_i}(t) \geq 0$ at all t, the right-hand side is negative. The result follows since $\beta_0 > 0$ and $S(t) > 0, I(t) > 0$ at all t (Lemma 7.1).

Strict monotonicity of $\psi^{N_i}(\cdot)$

$$\dot{\psi}^{N_i} = \frac{\partial}{\partial t}\psi^{N_i} = (\dot{\lambda}_I - \dot{\lambda}_S)\beta_0 IS + (\lambda_I - \lambda_S)\beta_0 \dot{I}S + (\lambda_I - \lambda_S)\beta_0 I\dot{S}$$

$$\Rightarrow \frac{\dot{\psi}^{N_i}}{\beta_0 I} = \kappa_I\beta_2 + \beta_2 u^M \lambda_D + \beta_0\beta_1 Su^{Nr}(\lambda_I - \lambda_S).$$

The R.H.S is positive from Lemma 7.2 and since $\kappa_I > 0, \beta_0 > 0$. Thus, $\dot{\psi}^{N_i} > 0$ since $\beta_0 > 0$ and $I(t) > 0$ at all t (Lemma 7.1).   □

In the next theorem, we show that under a sufficient condition, the quarantining period (by reduction of communication rates) ends before the immunization/healing effort is stopped. This is in accordance with our intuition that the primary use of quarantining is *buying time* for the recovery process in the network.

Theorem 7.3. *Let t_1 and t_2 be as defined in Theorem 7.2. If $\kappa_r \geq \dfrac{\beta_0 S_0}{\beta_2}$, then either $t_2 = 0$ or $t_1 < t_2$.*

Note that the condition of the theorem is quite intuitive. For instance, if $\beta_2 = \beta_0$, this condition is satisfied when $\kappa_r \geq 1$. Recall that the coefficients of the costs were rescaled so that the coefficient of the cost of patching is normalized. Thus, this means when the instantaneous cost of per unit reduction of commutation rates (quarantining) is no less than per unit patching. The other parameters in $\kappa_r \geq \dfrac{\beta_2 S_0}{\beta_0}$ refer to the cases of relatively small initial susceptible pool, and a relatively fast healing rate.

Proof. Recall from the proof of Theorem 7.2 that $\psi^{Nr}(t)$ and $\psi^{N_i}(t)$ each have at most one zero-crossing point and $\psi^{Nr}(t)$ terminates in a negative, and $\psi^{N_i}(t)$ terminates in a positive value at T. Now we show that at a *potential* zero-crossing point of $\psi^{Nr}(t)$, $\psi^{N_i}(t)$ is strictly negative.

$$\psi^{Nr} = (\lambda_I - \lambda_S)\beta_0 IS - \kappa_r = 0 \Rightarrow \lambda_I = \frac{\kappa_r}{\beta_0 IS} + \lambda_S, \text{ then,}$$

$$\psi^{N_i} = R_0 - \lambda_S \beta_1 R_0 S - \lambda_I \beta_2 R_0 I = R_0 - \lambda_S \beta_1 R_0 S - \beta_2 R_0 I \left(\frac{\kappa_r}{\beta_0 I S} + \lambda_S \right)$$

$$= R_0 \left(1 - \frac{\beta_2}{\beta_0} \frac{\kappa_r}{S} \right) - \lambda_S \beta_1 R_0 S - \lambda_S \beta_2 R_0 I.$$

According to Lemmas 7.1 and 7.2, the last two terms are strictly negative. The first term is negative following the condition of the theorem (i.e. $\kappa_r \geq \frac{\beta_2 S_0}{\beta_0}$) and the fact that S is a nonincreasing function of time. Similarly, one can show that given $\kappa_r \geq \frac{\beta_2 S_0}{\beta_0}$, then at a potential zero-crossing point of $\psi^{N_i}(t)$, $\psi^{N_r}(t) > 0$. The theorem follows from the continuity of $\psi^{N_i}(t)$ and $\psi^{N_r}(t)$ and by referring to (7.8) and (7.9). $\qquad\square$

Note that the case of $t_2 = 0$ occurs when ψ^{N_i} does not have a zero-crossing point and is hence non-negative throughout $[0,T]$. In this case, the immunization/healing effort is never launched.

7.3.4 STRUCTURE OF THE SADDLE-POINT ATTACK STRATEGY

The saddle-point attack has a simple *first-amass, then slaughter* structure in the special case that the worm benefits from killing only through the final tally of the dead (i.e. $\kappa_D = 0$), and the patches can only immunize the susceptibles, but cannot heal the infectives (i.e. $\beta_2 = 0$). Specifically, we have the following.

Theorem 7.4. When $\beta_2 = \kappa_D = 0$, for the saddle-point attack strategy $u^M(\cdot)$, there exists a time t_3, $0 \leq t_3 < T$ such that $u^M(t) = 0$ for $0 < t < t_3$, and $u^M(t) = u_{\max}^M$ for $t_3 < t < T$.

Thus, the worm does not kill any infective during the initial amass period of $(0, t_1)$ when it uses them to spread the infection; it slaughters them at the maximum rate subsequently. The intuition behind this structure is as follows. Once the worm infects a host, it never loses it to the recovery process, and thus, since it benefits from killing a host only because this enhances the final tally of the dead, it ought to kill hosts toward the end and utilize them before. The proof follows.

Proof. Note that $\psi^M(T) = \kappa_I I(T) > 0$ (because of Lemma 7.1). Thus, as in the proof of Theorem 7.2, the result follows if we can show that $\psi^M(t)$ crosses zero at most once. We establish this slightly differently: we show that $\dot{\psi}^M$ is strictly positive at its zero-crossing point (as opposed to showing it for all t). But this is also sufficient to conclude ψ^M has at most one zero-crossing point,

$$\dot{\psi}^M = I(\lambda_D - \lambda_I) + \dot{I}\psi^M$$

$$= \kappa_I - \kappa_D + u^M(\lambda_D - \lambda_I) - \beta_2 \lambda_I u^{N_i} + S\beta_0 u^{N_r}(\lambda_I - \lambda_S) + \dot{I}\psi^M.$$

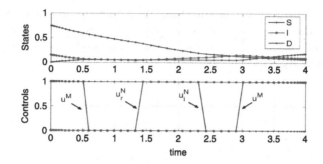

FIGURE 7.2

State evolution and saddle-point strategies. The parameters of the game are as follows: $\kappa_I = 10$, $\kappa_D = 13$, $\kappa_u = 10$, $\kappa_r = 5$, $K_I = K_D = 0$, $\beta_2 = \beta_1 = \beta_0 = 4.47$, $\pi = 1$, and initial fractions $I_0 = 0.15$, $R_0 = 0.1$, $D_0 = 0$, and $T = 4$.

-From © 2012 IEEE. Reprinted, with permission, from Khouzani MHR, Sarkar S, Altman E. Saddle-point strategies in malware attack. IEEE J Sel Areas Commun 2012;30(1):31–43.

At a zero-crossing point of ψ^M, the last term vanishes. Also, $\kappa_D = \beta_2 = 0$. The third and fifth terms are non-negative because of (7.10) and Lemma 7.2, respectively. The result now follows since $\kappa_I > 0$.   □

The saddle-point attack strategy may however be more involved when either $\beta_2 > 0$ or $\kappa_D > 0$. For example, Fig. 7.2 depicts the saddle-point strategies and the state evolution in an example scenario where $\kappa_D = 13$, $\beta_2 = 4.47$. The malware starts killing the nodes from the beginning, but then it stops the killing and infects the newly accessible susceptible nodes, boosting the fraction of the infective and toward the end, starts to kill them all again.

SUMMARY

In this chapter, we discussed how to model and analyze settings in which multiple decision-makers with misaligned interests can dynamically affect the spread of a malware in a large network. Specifically, we presented a framework based on differential game theory and introduced the notion of saddle-point strategies as mutually optimal dynamic controls in a game between a malware and a network defender. Through a stylized example, we showed how similar techniques from optimal control theory as in the previous chapter can help discover structural characteristics of such strategies in the absence of a closed-form solution. In particular, we showed that a killing worm facing a dynamic and strategic defense is still best off to have an initial phase of maximum intensity with no killing even while losing some of the infected nodes to be recovered by the system until a certain time after which it should start killing with maximum intensity. On the other hand, the defender should aggressively reduce the transmission range of the nodes and patch intensely right from the beginning until threshold times, after which it should relax the network operation to normal and subsequently stop patching.

Qualitative comparison

8.1 INTRODUCTION

In the previous four chapters of Part 2, namely, Chapters 4–7, we presented and analyzed in detail four generic modeling frameworks, developed for studying, and analyzing the malware diffusion processes and further controlling the dynamics of malware diffusion in various types of complex communication networks and different operational scenarios. For all these four frameworks, namely, the queuing, MRF, optimal control, and game-theoretic based frameworks, respectively, the corresponding techniques were developed mainly for one or more types of wireless complex networks. However, as also mentioned before, all of these four frameworks can be in principle generalized to arbitrary types of complex communications networks, e.g. wired and hybrid, in a straightforward manner.

Given these frameworks, their analysis, and provided numerical/simulation results, in this chapter, we attempt a qualitative comparison of the four generic approaches, in order to increase intuition and the potential for their practical exploitation. Considering prior experience and pitfalls identified while developing these frameworks, useful highlights, intuitions, and suggestions are shortlisted and presented in this chapter. We provide suggestions with respect to a number of important facets, such as computational complexity, implementation efficiency, practical value of the approaches, and others, by comparing the four approaches within an as common as possible set of assumptions. Thus, we provide in a codified manner, educated suggestions for applying the more suitable approach in theoretical and practical problems, given the network, operational, and other factors involved.

Consequently, this chapter further complements Part 2 of the book, toward creating longer-term value for the corresponding state-of-the-art techniques. It further aspires to highlight the most important aspects that each of these frameworks is characterized from, which can be exploited in the future in a more intense basis for modeling and analyzing other types of diffusion processes in similar or different types of complex networks. In other words, Chapter 8 further aids toward one of the initial goals of this book namely contributing to the evolution of Network Science, with respect to the diffusion of malicious information, and in general, in the study of information diffusion processes over various types of complex networks.

In the following, we first compare the four frameworks with respect to various designs and operational aspects, mentioned above, and then summarize the comparison in a tabular form.

8.2 COMPUTATIONAL COMPLEXITY COMPARISON

In general, computational complexity characterizes the class of computational problems according to their inherent difficulty, and relating those classes to each other [56]. A problem is regarded as inherently difficult if its solution requires significant resources, whatever the algorithm used. Computational complexity formalizes this intuition by introducing mathematical models of computation to study these problems and quantifying the amount of resources needed to solve them, such as time and storage. Other complexity measures are also used, such as the amount of communication (communication complexity) and the number of processors.

In terms of the previously presented malware diffusion modeling frameworks, computational complexity refers to the difficulty posed by each framework in modeling malware diffusion and providing the desired solutions. Each of the four frameworks is based on a mathematical approach, i.e. queuing systems, MRFs, stochastic optimal control, and game theory, each providing various algorithms and solution techniques. In the framework of this study, computational complexity formalizes cumulatively the inherent processing (computational) load, communication (synchronization) load, etc., imposed by the corresponding decision or computational algorithms operating within each framework for obtaining the desired solutions.

Consequently, the computational comparison in this section will indicate roughly which approach requires more computational resources for obtaining a representation of malware diffusion and studying its impact on various types of networks. Computational resources of interest include processing resources, memory requirements, and communication, namely, signaling required to exchange between distributed agents, if so the case, by the corresponding algorithms.

In terms of computational complexity, the queuing-based and the MRF-based approaches have, in general, lower overall complexity than the other two. The first framework based on queuing systems does not employ any distributed or iterative algorithm. In order to acquire results on the network behavior, analytic expressions obtained can be directly applied to yield the desired evaluation metrics. At the same time, if simulation is involved, discrete-event simulation [63] can be applied by taking into account that two events are possible, infection or recovery. The corresponding simulator then determines which nodes are involved in the determined event and performs the required updates in the system state. Computationally, such algorithm is very simple, with very small computational complexity, namely, that of a for loop. However, the simulation should be run for sufficient time to ensure a large number of events take place, and this imposes CPU time requirements. Furthermore, if large networks are simulated, memory requirements also increase given the simulation time required. Thus, the first framework has very low computational complexity, but typically imposes demanding CPU and memory requirements.

On the other hand, the MRF approach is based on an iterative algorithm (Simulated Annealing (SA) combined with Gibbs sampling). However, this is also characterized by low complexity in the order of $O(N)$,[1] N being the size of the network, for the case of the sequential implementation. The parallel and hybrid ones may have the same or smaller computational complexity, thus making this approach the best in terms of theoretical performance. Furthermore, the memory requirements scale in the order of $O(N)$ as well, requiring maintaining a simple vector with current and possibly previous state information. Regarding CPU time, again the number of iterations (sweeps) performed is usually in the range 1000–10,000, a load which is not of significant magnitude for modern machines. When a parallel implementation is employed, the computational complexity is even lower, but some signaling requirements emerge for the synchronization of the process. However, even in this case, the associated message passing is minor, scales well with a growing network size, and does not impose noteworthy requirements on the involved network. Overall, this approach emerges as the one with the lowest computational, communications and memory complexity requirements.

The malware diffusion control framework and the game-theoretic techniques, unless in cases where the solution can be shown to have simple (e.g. threshold-based) structures, lie on the more computationally demanding end. Both again provide analytic solutions, however, especially with the game-theoretic ones, solutions are typically computed iteratively. Convergence of the computational algorithms to the desired solutions can be an issue, especially when distributed approaches are required. In such cases, the selection of solution approach can have a significant impact on the speed of convergence to the corresponding solutions and even on the accuracy of the solution itself if time constraints do not permit an adequately ample amount of time to accommodate a sufficient number of iterations.

Thus, among the four frameworks, the MRF-based emerges as the most convenient in terms of computational complexity. Even though solutions are computed iteratively, and in principle infinite time is required for obtaining the desired solution, acceptable approximations are obtained rather fast compared to, e.g. game-theoretic solutions, where typically convergence of the corresponding iterative algorithms takes longer times. The more interested reader may consult more specialized treatments regarding the convergence behavior of the iterative algorithms typically employed in the four presented frameworks. Some relevant references include [228] for MRFs, [137] for optimal control based approaches, and [79] for game-theoretic based ones. The queuing-based framework is also very light in terms of

---

[1]The Big-O notation employed here is defined as follows. Let f, g be two functions defined on some subset of the real numbers. Then $f(x) = O(g(x))$ as $x \to \infty$, if and only if there is a positive constant M such that for all sufficiently large values of x, the absolute value of $f(x)$ is at most M multiplied by the absolute value of $g(x)$, i.e. if and only if there exists a positive real number M and a real number x_0 such that $|f(x)| \leq M|g(x)|$ for all $x \geq x_0$.

computational complexity, but it is characterized by stricter operational assumptions and conditions that might not be always possible to meet in practical scenarios compared to the MRF or game-theoretic frameworks.

8.3 IMPLEMENTATION EFFICIENCY COMPARISON

A significant aspect of each framework relevant to the computational complexity explained previously is the efficiency of implementation of each approach and its solutions. The efficiency of implementation refers to the total effort required to develop, deploy, test, and fully exploit the software implementing each framework, which can vary significantly if the framework is fully distributed, asynchronous, etc. The efficiency of implementation of each of the four presented frameworks can be very critical for different operational, network, and environmental conditions on a per case basis, and especially for their practical application. Furthermore, the implementation efficiency could improve or degrade computational complexity performance, by enabling obtaining solutions easier.

The queuing-based approach is perhaps the most agnostic of the four, since it does not require the computation of solutions iteratively and it only provides analytic results with respect to the parameters of the system analyzed. The implementation cost is minor requiring performing the corresponding computations with a suitable computational package, or in an ordinary programming language. Parameters can be modified easily and the results can be obtained in very fast-response scales. No signaling overhead exists and the exploitation of the results is direct. However, simulation of the network behavior can be tricky, as it depends on discrete-event techniques that typically take a significant amount of development and validation time.

The MRF framework for modeling malware diffusion also bears a light implementation load. Apart from the network/neighborhood implementation, which is necessarily common for all four frameworks and other traditional and emerging ones, this approach requires the implementation of the Gibbs sampler. If the Gibbs sampler is sequential, it is only required to compute implicitly the partition function based on the distributed potential function given by (5.4) for all sites in order to successfully compute the probability that each site will change state. A simple for loop is sufficient for this purpose. This can be easily extended for multiple states as well, if the infection model requires so. Furthermore, the same approach will be used for implementing a parallel Gibbs sampler. A minor complication might arise in the case of a synchronous parallel operation. In the event of an asynchronous one, sites do not have to exchange messages and thus implementation remains as simple as the sequential version of the approach. In the synchronous case, some signaling will be required to ensure that all agents (one for each site) have the same picture of system state, which will require message exchanges and thus slightly complicate implementation, especially if the underlying network substrate is wireless. In any case, piecemeal solutions for distributed agents communication available in the literature [12, 228] can be employed, ensuring as little signaling overhead as possible.

The implementation of the optimal control framework is heavily dependent on the computation of solutions using Pontryagin's maximum principle, which involve mixed boundary differential equations, i.e. differential equations for the states with initial conditions and differential equations for the costates with final conditions. Such implementations are readily available for various applications in the control systems domains, e.g. robots and vehicles. These implementations can be easily modified and used for obtaining solutions for malware modeling. Furthermore, the implementations are not difficult to achieve, even though the computational complexity could be in general high.

The implementation of the game-theoretic approaches is comparable but slightly more challenging than the optimal control framework. Specifically, the information structure of each player has to be carefully identified: for instance, whether players have no feedback information about the history of actions and states of the network (hence an open-loop scenario), or they have perfect or imperfect information of the current state, etc. Convergence issues and stability properties of the targeted solutions have to be considered carefully and evaluated within each specific network topology analyzed and attack scenario considered.

Overall, the queuing-based framework emerges as the more efficient for implementation, followed by the MRF-based approaches. However, given the fact that if the studied network changes, MRF can be applied with a minor only change, namely, the correct definition of new neighborhoods, while the queuing-based framework requires more complicated changes, MRF seems again the best "value-for-money" choice in terms of implementation efficiency.

8.4 SENSITIVITY COMPARISON

Sensitivity analysis can be, in principle, defined as the study of how the uncertainty in the output of a mathematical model or system (numerical or otherwise) can be apportioned to different sources of uncertainty in its inputs. In plain terms, in sensitivity analysis, one studies the variation of outcomes of a model for varying input parameters. Another form of sensitivity analysis is to explore the validity/accuracy of outcomes of a model, when the assumptions are slightly varied, as well as the impact that each such variation of assumptions might have on the quality of solutions, speed of convergence, etc.

In terms of the above, the queuing systems based framework is rather sensitive to the provided assumptions. Especially those regarding the ergodicity of the queuing system are very critical for the validity of the underlying Markov process that is employed for obtaining results on the evolution of malware diffusion. Furthermore, environmental factors affecting the topology, e.g. channel quality, mobility and operational variations, can easily throw systems off their steady-state and thus affect ergodicity of the system. In those cases, careful consideration and potential modification, if possible, of the Markovian setting to ensure that the system remains ergodic in the new steady-state are critical.

The MRF framework is far less sensitive to the above considerations. First of all, the framework is fairly transparent to the type of network topology. As long as local interactions take place, which is the case in all communication networks at the physical layer, the spatial Markov property of the model can be exploited for studying the system. The only modification needed is in the definition of the neighborhoods of nodes. Thus, from the generic perspective, the framework is fairly robust. However, it appears more sensitive to all factors affecting the convergence of the Gibbs sampler. Depending on the type of sampler employed, sequential or parallel, synchronization aspects might have an impact on the speed of convergence. Similarly, topological variations, e.g. network churn and mobility, can affect the speed of convergence of the Gibbs sampler, and even convergence of the sampler itself if the topology variations are extremely frequent. Of course, in any case, the system is capable of following up with the variations, driving the solution search toward the correct convergence points. However, it might be the case that variations take place at such a rate that the sampler cannot follow up. This is a form of "dynamic convergence," which given the extremely varying nature of this system can be acceptable to some degree.

The optimal control and game-theoretic frameworks more or less follow the same lines. Depending on the spreading model governing the interaction between susceptible and infected nodes, as well as the cost and reward functions, the changes in the obtained solutions for a variation in the parameters of the system can be considerable. Hence, convergence to the desired solution requires careful reconsideration when extrapolating the obtained solutions for obtaining similar ones with varied input parameters.

8.5 PRACTICAL VALUE COMPARISON

The four frameworks presented exhibit several overlaps in terms of their modeling objectives and features, e.g. focus on the macroscopic evolution of the systems analyzed and application of stochastic techniques. However, at the same time, they also complement each other in various facets of malware diffusion. One such aspect is the practical value of each framework, and the outcomes that can be obtained by each one of them.

The queuing-based framework aims toward a more theoretic indication of robustness capabilities of an attacked complex network. Thus, it allows assessing in a compact fashion the overall and macroscopic capabilities of a network under attack, by identifying which parameters, e.g. number of network nodes and transmission radius, are prone to make the network more vulnerable. The framework can provide ample information on such longer-term aspects of the malware diffusion dynamics and especially the network topology parameters. However, it cannot be used in, e.g. real-time diffusion control. Any control policies obtained using the corresponding analytic results will be necessarily of open-loop character, i.e. control policies will be equivalent to threshold-based decisions on how to react to given circumstances and system states.

This is not the case with the optimal control framework, which allows for a broader span of control scenarios to be accommodated and various parameterized control policies to be obtained. The main objective of this approach is centered around controlling the dynamics of the malware diffusion in a generic fashion, and thus, it seems straightforward that one should prefer this modeling substrate when control is the main objective of the imposed problem. Of course, the MRF and game-theoretic based frameworks enable developing optimization formulations that allow control of malware dynamics as well. However, since both operate iteratively, this can have an impact on real-time controls, complicating their implementation and quality of obtained results.

The MRF-based framework on its part is a very simple to implement and use approach, which can accommodate all types of network topologies and model generically various types of infection models (with many different states, rather than only binary). Thus, this modeling substrate emerges as the most general and "handy" for complex networks, further allowing the study of malware diffusion under complex conditions, e.g. mobility and network churn.

Finally, the game-theoretic framework, which is capable of more accurately modeling the dynamic interactions between pairs of susceptible-infected/malicious nodes, can be employed for those purposes where the impact of the interaction and node response is analyzed. For instance, when the focus of the analysis is on protocol-user interactions, the behavior of legitimate users, e.g. quarantine, and the strategies of attackers, e.g. which network parameter to exploit more, the game-theoretic framework allows more flexibility than, e.g. the queuing framework (an example of attack optimization based on the queuing models is presented in Chapter 9, Section 9.1). The MRF-based framework also models susceptible-infected interactions, especially in networks where these interactions are of local nature via the clique potential functions. However, such interactions are restricted on a physical connectivity scope, contrary to the game-theoretic framework, where arbitrary types of interactions between susceptible-infected pairs can be analyzed and studied with respect to their impact on malware diffusion.

One significant aspect of all the presented frameworks and their parameters is the extent to which all such model parameters are associated with their real counterparts. In most of them, e.g. average/instantaneous infection/recovery rate, node/edge churn rates, and link infection probability, it is possible to obtain them accurately by measurement and statical processing. This is possible for all model parameters that can be observed macroscopically in a network through a simple counting process (either in the long-term yielding average quantities, or instantaneously yielding infinitesimal quantities). An example of such case is the average infection/recovery rate, which can be measured by a distributed process running over a legitimate network and updating two counters after each infection/recovery event.

On the other hand, there are model parameters/assumptions that are more complicated to observe/measure and validate in real systems. For instance, in the queuing and optimal control approaches, it is assumed that infections arrive according to Poisson processes in the network. This a very tough assumption to validate in the case of malware diffusion applications, because very specialized software is required to collect the involved data, in addition to the willingness of users to collaborate. However, it seems more viable to validate such assumption in information dissemination applications that resemble malware diffusion (Chapter 9, Section 9.2), where data from online social networks can be collected (see, for instance, relevant datasets provided in [54]) and processed to reconstruct the sequence of events taking place. For all these model parameters and assumptions, great caution is required by the interested researchers, first in validating them, if possible, and then in the application of the provided frameworks and the extent of accuracy of the obtained results.

8.6 MODELING DIFFERENCES

The malware diffusion modeling frameworks presented in Chapters 4–7 can be grouped into two major classes, the first group (queuing and MRF-based) mainly deals with SIS types of networks via probabilistic and Markovian tools, while the second (optimal control and game-theoretic techniques) mainly addresses epidemics models (SIRD type). By an overall inspection of the mathematical techniques involved in the basis of each framework, it can be readily observed that there exist some commonalities between them, but also fundamental differences as well. They are concisely explained in the following.

Regarding the employed node infection model, the queuing and MRF-based approaches assume both a SIS type of malware diffusion. This means that both study the long-term behavior of the attacked network, when possibly multiple threats propagate. Thus, both frameworks aim at characterizing the long-term behavior and robustness capabilities of the network and can be used extensively as benchmarks for more specific attacks. Moreover, they can be used for design purposes, analyzing the inherent defensive capabilities of networks, and allowing the design of efficient countermeasures. On the contrary, the optimal control and game-theoretic frameworks assume a SIRD type of malware diffusion, which is more suitable for the study of specific single attacks. Thus, these approaches can provide useful outcomes for further assessing and securing attacked networks, but in a more targeted manner. This means that the two emerging groups of approaches can be used complementary for the study of malware diffusion.

In both optimal control and game-theoretic frameworks in Chapters 6 and 7, the dynamics of the epidemic is subject to dynamic decisions of rational entities, in order to achieve desired objectives. The difference between the two frameworks is that in optimal control framework, only one agent, either the attacker or the defender of the network, is assumed to be able to dynamically manipulate the dynamics of the spread, while in the game-theoretic framework, both of them can. For instance, while in both Eqs. (6.1) and (7.1) the worm is assumed to be able to dynamically decide the rate of killing the infective nodes, in (6.1), the contact rate of nodes as well as the rate of patch dissemination is assumed as given exogenous functions, while in

Table 8.1 Qualitative Comparison of State-of-the-Art Malware Modeling Frameworks

Features	Queuing	MRF	Optimal Control	Game Theory
Computational complexity	Low	Low	High	High
Implementation cost	Low	Low/average	Average	High
Sensitivity	High	Low	Medium	Medium
Practical value	Medium	High	High	Medium

(7.1), these rates can be dynamically decided by the defender, and the attacker can only reasonably assume they will be chosen optimally in response to its choice of the dynamic killing rate, and likewise for the defender. In an optimal control framework, the decision of the controlling entity should take into account the instantaneous cost of a control at a time as well as its effect on future costs by affecting the evolution of the spread. In a differential game framework, each player, in addition, needs to take into account the strategic effect of the control decisions at any given time as well.

A very important common assumption among all frameworks is homogeneity and uniformity. Homogeneous mixing and uniform infection/recovery rates have been considered for all legitimate network nodes. Regarding the second, it is straightforward to extend the analysis in all frameworks to include nonuniform infection/recovery rates. However, that would involve more complicated algebraic computations for obtaining analytical results. Mathematically, it seems easier to do so for the MRF-based framework, due to the simpler analysis and respective expressions emerging. With respect to the first, this is a broader open problem, which is further discussed in Chapter 10.

Regarding the node states, the first two approaches consider mainly two states (S,I) and the second two approaches four states (S,I,R,D). However, extending to cases with more states is straightforward for the queuing and MRF frameworks. Especially for the first aspect, this was shown for networks with churn, while extension of states in the MRF models is rather easy (extension of the phase space). This is not the case with the optimal control and game theory based approaches, where modeling is tightly related to the employed SIRD model and associated states.

8.7 OVERALL COMPARISON

In Table 8.1, we present cumulatively the main features of the comparison of the four state-of-the-art frameworks presented previously, modeling the diffusion of malware and information in general. Their main performance indices are provided in a codified manner in order to provide an overall picture of their qualities and drawbacks in exploiting them in future research and industrial applications.

In addition, in Chapter 10, an overview of open problems is provided for each of the approaches presented in Part 2 of the book. Such open problems also constitute a comparative basis, in the sense that they indicate the maturity of each approach, tough constraints that need to be currently removed in order to obtain even more accurate and generic solutions, as well as the complete potential of each modeling framework for successfully describing malware diffusion in complex wireless and possibly wired networks.

Applications and the road ahead

3

Applications of state-of-the-art malware modeling frameworks

9.1 NETWORK ROBUSTNESS

9.1.1 INTRODUCTION AND OBJECTIVES

The queuing framework presented in Chapter 4 can be exploited for studying and analyzing the potentials of the overall network robustness against generic malware attacks for the corresponding networks. In this section, we demonstrate this possibility by presenting an optimization methodology that allows for assessing the maximum damage capabilities of generic attacks in wireless multihop networks. By exploiting the framework of Chapter 4, an optimization problem can be formulated and the obtained results can be used for designing efficient countermeasures. The methodology is demonstrated for wireless multihop networks; however, other topologies can be also accommodated by proper modifications, since the underlying queuing-based framework can be also extended to other types of wired and wireless networks.

The malware spreading regime is considered, where infected legitimate nodes cannot infect other susceptible peers. Furthermore, the attackers are allowed to develop various attack strategies by varying their transmission radius prior to the beginning of their operation, namely, exploit topology control capabilities that are feasible with modern wireless networks [191]. The application presented in this chapter takes on the attackers' perspective in order to identify the critical network parameters that may constitute a multihop network vulnerable, as well as the extent of the corresponding network damage when attackers exploit them. Availability of the aggregated steady-state behavior of the system from Chapter 4 allows the formulation of an optimization problem in order to identify the optimal values of the network parameters that can be used for the aforementioned purposes.

9.1.2 QUEUING MODEL FOR THE AGGREGATED NETWORK BEHAVIOR UNDER ATTACK

In the considered model, multiple attackers spread throughout a network of legitimate nodes. Within this setting, M attackers, all having a common transmission radius r, are assumed over a multihop network of N legitimate nodes. The latter also have a common transmission radius, each with value R. Under the topology control regime, attackers are assumed more capable and the value of their transmission radius r can be varied at the beginning of the attackers' operation. Note the different transmission radii employed for attackers and legitimate users in this formulation, contrary to a common radius employed in the analysis in Chapter 4. All nodes were assumed initially randomly and uniformly spread and following the random walk mobility model with wrapping [37]. Two nodes are assumed connected whenever each one lies within the other's transmission radius, while no energy constraints are considered.

Legitimate nodes follow the SIS infection paradigm [178] as in Chapter 4, but only the spreading case. A legitimate node having multiple malicious nodes in its (one-hop) neighborhood is prone to receive infections from any one of them. It is assumed that the arrival process of infections in each link of a legitimate node is Poisson, while the recovery process of each network node is exponentially distributed as in Chapter 4. We denote by λ_k the infection arrival rate of link with node k and by μ_m the recovery rate of node m. Thus, the described node model can be mapped to that of an $M/M/1$ queue [117, 118, 124], where an infection corresponds to an arrival that requires service, and the recovery to the service itself.

Network nodes are separated at each time instant in two groups, namely infected and susceptible. According to the methodology described in Chapter 4, the whole network can be represented by a closed multiqueue network, as shown in Fig. 4.2(a), so that each group of nodes (infected and noninfected) corresponds to an $M/M/N$ sub-system. Chapter 4 (Sections 4.2.2 and 4.3) explained in detail the transition from a single queue representing one node to the multiqueue network modeling the malware diffusion over the communication network shown in Fig. 4.2(a).

Dealing with the closed queuing network of $M/M/N$ queues directly is not advised and once more, the Norton equivalent is developed and solved instead [123]. In the Norton equivalent, each of the $M/M/N$ subsystems is substituted with an $M/M/1$ subsystem with equivalent rate. In our case, the equivalent rate will be the sum of the corresponding rates. Consequently, the Norton equivalent of the original system will have the form shown in Fig. 4.3. This transformation of the queuing network is agnostic to the actual topology of the underlying communication network, i.e. it can take place irrespective of the specific type of the topology considered. The specific structure of the topology will determine the equivalent rates of the Norton equivalent later.

The Norton equivalent is a two-queue closed queuing network, where N customers (legitimate nodes) circulate. Since the more the susceptible nodes in the network, the more the total infection rate that can be achieved by the attackers in the network, the infection rate of the queue of susceptible nodes will be state-dependent. Similarly, since the total recovery rate (service rate of queue of infected nodes) will be greater for a more occupied queue, the total recovery rate in the network is state-dependent as well. The overall methodology to be followed is depicted in Fig. 9.1.

9.1.3 STEADY-STATE BEHAVIOR AND ANALYSIS

In this subsection, we provide the analysis of the corresponding closed queuing network from the perspective of the attacker, while in Chapter 4, the problem was

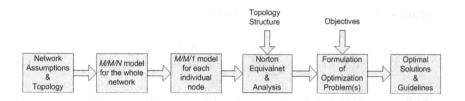

FIGURE 9.1

Methodology for studying optimal attacks.

solved from the legitimate node's perspective. In this application for the study of optimal attack strategies and robustness assessment, the attacker's perspective is adopted in order to analyze their potentials with respect to the parameters the attackers can control.

The analysis follows the same lines and focuses on the infected queue. Using balance equations for the underlying Markov chain, as shown in more detail in Appendix B.6.1, the expression for the steady-state distribution, i.e. the probability for i infected nodes in the network, can be obtained as

$$\pi_I(i) = \pi_I(0) \cdot \prod_{j=1}^{i} \frac{\lambda(N - i + j)}{\mu(j)},$$ (9.1)

with $\pi_I(0)$ the probability that no infected nodes exist in the network. By applying the standard normalization condition $\sum_{k=0}^{N} \pi_I(k) = 1$, we get

$$\pi_I(0) \cdot \sum_{i=0}^{N} \prod_{j=1}^{i} \frac{\lambda(N - i + j)}{\mu(j)} = 1.$$ (9.2)

As in Chapter 4 and without loss of generality, it is assumed that all legitimate network nodes have the same link infection rate λ, i.e. $\lambda_k = \lambda, \forall k \in \{1, 2, \ldots, N\}$, and a common recovery rate μ, i.e. $\mu_k = \mu, \forall k \in \{1, 2, \ldots, N\}$. With the above assumptions, the total recovery rate, i.e. the total service rate of the queue of infected nodes can be computed as

$$\mu(i) = \sum_{k=1}^{i} \mu_k = i \cdot \mu,$$ (9.3)

since all the corresponding procedures follow the exponential distribution and thus the whole process will be exponentially distributed as well.

Now, the total infection rate, i.e. the total service rate of the queue with noninfected nodes, as perceived from the attackers' perspective when computed for the distribution $\pi_I(i)$ in (9.1) can be obtained as follows:

$$\lambda(N - i + j) = \sum_{m=1}^{M} k'_m \lambda_m = \lambda \frac{\sum_{m=1}^{M} k'_m}{M} M = \lambda K' M, \qquad (9.4)$$

where it is assumed that for the specific malware diffusion instance $N - i + j$ noninfected legitimate nodes exist, and k'_m is the number of susceptible nodes within the neighborhood of an attacking node indexed by m. This is an important difference from the analysis in Chapter 4, where in Chapter 4 parameter k_m represented the number of attackers in the neighborhood of a susceptible node. Observing that the sum of all k'_m's divided by the number of summands yields the average number K' of legitimate noninfected neighbors of an attacker, given by

$$K' = \frac{\pi r^2}{L^2} \cdot (N - i + j)$$

if a square deployment region with side length L is considered, allows Eq. (9.4) to yield a total infection rate of

$$\lambda(N - i + j) = \lambda \pi M \left(\frac{r}{L}\right)^2 (N - i + j) \qquad (9.5)$$

for an instant with $N - i + j$ noninfected legitimate nodes at the time.

Eq. (9.5) should be compared with the corresponding one provided in Chapter 4, where instead of the attacker's transmission radius r, the legitimate node's attack radius R is used. It also reveals the time-dependent nature of the corresponding queue.

Using Eq. (9.5) for the total infection rate, Eq. (9.2) becomes

$$\pi_I(0) \cdot \sum_{i=0}^{N} \left[\binom{N}{i} \left(\frac{\lambda \pi M}{\mu} \cdot \left(\frac{r}{L}\right)^2 \right)^i \right] = 1. \qquad (9.6)$$

In order to obtain the solution of this equation, the binomial theorem is applied in expression (9.6), yielding the corresponding analytical equation for $\pi_I(0)$,

$$\pi_I(0) = \frac{1}{\left(1 + \frac{\lambda \pi M}{\mu} \left(\frac{r}{L}\right)^2\right)^N}. \qquad (9.7)$$

Thus, the steady-state distribution can be derived as

$$\pi_I(i) = \frac{1}{\left(1 + \frac{\lambda \pi M}{\mu} \left(\frac{r}{L}\right)^2\right)^N} \binom{N}{i} \left(\frac{\lambda \pi M}{\mu} \left(\frac{r}{L}\right)^2 \right)^i. \qquad (9.8)$$

Using Eq. (9.8), the probability of a completely infected network is obtained as

$$\pi_I(N) = \frac{\left(\frac{\lambda \pi M}{\mu} \left(\frac{r}{L}\right)^2 \right)^N}{\left(1 + \frac{\lambda \pi M}{\mu} \left(\frac{r}{L}\right)^2\right)^N}. \qquad (9.9)$$

Based on (9.8), the average number of infected nodes (expected number of customers in the queue), obtained by $E[L_I] = \sum_{i=1}^{N} i \cdot \pi_I(i)$, can be obtained as

$$E[L_I] = \frac{\lambda \pi M N \left(\frac{r}{L}\right)^2}{\mu + \lambda \pi M \left(\frac{r}{L}\right)^2}. \tag{9.10}$$

An additional metric that can be obtained and can be indicative of the effectiveness of an attack strategy is the average throughput of the noninfected queue $E[\gamma_S] = \sum_{i=1}^{N} \lambda(i) \cdot \pi_S(i)$, where $\pi_S(i) = 1 - \pi_I(i)$ is the steady-state distribution of the noninfected queue, i.e. the expected number of susceptible nodes. With respect to this queue, its throughput ($E[\gamma_S]$) denotes the average rate at which noninfected nodes become infected, and thus can be very "useful" for monitoring the effectiveness of an attack strategy. Thus, in the following, we use interchangeably the terms *average node infection rate* and *average throughput of the noninfected queue*. If $E[L_S]$ denotes the average number of packets in the noninfected queue, then the average node infection rate is given by

$$E[\gamma_S] = \lambda \pi M \left(\frac{r}{L}\right)^2 \sum_{i=1}^{N} i \cdot \pi_S(i)$$
$$= \lambda \pi M \left(\frac{r}{L}\right)^2 E[L_S] = \lambda \pi M \left(\frac{r}{L}\right)^2 (N - E[L_I])$$

since $E[L_I] + E[L_S] = N$. After some calculations, the above equation yields the following closed-form expression:

$$E[\gamma_S] = \frac{\lambda \mu \pi M N \left(\frac{r}{L}\right)^2}{\mu + \lambda \pi M \left(\frac{r}{L}\right)^2}. \tag{9.11}$$

9.1.4 OPTIMAL ATTACK STRATEGIES

In the attempt to study network robustness by identifying the worse possible attack potentials, malicious nodes were assumed more capable, having the ability to vary their transmission radius prior to their operation. Similarly, the number of attacking nodes can also be chosen by the attacker in conjunction with the adopted topology control capability, in order to maximize the desired level of damage. Naturally, different combinations of these two parameters yield different attack strategies and in the following robustness analysis, we focus on different strategies of this type.

In the following demonstration of application of the queuing framework on the study of network robustness, we consider the quantities $E[L_I] = \frac{\lambda \pi M N \left(\frac{r}{L}\right)^2}{\mu + \lambda \pi M \left(\frac{r}{L}\right)^2}$ and $E[\gamma_S] = \frac{\lambda \mu \pi M N \left(\frac{r}{L}\right)^2}{\mu + \lambda \pi M \left(\frac{r}{L}\right)^2}$ as the objectives to be maximized. The rest of the system parameters are assumed to have constant values. A maximization problem on each of these two objective functions can be formulated as follows. Both optimization problems will have the generic form

$$\max \frac{cx_1x_2{}^2}{1 + dx_1x_2{}^2}, \tag{9.12a}$$

$$s.t. 1 \leq x_1 \leq M_{max}, \tag{9.12b}$$

$$r_{min} \leq x_2 \leq r_{max}, \tag{9.12c}$$

$$x_1, x_2 > 0, \tag{9.12d}$$

with respect to optimization variables x_1, x_2. Variable x_1 corresponds to the number of malicious nodes M, x_2 corresponds to the transmission radius r of each attacker and the M_{max}, r_{min} and r_{max} are constants, indicating the maximum number of malicious nodes, the minimum and maximum values of a malicious node's transmission radius, respectively. Parameter d is the same ($d = \frac{\lambda\pi}{\mu L^2}$) for both problems, while parameter c equals $c = \frac{\lambda\pi N}{\mu L^2}$ for the optimization problem involving $E[L_I]$ and $c = \frac{\lambda\pi N}{L^2}$ for the problem involving $E[\gamma_S]$. Intuitively, the form of the objective function indicates a linear dependence on the number of attackers and a power dependence on their radius, signifying that the impact of a small increase in the radius can be greater than a bigger increase in the number of attackers.

In the more general case, where nodes are energy constrained, the optimization objective can be different; however, similar optimization problems can be formulated, as explained in Chapter 10.

The objective function $f(x_1, x_2) = \frac{cx_1x_2{}^2}{1+dx_1x_2{}^2}$ of the generic optimization problem is twice differentiable in $\mathcal{R}^2$ and thus, its Hessian matrix $H(x_1, x_2)$ may be obtained through standard calculations. The Hessian $H(x_1, x_2)$ of $f(x_1, x_2)$ is negative definite if condition $3dx_1x_2^2 > 1$ is satisfied, in which case the objective function is a concave function. In terms of network parameters, the condition for concavity of $f(x_1, x_2)$ is given by the following expression:

$$3\left(\frac{\lambda}{\mu}\right)\pi M > \left(\frac{L}{r}\right)^2. \tag{9.13}$$

In Fig. 9.2, we provide the zero-level contours of the function $g(x_1, x_2) = 3\left(\frac{\lambda}{\mu}\right)\pi x_1 - \left(\frac{L}{x_2}\right)^2$ for $L = 1500$ m and different values of $\frac{\lambda}{\mu}$. The regions of the $[x_1, x_2]$ plane that $g(x_1, x_2) > 0$ holds represent combinations of the optimization parameters for which condition (9.13) is satisfied, and thus for which $f(x_1, x_2)$ is concave. In Fig. 9.2, these regions lie above the corresponding contour curves. It can be observed that for values of the attackers' transmission radius usually employed in wireless networks, i.e. 200 m $\leq r \leq$ 300 m, the objective function can be safely considered concave. This can be more clearly seen if (9.13) is solved with respect to r and the smallest possible value $M = 1$ is employed, namely,

$$r > \sqrt{\frac{L^2}{3\left(\frac{\lambda}{\mu}\right)\pi}}.$$

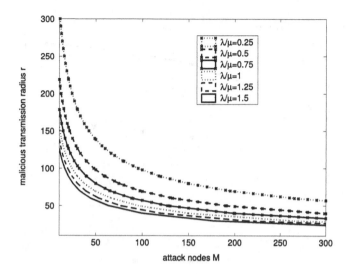

FIGURE 9.2

Zero-level contours of $g(x_1, x_2)$.

-From Karyotis V, Papavassiliou S. On the malware spreading over non-propagative wireless ad hoc networks: the attacker's perspective. In: Proceedings of the 3rd ACM workshop on QoS and security for wireless and mobile networks. p. 156–9. © 2007 Association for Computing Machinery, Inc. Reprinted by permission.

Of course concavity depends also on the value of L. The value $L = 1500$ m has been considered fixed, so that the impact of the parameter r/L is studied only through the parameter r. The value of $L = 1500$ m can be also considered a reasonable one for applications, given the typically employed r/L ratios in the literature [29–32].

9.1.5 ROBUSTNESS ANALYSIS FOR WIRELESS MULTIHOP NETWORKS

In this subsection, based on the previous analysis of the queuing-based model for malware diffusion modeling, we provide some results that can be used to characterize generic topology control based attacks and thus assess network robustness. The results are indicative and focus only on wireless multihop networks. However, extension to other types of network topology is straightforward.

In the following, for presentation purposes, we choose the maximum number of available attackers to be $M_{max} = 50$, while the range of the malicious transmission radius is within $[r_{min}, r_{max}] = [200, 300]$. The transmission radius of legitimate nodes is assumed fixed and equal to $R = 200$ m. The rest of parameters of the optimization problem, i.e. in c and d are provided in each figure separately.

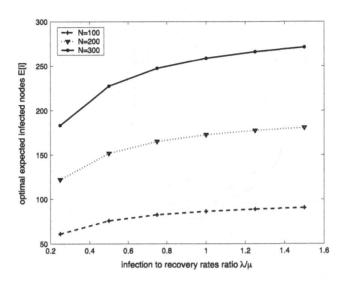

FIGURE 9.3

Optimal $E[L_I]$ versus λ/μ.

-From Karyotis V, Papavassiliou S. On the malware spreading over non-propagative wireless ad hoc networks: the attacker's perspective. In: Proceedings of the 3rd ACM workshop on QoS and security for wireless and mobile networks. p. 156–9. © 2007 Association for Computing Machinery, Inc. Reprinted by permission.

The generic objective function, as explained in the previous subsection, is concave in the optimization interval defined by the above selections of the optimization variables, and the search of the optimal solution is constrained in a closed interval defined by linear constraints. Thus, the optimal value will lie on the stationary points of the function or at the boundary region of the corresponding search space. In this case, since the gradient of this function is nonzero at all points in the admissible region, the objective function has no stationary points in the corresponding interval, and consequently, the optimal value will lie at the boundary region. The two objective functions presented (yielded by Eq. (9.12a) for the two different values of c mentioned before) differ only in a multiplicative constant that only affects the respective optimal values of each optimization problem at the optimal solution and not the form of solutions themselves.

As expected by the form of optimization problem (9.12), the optimal value of the objective function in each of the above problems is attained when the two optimization variables satisfy their upper bound constraints, i.e. $x_1 = 50, x_2 = 300$, namely, at the boundary of the search space. In this case, the optimal strategy that maximizes network damage is $\{M = 50, r = 300\}$. Each attacker should fully utilize its transmission radius and all available attackers should be used to maximize the expected network damage. However, that would not be the case in the event that

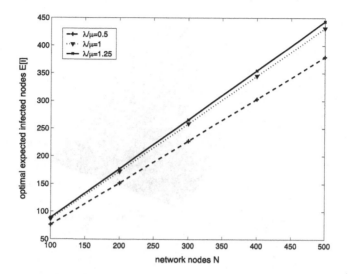

FIGURE 9.4

Optimal $E[L_I]$ versus N.

-From Karyotis V, Papavassiliou S. On the malware spreading over non-propagative wireless ad hoc networks: the attacker's perspective. In: Proceedings of the 3rd ACM workshop on QoS and security for wireless and mobile networks. p. 156–9. © 2007 Association for Computing Machinery, Inc. Reprinted by permission.

the optimization objective was a utility function that would take into account the available energy resources, in which case each attacker would have to constantly adapt its transmission radius in a clever way, in order to save energy resources without sacrificing potential network damage. It should be noted that even though this result seems too obvious to require an optimization problem to be solved, this is not always the case if other requirements and operational aspects are posed. However, the methodology and solution approach will be exactly the same and it can be directly extended to cover those cases as well. The above example motivates such methodology in a simple and straightforward manner.

In Fig. 9.3, we provide the optimal values of the expected number of infected nodes for various values of λ/μ and numbers of network nodes N. As expected, higher values of λ/μ, as well as denser topologies, yield greater network damage, meaning that networks where the ratio λ/μ works in favor of the attacker further improve the result of an attack strategy. However, as the ratio λ/μ grows beyond a threshold value, no further significant damage is caused. This means that after this point a chosen attack strategy remains relatively unaffected by λ/μ, and thus no effort needs to be put to improve the attacker's side (i.e. possibly increase the link infection rate). Consequently, a group of attackers applying the aforementioned attack strategy

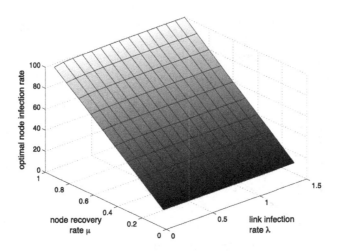

FIGURE 9.5

Optimal $E[\gamma_S]$ versus (λ,μ).

-From Karyotis V, Papavassiliou S. On the malware spreading over non-propagative wireless ad hoc networks: the attacker's perspective. In: Proceedings of the 3rd ACM workshop on QoS and security for wireless and mobile networks. p. 156–9. © *2007 Association for Computing Machinery, Inc. Reprinted by permission.*

should only improve its one-hop operation (namely, λ) if the attacked network is either dense or the ratio λ/μ is below unity.

In Fig. 9.4, the objective function value depicted against the number of network nodes, for various values of λ/μ. The optimal strategy maximizing network damage is again $\{M = 50, r = 300\}$ and the optimal value scales linearly with the number of total network nodes. As λ/μ increases, so does optimal $E[L_I]$, but there exists a threshold value for λ/μ, above which additional increase in the rate of link infection to the recovery rate ratio does not offer significant profit in terms of network damage, as shown in Fig. 9.4. It can also be observed that for sparse topologies (i.e. fewer network nodes for the same network deployment area), different values of λ/μ do not yield different results, as the sparsity of the network does not allow many infections and recoveries to take place and thus the results appear similar. On the contrary, such differences are evident for denser topologies.

As mentioned before, both optimization problems have the same form. Consequently, the optimal strategy that maximizes network damage is again $\{M = 50, r = 300\}$. It should be noted that this strategy corresponds to a bang-bang control policy (bang-bang policies arise in minimum-time problems or as application of Pontryagin's minimum or maximum principle, as explained in more detail in Appendix C) if one assumes that the noninfected queue is the controller of the

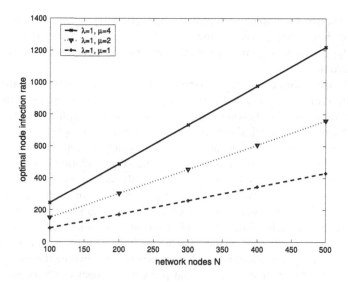

FIGURE 9.6

Optimal $E[\gamma_S]$ versus N.

-From Karyotis V, Papavassiliou S. On the malware spreading over non-propagative wireless ad hoc networks: the attacker's perspective. In: Proceedings of the 3rd ACM workshop on QoS and security for wireless and mobile networks. p. 156–9. © 2007 Association for Computing Machinery, Inc. Reprinted by permission.

infected queue and the aggregated service rate of the noninfected queue the control signal of the infected queue, [194].

Fig. 9.5 depicts the optimal values of the average node infection rate with respect to λ and μ for a network of $N = 100$ nodes. A similar form (properly scaled) would be obtained for different network densities. The optimal values of $E[\gamma_S]$ scale linearly with λ and μ, but the rate of increase is different, with μ having a slightly greater rate between the two.

Fig. 9.6 depicts the dependence of the optimal value on the number of network nodes, for various values of λ and μ. Similar observations, as those mentioned above for the maximization of the average number of infected nodes, apply. Linearity underlies the dependence of $E[\gamma_S]$ with network density and for sparser topologies the results do not bear significant differences. However, as the network density increases, differences between various network cases (i.e. different $\{\lambda, \mu\}$ combinations) become more prominent.

9.1.6 CONCLUSIONS

The above framework and provided results depict the potentials for studying the behavior and outcomes of various generic attack strategies that are based on special features of network topologies. In the above subsections, these features were topology control capabilities of attackers in wireless multihop networks. Other features of other network types can be exploited in similar studies and obtain worse-case scenario attack outcomes.

In any case of attack strategies capabilities/network topology features, the concept behind the previously presented approach is that initially the queuing framework allows obtaining analytical expressions of the parameters controlling the quantities of interest. Then, these analytical expressions allow formulating optimization problems according to the objectives of the attackers, which can vary considerably from attack strategy scenario to scenario and/or attack type. Then solutions over the various optimization problems can be obtained via analytical or numerical techniques, pending on the complexity and type of each optimization problem. The results obtained in each case are indicative of the worst-case scenario damage that an attack strategy can have on a legitimate network, while also allowing studying this damage over the various parameters of the system involved. The latter can be employed from network designers, administrators, and users for anticipating the worse possible attacks and develop countermeasures that prevent the attacks fully deploying their capabilities. Thus, they can be used for designing efficient countermeasures that cope against the worse possible attack strategies.

In the scenario demonstrated above, the straightforward guideline for countermeasures would be to restrict by as much as possible the number of attackers with suitable detection methods, or reduce interactions when an attacker is identified within range of a legitimate node, driving them to increase their transmission radius as much as possible and thus expediting faster their energy resources.

9.2 DYNAMICS OF INFORMATION DISSEMINATION

9.2.1 INTRODUCTION TO INFORMATION DISSEMINATION

Information dissemination (diffusion) has been a key social process, but especially in modern information-centric societies, it has become one of the most critical ones [46–48, 59]. Furthermore, it can be observed that most of the commercial communication infrastructures have been initially developed in the last thirty years mainly to allow transferring diverse types of information. There are different types of information disseminating in human societies and especially through the computer and communications networks available. These different types range in scope (e.g. academic, educational, healthcare, gossip, financial, and military to name a few), criticality (e.g. confidential, noise-sensitive, and public information), value (e.g. low/medium/high cost and invaluable), and overall desire by human consumers (e.g. useful, harmful, and indifferent).

Recent advances in networking infrastructures and services developed have been partially stimulated in order to accommodate the emerging trends of increasing volumes and service demands of disseminated information, in conjunction with the diversity of information types, as explained above. In general, with respect to the manner that each piece of information is received by human users, information may be distinguished in three types, characterized of *useful*, *malicious*, or *indifferent* content and denoted accordingly. Each type is analyzed in more detail in the following.

Useful information consists of many diverse types of data, all of which are expected to be of some immediate or later use by the end-users. It may consist of news, multimedia content, financial, healthcare, educational data, etc. People are willing to accept such types of information, and usually store it for further use, e.g. e-books and health examinations. Consequently, with respect to useful information, a single state transition takes place for a user, namely, from the state of not having it to the state of having received and stored the information.

On the other hand, users may receive *malicious information*, most prominently malware. The users are, in principle, reluctant to accept and/or use this type of information. Frequently, the producers and distributors of malicious information devise ways to have their information accepted by some users, e.g. by hiding it in other types of useful or indifferent information, such as email viruses. Furthermore, the sources of malicious information also manage to devise new ways of spreading such data, by using hybrid diffusion means, alternating transmission channels they distribute their threats, and in general discovering new well-hidden methods for entrapping their victims. The net result of this trend is that malicious content is characterized by recurrent behavior, i.e. by a SIS diffusion model explained in the previous chapters. Users will eventually find a way to remove the currently diffusing malware. However, and depending on their level of acquaintance, they will become prone to become infected by the same or new malware that will disseminate in the network at a future time. This of course costs in time and money to the end-users that become victims.

Finally, *indifferent information* describes cumulatively multiple and diverse types of information that disseminates in societies, nowadays mostly through the Internet and social networks, that the user is not willing to follow, but at the same time the user does not consider harmful. Characteristic cases of indifferent information include spam email (typically unwanted advertisements, etc.), pop-up advertisements in various websites, leaflets in the general case, and other email targeted for general promotional purposes. The user typically discards such information, but relevant messages are of recurrent nature, namely, they are repeated at frequent rates in order to achieve their goal. Sometimes users can be bothered by very frequent recurring indifferent information; however, most of the times the user cannot do much to restrict or discard them. Characteristic examples include phishing email messages and promotional/discount pop-ups are various electronic shopping websites.

We should note that although malicious and indifferent types of information are essentially of different nature in terms of context, in many occasions, their diffusion nature is identical, or at least very similar. Thus, later in this chapter, we use the same SIS infection model to study their propagation. The reader should keep in mind their diverse contextual nature and use the corresponding models appropriately given the specific application framework they emerge.

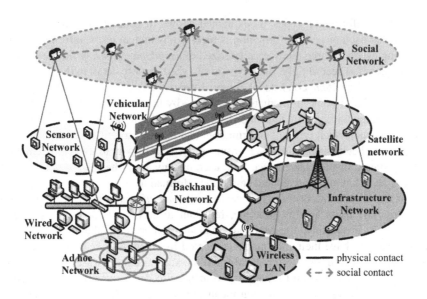

FIGURE 9.7

Contemporary wireless complex communication network architecture depicting all the considered and converged types of networks, including interconnections to wired backhauls.

- From Cheng S-M, Karyotis V, Chen P-Y, Chen K-C, Papavassiliou S. Diffusion models for information dissemination dynamics in wireless complex communication networks. J Complex Syst 2013;2013: 972352.

Thus, it may be observed that the various types of information disseminated through networks and humans resemble different cases of malware and epidemic diffusion. Consequently, given specific assumptions, it may be possible to model the information dissemination via malware diffusion modeling techniques. In this section, we focus especially on these cases of application of the previously proposed malware diffusion modeling frameworks on information dissemination over wireless complex communication networks. As already explained in Part 1 of the book, modern networks typically consist of various complex subnetworks [164], which eventually merge as a heterogeneous large-scale network, possibly connecting to a wired infrastructure (Fig. 9.7). This is the trend in future 5G and heterogeneous networks, where contemporary users, equipped with devices having various radio interfaces, can communicate with each other in more complicated ways than in the past and utilizing multiple subnetworks. For example, a user may communicate with another user via mobile phone over cellular networks, while also communicating with a different user in his/her geographic vicinity via WiFi [107] or Bluetooth [230]. In order to evaluate the dynamics of information dissemination (information dissemination dynamics — IDD) in such heterogeneous complex communication networks, where communications are affected by both social relations (at overlay networks, e.g.

social network) and physical proximity (contacts of network nodes by exchanging packets point-to-point), novel analytical models exhibiting parameters representing IDD for different types of underlying communication networks are required. In this section, as already explained, a radical approach of using a mapping between malware diffusion and information dissemination is presented and exploited for studying the dynamics of IDD, and moreover for obtaining means to control the IDD.

In the following, we present a generic modeling framework for IDD and discuss the analogy to epidemics models for describing such dynamic operation. Within this framework, specific IDD models for different types of complex networks and assessment metrics are developed. Different analytical approaches for describing useful-information dissemination malicious/indifferent-information spreading are presented and briefly analyzed, followed by indicative evaluation results.

9.2.2 PREVIOUS WORKS ON INFORMATION DISSEMINATION

Information dissemination modeling has attracted significant attention the past years, with numerous works attempting to provide accurate and effective frameworks and techniques for modeling the spreading of information, many of which have focused mainly on wireless networks from as early as 1999 [97]. An indicative comparison of such protocols may be cumulatively found in [4]. However, the approaches in [97] are not based on traditional epidemic models, such as those presented in Chapter 3, e.g. [8] and [9]. Both [8] and [9] are involved in the development of adaptive bio-inspired information dissemination models for wireless sensor networks. However, such models consider only this specific type of networks and in addition they consider one type of information propagating in the network. Another type of approaches are probabilistic ones, such as those described in [131], and the family of gossip methods, such as those in [129], [91], [41], which are based on a combination of random walk protocol variations and epidemic routing techniques. These randomized techniques have additionally stirred novel research in information dissemination for delay-tolerant (intermittently connected) networks, such as those presented in [156] and [95], where combinations of probabilistic methods, gossip algorithms, or other epidemics based approaches can be used.

The main problem with all the above families of approaches and individual techniques is that typically, in most of these approaches only one type of information is considered over a specific network topology. The objective of each such approach is to spread the information to as many users as possible. However, this is not always the case in arbitrary networks, where different types of information may propagate.

Substantial works [46, 60, 68, 231, 236] have employed an analogy of IDD in a single network as infectious spreading diseases by using ODE in the field of epidemics [58, 219], which could act as a quick reference to efficiently gather approximate knowledge of information dissemination speed and status with various settings of average node degrees and attain further control [47] in those networks. A more thorough summary of such protocols can be found in [4]. However,

Table 9.1 Types of Diffusing Information and their Features

Features	Types of Information		
	Desired	*Malicious*	*Indifferent*
Nature	Permanent	Recurrent	Recurrent
Infection model	SI	SIS	SIS

IDD are further complicated in wireless complex communication networks due to the presence of dynamic behaviors regarding the variability of topology and user behavior. Epidemic routing protocols are characterized as those randomized approaches that are based on both stochastic processes and epidemics models, such as the ones presented in [41, 91, 129] and references therein.

However, all such works mainly focus on specific types of information and network topology. The framework that will be presented in the following serves as one of the early attempts to systematically analyze IDD in a generic manner and allow assessing the dissemination and robustness performance in both current and future wireless complex networks and different types of information.

The main feature of this application, which is a joint application of the epidemics and queuing frameworks presented in Parts 1 and 2, is in realizing that different dynamics are governing the propagation of different types of information and provide appropriate models for each case. Specifically, with regard to useful data, the objective is to spread such information to as many users as possible, so that useful information reaches potentially all nodes of a network. On the contrary, in the event of malicious/indifferent-information spreading, the objective is to study the robustness of various network types against harmful or indifferent, but nevertheless network-stressing content. Consequently, there are two broad types of information, short-lived, i.e. useful information, with a short circulation lifetime and long-lived information, such as malicious information, with a longer circulation lifetime. We summarize the features of these types of information that we consider most important in Table 9.1.

To address the above, two different paradigms have been proposed that describe the behavior of systems, a SI for the first and SIS for the second type of information. Both paradigms were inspired by drawing analogies to the field of epidemics [48]. Two specific approaches have been analyzed, and the behavior of each case is demonstrated, showing how they can be used in order to identify the parameters and properties of the underlying wireless complex communication networks that govern IDD [48].

9.2.3 EPIDEMIC-BASED MODELING FRAMEWORK FOR IDD IN WIRELESS COMPLEX COMMUNICATION NETWORKS

As already explained, considerably different dynamics govern the propagation of useful and malicious/indifferent types of information. Regarding the spreading process, the dissemination of useful information resembles that of an epidemic disease, in which population members that get infected by a virus permanently transit to immunization (after recovery) or termination (if the infection is lethal). In epidemics, such models are readily referred to as Susceptible-Infected (SI) [58] or susceptible-removed (SR) [177]. Such behavior is effectively described by a two-state model, where nodes are initially prone to receive epidemics (susceptible) and then permanently transit to the infected state, once they receive the epidemic (SI model). The SR model essentially expresses the same behavior when nodes are removed from the network in cases of lethal epidemics. The SI (SR) model captures the characteristic behavior, where for each node, only a single and permanent transition takes place from the susceptible to the infected (removed) state.

Recognizing the similarities between IDD and the spread of infectious diseases, ODE models in the field of epidemic [58, 219] have been widely adopted as tools to analyze IDD in communication networks [46, 60, 68, 204, 231, 236]. Since ODE models provide closed-form formulas for the performance metrics of IDD, a wide range of effects can be encompassed by aggregating individuals in the network into two states (i.e. infected and susceptible), and thus reducing computation time and required resources. In contrast, both agent-based emulation model presented in [230] and simulation experiments presented in [221] try to precisely capture attributes of individuals in the network and the interactions among them via massive experiments; thus, the appeals of analytical tractability are neglected. Obviously, the complexity of modeling individual-level details significantly increases with the number of individuals, and thus the ODE model is suitable to act as a quick tool to identify IDD in large complex networks.

The dissemination of useful information resembles the SI process in epidemics, since a rational user waits to receive desired or useful information and then stores it for further use. If the information is indeed useful and valid, no further transaction for this information will be required/take place. Thus, the user experiences a single state transition when (s)he receives useful information from the noninformed state (node can potentially receive data, i.e. susceptible) to the informed state (i.e. infected). Under this regime, users obtain information either from information sources, or from legitimate users that have already stored the information (already in the "I" state). Several works have provided epidemic-based models for such type of information dissemination in the Internet [218], and in wireless networks [9, 156].

This is not the case however in malicious/indifferent-information dissemination. Regarding malicious-information spreading, users are reluctant to retain the malicious information they receive. However, malicious information might return (if the user is not properly protected) or adversaries are capable of devising a new malware type or package for the malicious information. With respect to indifferent information, even tough it is usually of no interest to users, it might further load an already stressed infrastructure, and especially in the case of spam, its repetitive nature might eventually become disturbing, similarly to malware. Consequently, one may consider indifferent information as a special case of malware with respect to the users' macroscopic behavior. In our treatment, we will employ this observation, due to the fact that we will focus on modeling the macroscopic behavior of IDD. For the rest of this chapter, we will focus on "malicious" information, while keeping in mind that similar models and behaviors can be obtained for "indifferent information" as well.

In principle, a user is able to recover (i.e. dispose malicious information) and return to the previous state, where it is possible to become infected again. As will be seen later from the obtained results, the model and employed methodology is similar to that of Chapters 3 and 4. The main difference is that now the topology corresponds to a potentially mesh-like network, and thus the network graph can be now any of small-world (SW), scale-free (SF), regular, or other type, compared to the random geometric graph (RGG) of the previously studied distributed wireless networks case (Chapter 4).

This behavior resembles epidemics, where individuals become infected, then recover from a disease, and become susceptible again to a different or the same disease (if they do not properly "vaccinate") — SIS model. Contrary to useful information, we notice that the SIS model defines two potential transitions between the two possible states of a user, namely, switch from susceptible to infected and then switch back to the susceptible state. This model has been also adopted in [190] for information dissemination, but again it is targeted toward only a specific type of information and network topology, as the SI epidemic models mentioned above.

9.2.4 WIRELESS COMPLEX NETWORKS ANALYZED AND ASSESSMENT METRICS

As already explained in Part 1 of the book, in Network Science (complex network theory) [22, 125, 155], a network represents a system of interactions. We consider a graph $G(V, E)$ consisting of a set V of nodes (i.e. wireless terminals or Internet users) and a set E of undirected edges (i.e. physical channels). The number of nodes is denoted by N. In this consideration of complex networks and without loss of generality, in order to focus on the IDD rather than on the wireless propagation details, two nodes connected by a link are called neighbors and the number of neighbors for a node is defined as the degree k. We consider the degree distribution $P(k)$, as the probability of having k channels for a node and $\bar{k}$ as the mean value of k. Also, it is assumed that there is at most one edge between any node pair and no self loops in $G(V, E)$.

By employing the aforementioned analogy to epidemics and malware diffusion, IDD can be effectively described and mathematically analyzed for different types of complex communication networks and their topologies. Each network type is characterized by different topological features and the proposed approach allows in both cases of useful and malicious information to identify the impact of each topology on the IDD performance and robustness. The classification of these models will be based on the following critical parameters involved:

- **Degree distribution**: It provides a suitable representation of the structure of a network, especially for social ones (networks that their topology is of relational

Table 9.2 Complex Network Classification

Connectivity Mixing Type	Partially Connectible	Equally Connectible	Unequally Connectible
Homogeneous mixing	Lattice network wireless sensor network	ER network	Small-world network
Heterogeneous mixing	Machine-to-machine network wireless mesh, smart grid		Scale-free network

nature, rather than spatial). Based on the degree distribution, a network is of homogeneous mixing if the degree distribution of each node is centered around $\bar{k}$ and the degree variance $\sigma_k^2 \leq \epsilon$, i.e. that the degree variance is smaller than some threshold level. Otherwise, it is of heterogeneous mixing.

- **Link connectivity**: The finite transmission range of a wireless node determines its neighborhood. Moreover, wireless channel quality affects the success of transmissions. The transmission range and channel quality jointly affect link connectivity.

Connectible networks: For a network with variable connectivity, the network is *(partially) connectible*, if each node has a positive probability to establish an undirected connection to any other node in the network. A network with (different) same connection probability in each node is defined as *(un)equally connectible*. Regarding connectible networks, the probability to establish an undirected connection to any other node in the network involves a partial set of nodes in the network, not all of them.

Additional properties may be identified, capturing topological properties of the underlying complex networking infrastructures, and could be exploited for analyzing IDD in these types of networks. The clustering coefficient (indicative of the clusters building up due to social or other types of interaction) and the average path length between randomly selected node pairs are appropriate quantities [164]. Based on the specific metrics, the complex networks of interest for the analysis of the IDD via malware diffusion techniques can be classified in five main categories, cumulatively depicted in Table 9.2.

Homogeneous Mixing and Partially Connectible (HoMPC): In a (regular) lattice with degree k every node connects to its k nearest neighbors [23]. A lattice is a HoMPC network since the degree distribution is a delta function with magnitude equal to 1 located at k [23]. Another example of HoMPC networks is wireless sensor networks, where sensors are uniformly and randomly distributed on a plane [202]. In the case of grid-based sensor networks, the number of neighbors in the communication radius of a sensor is fixed, which behaves exactly as a HoMPC network.

Homogeneous Mixing and Equally Connectible (HoMEC): ER random graphs [67] assume that an edge is present with probability p_e for $N(N-1)/2$ possible edges.

The degree distribution of an ER network in relatively large networks is

$$P(k) = \frac{(e^{-\bar{k}_e} \bar{k}_e^k)}{k!}, \tag{9.14}$$

where $\bar{k}_e = Np_e$. The above is a Poisson distribution that is homogeneously mixing because the variance is $\sqrt{Np_e}$, which is orders of magnitude less than $\bar{k}_e$, for large N. It is observed that the degree distribution is centered around $\bar{k}_e$ and a node has equal probability p_e connecting to any other node in the network. Thus, an ER network is a HoMEC network. Some complex wireless networks can be effectively modeled by ER graphs, especially when considering the joint cyber-physical system forming from a social network overlaying a wireless one [198].

Homogeneous Mixing and Unequally Connectible (HoMUC): SW networks generated by the WS model [225] are constructed from a regular ring lattice[1] with $2j$ edges for each node (where j is an integer number), and randomly rewiring each edge of the lattice with probability p_w such that self-connections and duplicate edges are excluded. The degree distribution of a SW network is

$$P(k) = \begin{cases} \sum_{n=0}^{\min(k-j,j)} \binom{j}{n} (1 - p_w)^n p_w^{j-n} \frac{(p_w j)^{k-j-n}}{(k-j-n)!} e^{-p_w j}, & \text{for } k \geq j, \\ 0, & \text{otherwise.} \end{cases} \tag{9.15}$$

Since the degree distribution has a pronounced peak at $\bar{k}_w = 2j$, it decays exponentially for large k [164], and the connection probability is unequal except $p_w = 1$ (extreme randomness), we regard SW as HoMUC network. This implies that as $p_w \to 1$, a SW network becomes HoMEC. Many social network models are extensions of the WS model, since the HoMUC property explains well their clustering features [164]. SW topologies emerge very often in wireless complex networks, especially in cyber-physical systems with social networks overlaying the physical ones [198].

Heterogeneous Mixing and Partially Connectible (HeMPC): M2M networks [103], wireless mesh networks [5], or smart grids have potentially nonstructured degree distributions due to finite transmission range, heterogeneity in location, mobility, channel quality variations, and time-varying user behavior, eventually classifying the underlying topology as HeMPC.

Heterogeneous Mixing and Unequally Connectible (HeMUC): A HeMUC network is a power-law distributed network, i.e. having degree distribution $P(k) \sim k^{-r}$, with r in the range $2 \leq r \leq 3$, which is also called a SF network. Barabasi and Albert observe two essential factors of SF network: growth and preferential

---

[1] A regular ring lattice is just a ring network, i.e. a chain of nodes, where there exists a connection between the head-tail node as well, so that all nodes have degree exactly two.

attachment [1]. In this model, denoted by BA, every new node connects its m edges to existing nodes according to the preferential attachment rule, and $\bar{k}_b = 2m$. The power-law degree distribution suggests that most nodes have few neighbors, while some "supernodes" (hubs) tend to have a great amount of neighbors. Thus, the power-law distributed network can be treated as a HeMUC network.

Table 9.2 summarizes the aforementioned complex network categories. Note that the size of the largest connected component ("giant component") is also an important measure of the effectiveness of the network at information epidemics or cooperations. Regarding this, a network is saturated if the giant component size approaches the total number of nodes in the network. Otherwise, nodes may be isolated from the giant component (nonsaturated network). For instance, a giant component emerges almost surely in random graphs, if $\sum_k k(k-2)P(k) > 0$ [161].

9.2.5 USEFUL-INFORMATION DISSEMINATION EPIDEMIC MODELING

As it was explained in the previous subsections, the macroscopic behavior of useful-information dissemination can be described by the SI epidemic model, in which, nodes receive the designated data once in their lifetime. In this section, we present an analytical approach for quantifying the process of useful-information dissemination in the types of complex communication networks considered above.

SI epidemic spreading model

We adopt the SI model in order to describe the IDD of useful information. In the analogy we draw between IDD and epidemics, an uninfected node in epidemics corresponds to a noninformed node in IDD, i.e. a node not having received yet useful information. On the other hand, an infected node corresponds to a node that has received and stored useful information. Such model is appropriate for describing a specific type of useful-information spreading, e.g. handbook disseminated to a population, similarly to the SI model describing the dissemination of a single viral threat. In the sequel, we will use the terms noninfected and noninformed, as well as the infected and informed terms interchangeably. Assume that N nodes are initially all susceptible noninformed except for a small number that are infected and contagious (denoted as infectious). These "contagious" nodes represent the nodes that generate the useful content to be disseminated, as will be explained in more detail in the following paragraph. Thus, by the previous analogy, in this scenario, all nodes would like to eventually become infected (i.e. informed).

The infection parameter λ is adopted to characterize the rate of spreading between S-I pairs. Considering the volatile nature of wireless channels and MAC, protocol operation information transmission over a communication link may not be successful. The availability (or the successful transmission rate) of a link in wireless communication channels is typically modeled as a two-state Markov chain with on and off states [84], which is sufficient for the consideration of IDD (channel propagation effects can be considered in future extensions). For a dynamic topology, the set V of nodes is time-invariant, the set E of edges varies with time, while however, the network maintains its time-invariant statistical properties. The informed nodes adopt a "consistent broadcast" behavior, so that they tend to transmit the useful information to susceptible nodes in contact consistently, just as in the spread of biological viruses. Hence, the spreading rate (similar to infection rate) λ is equivalent to the probability of an available channel ("on" channel state), which is independently determined for all channels such that the long time behavior of channel availability is equal to λ.

Expressed quantitatively, if $I(t)$ and $S(t)$ are, respectively, the fraction of infected (informed) and susceptible (noninformed) nodes at time t, we have

$$\begin{cases} I(t) + S(t) = 1, \\ \dfrac{dI(t)}{dt} = \lambda \bar{k} I(t) S(t), \end{cases} \tag{9.16}$$

which has the same form as (3.1). Then, the simple analytic solution obtained identically as in Chapter 3 is

$$I(t) = \frac{I(0)}{I(0) + (1 - I(0))e^{-\bar{k}t}}. \tag{9.17}$$

From (9.17), it is clear that the informed density $I(t)$ approaches 1 as time evolves, as expected. In this study, we assume initially there is only one informed node, i.e. $I(0) = 1/N$. Note that the saturation[2] of any complex network should serve as one of the important sufficient conditions when adopting the SI model for IDD, since the first-order differential equation in expression (9.17) fails to distinguish the saturation of a network. In the sequel, the proposed SI framework for static and dynamic saturated complex networks is presented.

Static complex networks

The static network is regarded as a time-invariant graph $G(V, E)$ where the topology is unchanged in time. Given that, in Fig. 9.8, numerical results for a regular lattice (HoMPC), an ER (HoMEC), a SW/WS (HoMUC), and a SF under the BA model (HeMUC) networks are presented, based on ensemble averages, obtained by 100 simulations in saturated complex networks. Since an ER network can be totally characterized by parameters N and p_e, IDD in ER is described through the SI model by adopting $\bar{k}_e = Np_e$. However, compared to IDD in ER networks, the SI model amplifies the cumulative informed node fraction with time as it implicitly assumes the HoMEC property, resulting in inaccurate estimation. In Section 9.2.5, this discrepancy is addressed in order to obtain a more accurate IDD SI-based model.

---

[2]It should be noted that the term saturation refers to the tendency of a network to slow down the rate of further propagation/spreading of malware due to its topology.

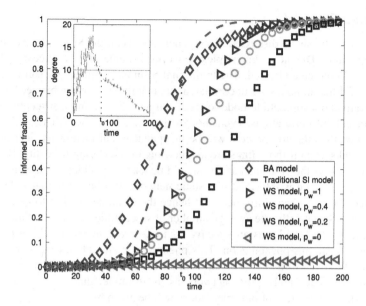

FIGURE 9.8

IDD in regular lattice (HoMPC; $p_w = 0$), ER (HoMEC; $p_w = 1$), SF (HeMUC), and SW (HoMUC) networks for different p_w with mean degree equal to 10, $\lambda = 0.01$, and $N = 2500$.
- From Cheng S-M, Karyotis V, Chen P-Y, Chen K-C, Papavassiliou S. Diffusion models for information dissemination dynamics in wireless complex communication networks. J Complex Syst 2013;2013: 972352.

An interesting phenomenon for the IDD in SF networks generated by the BA model can be observed. The corresponding IDD curve exceeds that of the SI model in the beginning, but at some instance t_0, the curve of the SI model transcends. This is in accordance with the fact that the information has higher possibility to be transmitted to supernodes than the nodes with low degrees at early stages. Once the information spreading begins, the number of informed nodes highly increases as supernodes eventually become informed. Then, the spreading rate decreases as information dissemination is now mainly propagated to nodes with lower degrees. Thus, although the SI model is not suitable for describing the IDD in SF networks, it still provides useful insights for better understanding the IDD in networks with the HeMUC property. Note that since it was assumed that homogeneity holds for nodes of the same degree [177], an enhanced model can be derived to accurately match the IDD curves of HeMUC networks.

Regarding SW networks, effects of different rewiring probabilities p_w are investigated (Fig. 9.8). Ranging from extreme regularity ($p_w = 0$) to extreme randomness ($p_w = 1$), IDD accelerates due to the enhancement of connectivity and the SW network transforms from HoMPC network ($p_w = 0$) to HoMUC network ($p_w \in (0, 1)$) and finally HoMEC network ($p_w = 1$).

Corrected SI model

To overcome the above discussed discrepancies between the SI model and the proposed epidemic IDD model in complex networks, the emerging "degree correlation" problems should be addressed. The traditional SI model implicitly assumes that the degrees of different nodes are uncorrelated, i.e. that the interaction between a source of malware and a susceptible node, or between an infected and a susceptible node, does not exhibit correlations, even though they can be close to each other. This implies that the infection process will be independent and identically distributed in different links, even if those links are neighboring or interfering in reality. However, this is not always the case and caution should be taken to validate the assumption in detail. When a node is informed in a static network, this suggests that at least one of its neighbors has been informed, and hence the mean degree has to be corrected accordingly. Specifically, the average number of susceptible neighbors of an infected node is less than $\bar{k}$ and thus an *effective mean degree* $\hat{k}$ metric has been proposed to accommodate this phenomenon. This phenomenon should be carefully modeled otherwise the overestimation problem [236] leads to significant deviation (see also Fig. 9.8). Considering HoMEC network as an example, due to its mean degree characterization, the rate of informed nodes will be given by

$$\frac{dI(t)}{dt} = \lambda \hat{k} I(t) S(t) = \lambda [\bar{k} - f(t)] I(t) S(t), \tag{9.18}$$

where $f(t)$ is a time-varying function accounting for the average number of informed neighbors of an informed node. By setting $f(t)$ as a constant, a tight upper bound on IDD can be obtained when $f(t) = 1$ for HoMEC networks since intuitively, a node that is infected implies that at least one of its neighbors is infected. Thus, the following observation is applicable:

Observation 9.1. *For a saturated and HoMEC network given the informed rate λ and mean degree $\bar{k}$ parameters, there exists a tight upper bound on IDD at any time instance.*

By setting appropriate values of $f(t)$ according to the features of different networks, the adjusted SI models successfully capture the corresponding IDD. This implies that upper bounds on IDD of HoMUC and HoMPC networks also exist. Considering IDD in sensor network (HoMPC) as an example, where sensors are uniformly and randomly distributed on a plane with node density σ, the communication radius of a sensor R and the radius of the circle containing the informed sensors $r(t)$, obviously, $\sigma \pi r(t)^2 = N I(t)$. This is approximated into the above model by having only the infected nodes that lie on the periphery of an informed circle in communication with the susceptible nodes located at a distance of at most R outside the infection circle, and thus have the potential to inform (infect) them. This is shown later in Fig. 9.10. In other words, the spatial broadcasting of the information is only contributed from the wavefronts of informed circles, while the infected nodes located in the interior of the informed circles are not engaged in further spatial dissemination. By using the general approximation of $(1 - x)^2$ by $1 - 2x$ for small x, since x^2 is negligible compared to x, function $f(t)$ can be calculated as

$$f(t) = \bar{k} \cdot \frac{\sigma \pi [r(t) - R]^2}{NI(t)} \cong \bar{k} \cdot (1 - c \sqrt{NI(t)}), \qquad (9.19)$$

where $c = 2 \sqrt{\sigma \pi} R$ and $\sigma \pi R^2$ is usually negligible compared with $NI(t)$. Applying $f(t)$ to the corrected SI model, leads to obtaining the same result derived in [60]. From this example, the following observation is emerging.

Observation 9.2. *For a saturated and homogeneous mixing network, the time T needed to inform a fraction s of nodes can be directly obtained from (9.18) as $T = I^{-1}(s)$, if $I^{-1}(\cdot)$ exists.*

As the cumulative informed fraction approaches 1 in a saturated network, Observation 9.2 serves as a more accurate benchmark to any broadcasting mechanisms in wireless complex networks for evaluating the performance of information dissemination. This can be justified by the fact that Observation 9.2 greatly mitigates the biased estimation due to degree correlations.

Dynamic complex networks

This section discusses the dynamic case where network topology changes with time, while maintaining the basic structure and properties. As it will be shown, a time-varying topology provides great chances for information dissemination to the entire network and therefore, it is suitable for describing complicated interactions within large-scale networks supporting mobility capabilities (e.g. routing protocols in MANET). Two nodes originally disconnected, might eventually establish a virtual link between them due to mobility, i.e. a link with short duration imposed by the mobility pattern that will last only while the nodes remain in proximity, thus yielding a virtual giant component. Given the condition that a network is originally nonsaturated (e.g. MANET in sparsely populated area), mobility may make the network *virtually saturated*, since all nodes can eventually receive the information due to change of connectivity in the dynamic sense.

Observation 9.3. *A dynamic network is virtually saturated in the sense that the virtual giant component size approaches the number of nodes.*

The mobility of nodes facilitates connectivity in homogeneous mixing networks originally not saturated and thus dynamic HoMEC, HoMPC, and HoMUC networks are virtually saturated. In the following, the focus shifts on IDD in such networks where benefits from mobility can be obtained in a more obvious manner.

Due to the virtual saturation property in Observation 9.3, the SI model that fails to describe the IDD of nonsaturated networks is nevertheless to characterize the

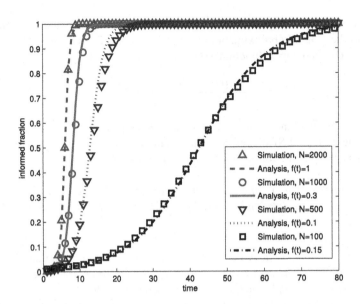

FIGURE 9.9

IDD in dynamic MANET with $R = 2m$, $\lambda = 1$, and $\bar{k} = 2.51, 1.26, 0.63$, and 0.13, respectively.
- From Cheng S-M, Karyotis V, Chen P-Y, Chen K-C, Papavassiliou S. Diffusion models for information dissemination dynamics in wireless complex communication networks. J Complex Syst 2013;2013: 972352.

IDD of dynamic homogeneous mixing networks. As we define $f(t)$ according to the properties of the networks, the IDD curves can be accurately matched by using Eq. (9.18).

Observation 9.4. *The IDD of dynamic homogeneous mixing networks can be characterized by the corrected SI model.*

To show the significance of these observations and the flexibility of the proposed model, Fig. 9.9 illustrates analytic and simulation plots depicting the epidemic routing via broadcasting in large-scale MANETs. Epidemic routing aiming at exploring the advantages of path diversity with concrete analysis [231] has been regarded as one practical way to achieve routing in dynamic HoMPC [172, 215]. In the simulation experiments, we assume that each node has the same transmission range R with uniform random deployment in a 100×100 m^2 plane with wrap-around condition. According to a stationary and ergodic mobility model, such as the truncated Lévy Walk model [185], the step length exponent $s = 1.5$ and pause time exponent $\varphi = 1.38$ have been set in order to fit the trace-based data of human mobility pattern collected in UCSD and Dartmouth [135]. The successful packet transmission rate to the spreading rate λ has been incorporated. Fig. 9.9 shows that the MANET is nonsaturated (however, virtually saturated), and our model captures the complicated interactions among numerous mobile nodes precisely.

Hybrid complex networks

When nodes are capable of communicating with each other using multiple heterogeneous connections, a hybrid complex network consisting of multiple complex subnetworks is built via heterogeneous links. As in the example we mentioned in Section 9.2.1, people could exploit mobile smart phones to communicate with individuals in their address books via traditional phone-call and short message, as well as the individuals in their geographic proximity via WiFi [107] or Bluetooth [230]. As shown in Fig. 9.10, the IDD in such complicated networks can be investigated by separately considering IDD in the social network constructed by contacts and IDD via broadcasting and then aggregating the results for the combined cyber-physical system.

According to the proposed categories, ER random network (HoMEC) and sensor network (HoMPC) are used to, respectively, model the delocalized and broadcasting dissemination patterns. Thus, the subpopulation function $I(t) = I_e(t) + I_s(t)$, where $I_e(t)$ and $I_s(t)$ are those that have been disseminated via ER and sensor networks at time t, respectively. The average degree $\bar{k}_e$ describing the social relationships between handsets provides the average number of contacts in the address book. According to Eq. (9.18), the basic differential equation that describes the dynamics of informed subpopulation is

$$\frac{dI_e(t)}{dt} = \lambda \frac{S(t)(\bar{k}_e - 1)}{N} I(t).$$ (9.20)

When an informed node intends to disseminate via broadcasting, it first scans to search the nearby nodes within its transmission range R and connects to a neighbor so as to determine the susceptible neighbors for propagation. In this case, the average number of neighbors $\bar{k}_s$ equals $\rho \pi R^2$. The behavior of such spontaneous spreading can be regarded as a ripple centered at the infected source node which grows with time. As shown in Fig. 9.10, the spatial spreading of the information here is only contributed from the wavefronts of informed circles, while the infected nodes located in the interior of the informed circles are not engaged in further spatial infections, where similar arguments, as the ones employed before for obtaining (9.19), apply.

Without loss of generality, it is assumed that a single informed circle is generated at time t_1 by a point source infected through and kept stretching for t_2 time units. Then, its incremental modified spatial infection at time $t_1 + t_2$ is

$$G'(t_1, t_2) \triangleq \frac{\partial G(t_1, t_2)}{\partial t_2} = \lambda \frac{S(t_1 + t_2) \cdot \frac{1}{2}\bar{k}_s}{N} c \sqrt{G(t_1, t_2)},$$ (9.21)

where $\frac{1}{2}\bar{k}_s$ accounts for the fact that for an infected node on a periphery, roughly half of neighbors outside the infection circle are susceptible, and $G(t_1, t_2)$ is the spatial

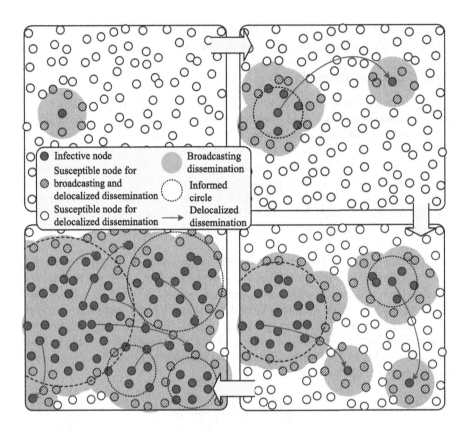

FIGURE 9.10

The IDD in wireless complex networks (cyber-physical systems) consisting of both long-range and broadcast dissemination patterns.

- From Cheng S-M, Karyotis V, Chen P-Y, Chen K-C, Papavassiliou S. Diffusion models for information dissemination dynamics in wireless complex communication networks. J Complex Syst 2013;2013: 972352.

infection within the same time interval prior to the rectification. The incremental spatial infection at time t of all infection circles is given by

$$\frac{dI_s(t)}{dt} = \int_0^t I_e'(\tau)G'(\tau, t - \tau)d\tau. \tag{9.22}$$

This means that there are $I_e'(\tau)d\tau$ point sources originated at time τ and each contributes $G'(\tau, t - \tau)$ incremental spatial infection at time t.

To validate the analytical model, experiments to simulate IDD in a hybrid network among 2000 individuals uniformly deployed in a 50×50 plane were developed. The constructions of social contact networks and setup of parameters (e.g. $\bar{k}_e = 6$) follow

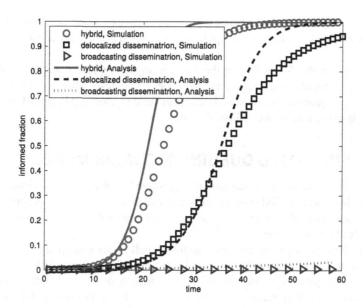

FIGURE 9.11

IDD in hybrid (HoMEC and HoMPC) complex networks of propagating information in both delocalized and broadcast fashions, where $\bar{k}_e = 6$, $\bar{k}_b = 3$, and $\lambda = 0.05$.

- From Cheng S-M, Karyotis V, Chen P-Y, Chen K-C, Papavassiliou S. Diffusion models for information dissemination dynamics in wireless complex communication networks. J Complex Syst 2013;2013: 972352.

the data set in [221]. Fig. 9.11 illustrates analytic and simulation plots depicting the IDD via both delocalized communication and broadcasting in hybrid (HoMEC and HoMPC) complex networks. It can be observed that the curves of propagation dynamics closely match the analytical model, where limited discrepancy exists, mainly due to the fact that information may propagate to individuals who have already been informed and uncertain boundary conditions could not be considered in the analysis. Comparing with the traditional SI in Fig. 9.8, the corrected SI could capture the IDD of ER network (HoMEC) more precisely.

9.3 MALICIOUS-INFORMATION PROPAGATION MODELING

In this section, we present an analytical model which is able to capture the behavior of malicious IDD (modeled as SIS epidemics) in wireless multihop networks. Contrary to the case of useful information, regarding malicious/indifferent information, one is interested in the robustness capabilities of the network to sustain such traffic, which in both cases, it is of no use (or it can even be harmful in case of malware). The presented analytical model is based on the approach exploiting queuing theory, but contrary to Chapter 4, here it was applied on the wireless complex topologies shown in Table 9.2.

Contrary to the SI model, in the SIS paradigm employed for the malicious-information spreading, average quantities are of interest, while in the SI model instantaneous quantities were considered, as in the long-run the SI network converges to a pandemic (all nodes are informed) state.

9.3.1 SIS CLOSED QUEUING NETWORK MODEL

In this subsection, some of the material presented in the previous subsections is repeated, but also modified appropriately. For clarity and completeness purposes, we present such material in its proper sequence, stressing on the emerging reusability of the involved methodologies.

In the SIS paradigm, susceptible (noninformed) nodes essentially wait until the arrival of malicious information, in which case they alter to the infected (informed) state. A propagative network is considered, where nodes spread further the malicious information they receive. Consequently, a node might become infected from malicious software either from an attacker or an already infected legitimate node. This holds for several types of viruses and worms that have appeared [133, 197]. Infections are assumed to arrive in a nondeterministic fashion. The recovery process (disposing malicious software) is of similar nature, but not necessarily of the same waiting behavior. Throughout the rest of this subsection, following current literature [80], it is assumed that the infection arrival process is Poisson, while the recovery process of each network node is exponentially distributed. The infection arrival rate in link i is denoted by λ and the recovery rate of node j by μ_j, as in Chapter 4.

Legitimate nodes are again separated in infected and susceptible. Their transition from state to state can be mapped to a closed two-queue packet network, as shown in Fig. 4.3 [48], where N customers, i.e. network nodes, circulate. At any instance, if i nodes are infected, then $N - i$ are susceptible. Both service rates are state-dependent according to the number of packets (user nodes) that exist in the corresponding queue at each time instance. Explicit definition of each queue's equivalent service rate, which is state-dependent, i.e. $\lambda(N-i+1)$ and $\mu(i)$ (for the corresponding system states with i infected and $N - i + 1$ susceptible nodes), depends on the underlying complex network and employed infection paradigm. Without loss of generality, we assume that the lower queue represents the group of infected legitimate nodes and denote it as "infected," while the upper queue represents the susceptible nodes and we denote the queue as "susceptible." The state-dependent service rate of the susceptible queue can be extended to the case of multiple malicious-information sources (i.e. attackers), in which case the service rate will have the form $\lambda(N - i + M)$, M being the number of attackers. Our analysis here is focused on the case of a single attacker (i.e. $M = 1$).

Standard methodology from Chapter 4 may be employed to analyze the two-queue closed network. The focus is on the infected queue. Its steady-state distribution, denoted by $\pi_I(i)$, represents the probability that there are i packets (nodes) in this queue. Using balance equations for the respective Markov chain, the explicit expression for the steady-state distribution can be obtained as

$$\pi_I(i) = \pi_I(0) \cdot \prod_{j=1}^{i} \frac{\lambda(N+1-i+j)}{\mu(j)}, \tag{9.23}$$

where $\pi(0)$ is the probability of no infected nodes in the network. Applying the normalization condition $\sum_{i=0}^{N} \pi_I(i) = 1$, and appropriately specifying the total infection and recovery rates (where $\lambda_k = \lambda$ and $\mu_k = \mu \; \forall k \in \{1,2,\dots,N\}$), and by setting $\alpha^{-1} = \frac{\lambda\pi}{\mu}\left(\frac{R}{L}\right)^2$ (following the reasoning presented in Section 9.2.4 and in Chapter 4) and considering a large number of legitimate nodes, the probability of no infected nodes can be approximated as

$$\pi_I(0) \cong \frac{\alpha^N}{N!} \cdot e^{-\alpha}, \tag{9.24}$$

and the steady-state distribution can be approximated as

$$\pi_I(i) = \frac{\alpha^{N-i}}{(N-i)!} \cdot e^{-\alpha} = \pi_S(N-i), \tag{9.25}$$

where $\pi_S(N-i)$ is the steady-state distribution for the noninfected queue. Using Eq. (9.25), the probability of a completely infected network equals $\pi_I(N) = e^{-a}$. It is noted that the error introduced by the above approximation is negligible for values of α and N commonly used in practice (in the order of less than 10^{-3}).

Eq. (9.25) clearly indicates the critical parameters that affect the behavior of the system. Assuming a fixed area of the network deployment region, the number of legitimate nodes (i.e. the density of the network) along with the common transmission radius and the ratio of the link infection rate to the node recovery rate are decisive factors regarding the overall behavior and stability of the system.

Based on such model, the expected number of infected (informed) nodes (corresponding to the expected number of packets in the lower queue) for different types of networks may be obtained. The general expression yielded has a Poisson form, due to approximation (9.25),

$$E[L_I] = \sum_{i=1}^{N} i \cdot \pi_I(i) = N - c, \tag{9.26}$$

where $\pi_I(i)$ is the steady-state distribution of the underlying Markov chain, and $c = \frac{\frac{\mu}{\lambda\pi}}{\left(\frac{R}{L}\right)^2}$ is the equivalent of the α parameter mentioned above for a HeMPC network (wireless multihop) over a square deployment region of side L and each node having a transmission radius R. For all other types of networks, $c = \frac{\mu N}{\lambda \bar{k}}$, where $\bar{k}$ is the average node degree for each network under discussion ($\lambda_i = \lambda$ and $\mu_j = \mu$ for all i, j without loss of generality).

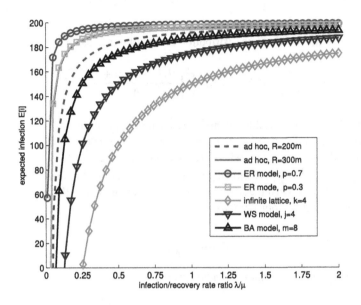

FIGURE 9.12

Average number of infected users of the legitimate network $E[L_I]$ as a function of λ/μ.

- *From Cheng S-M, Karyotis V, Chen P-Y, Chen K-C, Papavassiliou S. Diffusion models for information dissemination dynamics in wireless complex communication networks. J Complex Syst 2013;2013: 972352.*

Observing the analytic form of the average number of infected nodes for each network type, according to the specific expression of c, a major difference between HeMPC networks and the rest should be noted. More specifically, in a HeMPC, the spatial dependence among nodes (due to their multihop nature) is reflected by the fraction $\frac{\pi R^2}{L^2}$ representing the coverage percentage of each node with respect to the whole network area. On the contrary, in the expression of c for the rest of the networks for which the topology is mainly based on connectivity relations and not spatial dependence as in the multihop case, the corresponding quantity expressing the local neighbor impact is expressed by $\frac{\bar{k}}{N}$. It is evident, that for the rest of network types, quantity $\frac{\bar{k}}{N}$ expresses solely the special connectivity properties of the employed network through the value of $\bar{k}$, since for these networks, no spatial dependence is expressed in their connectivity graph and thus, no such spatial feature has impact on IDD.

The average throughput of the susceptible queue $E[\gamma_S]$, corresponding to the average rate that nodes receive malicious information, can be computed as

$$E[\gamma_S] = \sum_{i=1}^{N} \lambda(i) \cdot \pi_S(i) = \frac{\lambda \bar{k}}{N} \left[c(N+1) - (1 - e^{-c})(N+2) - c - c^2 \right]. \quad (9.27)$$

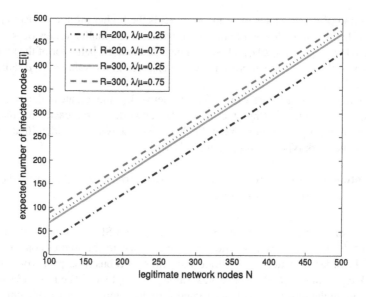

FIGURE 9.13

Average number of infected nodes $E[L_I]$ as a function of N (numerical result).

- From Cheng S-M, Karyotis V, Chen P-Y, Chen K-C, Papavassiliou S. Diffusion models for information dissemination dynamics in wireless complex communication networks. J Complex Syst 2013;2013: 972352.

Demonstration

As already mentioned, contrary to the SI model, in the SIS paradigm employed for the malicious-information spreading, average quantities are of interest, while in the SI model instantaneous quantities were considered, as in the long-run the SI network converges to a pandemic (all nodes are informed) state. Especially the average number of infected legitimate nodes is the most important quantity, since malicious information is of recurrent nature and if one observes the system macroscopically, nodes oscillate between the {S}, {I} states.

Fig. 9.12 shows the average number of infected nodes with respect to the infection/recovery ratio for various types of complex networks. It is evident that as the ratio λ/μ increases, $E[L_I]$ increases for all types of networks, denoting greater probability of users to get malicious information from a communication link. Parameters for each type of network are in accordance with the notation employed in the classification of Section 9.2.3.

With respect to *ad hoc* networks (HeMPC), the greatest the transmission radius of nodes, the denser the network, and thus the easier the malicious data to be spread. Especially for *ad hoc* networks, the dependence of the average number of infected nodes on the number of legitimate nodes is linear as shown in Fig. 9.13. The combination of larger values of R and λ/μ yields the higher number of $E[L_I]$ for all values of legitimate nodes, while the combination with smaller values of R and λ/μ yields the lower values of $E[L_I]$. It should be noted that the behavior of the spreading dynamics of the network is more sensitive to changes in the value of λ/μ rather than variations in R.

Similarly to HeMPC networks, in ER networks, the greatest the probability two nodes are connected (thus the denser the network is), the easier for malicious information to propagate on average. Such property may be also identified with the rest of the network types, leading to the following.

Observation 9.5. *The denser a network becomes, irrespective of its type, the easier for malicious information to spread.*

For regular, random (ER type), SW (WS type), and SF (BA type) networks with the same $\bar{k}$, the same result would be obtained, due to the expression of c given before. In order to better demonstrate the different robustness properties of these network types, in Fig. 9.12, a different $\bar{k}$ value was employed. The term robustness is meant here in the sense of resistance to malware diffusion so that a sufficient amount of susceptible nodes maintains the operation of the network. Regarding the different types of networks, HeMPC (*ad hoc*) is close to ER (HoMEC), exhibiting that in general, randomness aids the spreading of malicious information. This is a very useful outcome with significant practical value for designing efficient countermeasures for malign IDD. On the contrary, a regular lattice (HoMPC) makes the spread of malicious information more difficult, since each node is only connected to a typically small number of other nodes and it would take significant effort to quickly spread malicious information throughout the whole network. Similarly, WS (HoMUC) and BA (HeMUC) exhibit robustness closer to lattice networks, as their topologies are derived from such regular arrangements [164]. Among the latter three categories, a SF (BA) network may be more prone to spreading than a lattice (as shown in Fig. 9.12), because for the specific network instances, the given BA network is more dense than the lattice (the BA has mean degree $\bar{k} = 16$, while the lattice $\bar{k} = 4$ for the same number of network users). The following observations may be derived from the above analysis.

Observation 9.6. *Topological randomness favors the spreading of malicious software.*

Observation 9.7. *Among similar network types, the relative (local) density of each topology determines the robustness of the system against malicious-information spreading.*

The road ahead

10

10.1 INTRODUCTION

The malware modeling approaches presented in this book, legacy and state-of-the-art, cover a very broad spectrum of mathematical tools they employ and of applications. The frameworks presented in Part 2 describe malware diffusion dynamics accurately and holistically, both for SIS and SIRD infection models. Furthermore, one can capitalize on their analytic power and study more sophisticated aspects of malware diffusion, such as analysis of attack strategies, network robustness, and design of smart countermeasures.

The most important contribution of all the approaches presented in the previous parts of the book is that through the established results, they have paved the way for more advanced analysis of malware diffusion processes, stimulating multi- and cross-disciplinary research. This in turn has opened up new directions for research and practical considerations, as well as reconsidering several traditional problems from new perspectives. All these new problems present significant interest for both the academic and industrial communities.

However, at the same time, various less or more important open problems of already examined aspects of malware diffusion remain open. Even the presented state-of-the-art approaches were incapable of tackling the whole spectrum of emerging problems. Several mathematical limitations prevent state-of-the-art methodologies from providing piecemeal solutions, making the decision to use the more suitable approach in each network and application scenario an important one.

In this chapter, we focus on these aspects of malware diffusion research. We initially present some of the currently open minor or major problems for each state-of-the-art approach presented. Then we summarize some more general open problems of interest and suggestions on which approach would be more applicable or promising for tackling them in the future.

10.2 OPEN PROBLEMS FOR QUEUING-BASED APPROACHES

The proposed framework presented in Chapter 4, exploiting concepts of queuing theory, was demonstrated for wireless multihop networks (RGGs). Additionally, results were provided for static and dynamic networks, where for the case of networks with churn, the results involved other types of complex networks as well. However, a significant number of minor or major open problems remain open. In this section, we review the most noteworthy of these problems, accompanied with a small outline of the steps one might initially take in order to tackle them, or at least attempt a first approach.

- **Analytic solutions for nondynamic complex networks.** For the case of fixed (nondynamic) networks, the framework was demonstrated analytically for malware propagative and spreading random geometric (multihop) topologies. Similar results can be obtained for other complex networks, e.g. regular and scale-free. Obtaining analytic solutions for these topologies pends on the availability of closed-form expressions for the degree distribution of each network. Within Network Science some of these expressions are available, e.g. regular and scale-free, but for others, especially small-world, the characterization is based on rule-of-thumb definitions. Thus, for networks that the analytic expression of their connectivity is available, the methodology of Chapter 4 can be employed. On a per network type basis, the involved algebra might be cumbersome, but it seems viable to obtain at least sufficient analytic approximations in closed-form, depending on the special features of each topology.

- **Analytic solutions for complex networks with churn.** Malware diffusion models for dynamic networks with churn were explained in Chapter 4, and the results provided were obtained via simulations. An interesting extension would be to obtain analytic solutions for those types of complex networks, similarly to the nondynamic networks. The methodology will be similar, since due to the closed three-queue network model (Fig. 4.22(a)), an intermediate step is required to suppress the three-queue network into a two-queue Norton network before proceeding in its analysis as demonstrated in Chapter 4.

- **Malware-propagative networks with churn.** An equally interesting, and seemingly fairly straightforward, direction to extend the model presented for networks with churn is the malware propagation case. In Chapter 4, only the case of malware spreading was considered. Following the same lines as for malware-propagative nondynamic networks, similar results for malware-propagative networks with churn can be obtained.

- **Taking into account energy constraints.** In Chapter 4, the models presented for nondynamic networks do not take into account energy considerations. Incorporating directly such type of constraints on the model is a very challenging task. The main complication arises from the nodes depleting their energy reserves and being removed from the network. This impacts the ergodicity of the system. An alternative way to tackle this has been presented in the second half of Chapter 4, in Section 4.4. In the latter, modifying appropriately the churn processes to accurately describe the effect of energy depletion (node removal) and the addition of new recharged nodes allows to utilize the methodology of Section 4.4 to solve for the steady-state of the system. It allows also to obtain solutions for specific types of networks and their parameters, e.g. multihop or mobile (as currently this has not been achieved).

- **Heterogeneous infection/recovery rates.** In all of the models in Chapter 4, uniform infection/recovery rates were considered, i.e. $\lambda_m = \lambda, \mu_k = \mu$ for all links m, nodes k. Considering heterogeneous values is a straightforward but complicated task, with considerable practical merit, nevertheless, since in real scenarios such rates are expected to be heterogeneous. The analytic expressions are expected to be much more complicated.

- **Impact of mobility.** Currently, no mobility considerations have been taken into account, for the same purposes as with energy constraints. A possible way to tackle this is as with energy, via the model developed for dynamics networks with churn, where a node may be considered as disappearing in its original position and reappearing in the final predicted by the mobility model. However, this is a rather complicated approach. Searching for a seamless technique to address this within the queuing framework is currently an important open problem.

- **Comparison with traditional epidemics.** Comparing the results of the approaches in Chapter 4 with traditional epidemics problems presented in Chapter 3 will be a valuable task, revealing the degree to which generic malware assessing techniques based on queuing systems can accommodate the more specific epidemics-based results.

10.3 OPEN PROBLEMS FOR MRF-BASED APPROACHES

The framework presented in Chapter 5 exploited elements of the theory of MRFs and was applied to various types of complex networks, random, random geometric, scale-free, and small-world. However, for random, scale-free, and small-world no explicit solutions with respect to topological parameters were provided. In this section, we review noteworthy open problems, accompanied with a brief outline of the steps one might take to tackle them.

- **Analytic solutions for complex networks.** As already mentioned, the results presented in Chapter 5 for complex networks were not detailed for all networks explicitly. Taking Eq. (5.28) one step ahead to express it in terms of the topological parameters of each type of complex network is a very important and challenging extension of the framework that will enable many more results to be obtained further.

- **The case of dynamic complex networks.** Furthermore, all MRF-based models presented in Chapter 5 assumed a fixed topology. Considering the case of dynamic networks with churn is of critical importance for the practical realization of such framework. With respect to churn type, edge churn seems to be straightforward, since only the definition of neighborhoods will change. The results are not expected to be drastically modified. However, incorporating node churn is a very challenging task, since addition/deletion of nodes affects convergence of the Gibbs sampler employed, with very important outcomes if successful nevertheless.

- **Taking into account energy constraints and mobility.** As in Chapter 4, incorporating energy constraints and mobility in the presented MRF-based model is a very challenging task with many practical applications and interest. It is also related to the churn consideration problem explained above, and considering these directions jointly seems a viable possibility.
- **Multiple infection model states.** In Chapter 5, only two states were considered, modeling SIS malware dynamics. However, extension to multiple states, modeling other malware types, such as SIR and SIRD, is straightforward, by simply adding more values to the phase space.
- **Parallel/hybrid implementations of Gibbs sampler.** Chapter 5 presented results for the case of a sequential Gibbs sampler combined with simulated annealing. Developing parallel and hybrid samplers and studying their convergence behavior compared to the sequential would be a valuable extension with considerable interest for the practical adoption of the framework.
- **Attack strategy study.** Chapter 5 was devoted to the introduction and analysis of the MRF framework for modeling malware diffusion. However, this could be extended to study more intelligent attack strategies, e.g. topology control based, and study the long-term robustness of the network. This will be possible especially if analytic expressions with respect to each network's topology are obtained as explained in the first bullet above. Thus, these two extension directions can be jointly attained.

10.4 OPTIMAL CONTROL AND DYNAMIC GAME FRAMEWORKS

- **Relaxing the technical assumptions.** In the model presented in Chapters 6 and 7, we formally showed that optimal controls have a simple bang-bang structure with single or two jumps, which makes them conducive for computation and implementation purposes. However, we obtained these structures in the presence of some technical assumptions on the system dynamics and the cost functions. It will be interesting to investigate how far these assumptions can be relaxed. Moreover, even if these simple structures do not hold in the absence of these technical assumptions in general, it is useful to investigate the performance of such threshold-based strategies in such cases to quantify the loss in optimality, when they are adopted in practice.
- **Sensitivity to parameter estimation.** One of the assumptions in the optimal control problem is that the system is "identified" sufficiently accurately. For instance, the worm has a good estimate of the patching rate of the network, and the contact rate of the nodes. Accurate measurement of these parameters is not easy in practice; hence, sensitivity analysis of the optimal controls and saddle-point strategies to erroneous parameter estimations is another area of interest for future research. Modern techniques such as "H_∞ control" will be useful in this regard to guarantee a level of performance in the face of such measurement uncertainties.

- **Stochastic and nonhomogeneous systems.** Dynamic spreading models that take into account the underlying topological and spatial properties of the network have been developed, but largely not investigated how either defender or attacker on a network can take advantage of these spatiotemporal information in their favor. Specifically, developing defense mechanisms that are both stochastically and strategically robust and take optimal use of the local albeit noisy state information is an important subject of future research. Going from open-loop setting to other information structures in the differential game models of malware epidemics is another untouched area of research.

- **Exploiting cross-layer controls.** Another interesting direction of research is incorporating more detailed characteristics of the underlying network. Specifically, knowledge of specific states and available control at multiple layers of the network can be exploited by either the defender or attacker to their advantage, specially in dense and/or high traffic networks. Such knowledge can be at the physical/MAC layer such as the channel state and channel modulation and scheduling, and network layer such as routing and retransmission. For instance, a worm may avoid aggressive media access during its spreading period, in order not to "self-throttle" its propagation, only to then initiate a more effective jamming attack once a considerable number of nodes are compromised.

- **Mixture of attacks.** Considering dynamic defense and attack scenarios where different kinds of malwares are seeking to simultaneously infect the nodes, and where patching against one kind of malware does not provide immunity against others constitute another interesting problem.

- **Selfish defense.** Another promising line of research is designing optimal control and dynamic game strategies in the presence of the selfish response of the individual nodes. Specifically, nodes may not be oblivious to the state of their neighbors, and for instance, they may obtain potentially inaccurate estimates of the aggregate infection and recovery levels in their neighborhood by monitoring the media scanning activities, or through a globally announced message. The nodes then may selfishly choose their reaction potentially with bounded rationality. Both defender and attacker may be able to strategically affect and direct these reactions.

- **IoT and cloud computing.** The models developed for communication networks can be generalized to capture emerging application such as Internet of Things and Cloud Computing, where for instance in the latter, a large computational task is to be disseminated across many virtual machines and/or CPUs.

10.5 OPEN PROBLEMS FOR APPLICATIONS OF MALWARE DIFFUSION MODELING FRAMEWORKS

In Chapter 9, some applications of the presented malware modeling methodologies were presented, most notably for network robustness analysis and IDD. Directions for further extensions are summarized in the following nonexhaustive list.

- **Information spreading control and exploitation.** The proposed methodology for modeling IDD can be further exploited for defining more complex and application-oriented problems that arise in information diffusion processes in arbitrary complex networks. For instance, these could be used in order to optimize the spreading of useful information and designing more robust networks capable of sustaining large-scale attacks of malicious spreading information. Depending on the context of each application more effective dissemination campaigns can be designed and more robust infrastructures can be developed.
- **MRFs for IDD.** The MRF-based framework is also suitable for modeling IDD. Thus, applying this framework in place of the queuing-based one also allows for more general considerations of IDD. The corresponding extension requires identifying suitable potential functions and studying/showing the convergence of the corresponding MRF. Also, extending the MRF phase space can aid in studying more complex types of IDD.
- **Optimal control of spreading models for networks with churn.** Obtaining optimal controls for malware nonpropagative and propagative networks both from the attackers' and network's perspectives can be a challenging but fruitful exercise, with considerable practical merit. If different objective functions are employed, different settings and network operations can be modeled and studied.
- **Channel effects on IDD.** In all IDD study models contained in this book, no channel effects [214] have been considered. However, in practical scenarios, this could have an impact on the outcome of malware. Incorporating such considerations in the presented models is a challenging task, where prior experience from communications can be exploited.
- **New types of information models.** As technology evolves, new types of information and information dissemination means emerge. Modeling these new instances of information dissemination with the existing models, even state-of-the-art, can be a challenging task with considerable research interest.

10.6 GENERAL DIRECTIONS FOR FUTURE WORK

In this section, we focus on problems of broader interest. We briefly outline some of these more general considerations for possible future research directions in the broader field of malware diffusion modeling, as well as the role that the approaches presented in this book might have in addressing them.

- **Inhomogeneous mixing of populations.** In all the presented models, an explicit assumption was that the involved populations mix homogeneously. This is especially the case for epidemics-based models, but also for others. Exploring the case of inhomogeneous mixing is a very important research topic, as it will allow a direct comparison with previous results and reveal how realistic the homogeneous mixing assumption actually is. Such attempt requires radical, and potentially out-of-the-box thinking.

- **How mobility affects malware diffusion.** As already explained, mobility has not been considered explicitly in the presented models and only implicit ways to deal with it have been identified. However, developing solid methodologies addressing its impact on malware diffusion and for various types of wireless complex networks is a very important aspect with significant practical merit.
- **Generalized infection models.** Section 3.4 explained a more general epidemics model with generic states. Extending this model to an arbitrary number of states would allow capturing even more generic and complex malware in the future. This seems fairly viable for the queuing and MRF-based frameworks, especially the second, where extension of the phase space is straightforward.
- **Co-evolving malware.** The queuing and MRF-based approaches assume SIS malware, where potentially multiple attacks diffuse, paying attention only to whether infection happened, not the type of it. Extending these models in addition to the other two, to take into account multiple co-evolving attacks is a challenging effort with great research and industrial interest. Among others, this means studying hybrid propagative-spreading networks, where some nodes propagate malware, while others not.
- **Bulk infections.** An integral assumption of all approaches was that a single-event takes place, namely, at infinitesimal time no concurrent infections take place. A more pragmatic consideration would allow both bulk infections-recoveries and this needs again radical thinking in terms of mathematical tools employed.
- **Adaptive epidemics.** Until now, few works have addressed the possibility of feedback in malware dynamics. The optimal control and game-theoretic frameworks, as well as the optimal attack strategies application have done so. However, more complex epidemics, where, e.g. network links are adapted based on malware strategies reactively or proactively, are needed. Research toward this direction is still immature, but the outcomes are expected rather fascinating.
- **Cloud and IoT systems security.** Cloud systems have proliferated vastly the past five years and more users trust them for their daily operations, e.g. storage, computing, and web services. However, they are essentially complex cyber-physical systems [153, 157], namely, multitier topologies where each layer has a different type of topology, and as such they can also suffer from various types of malware attacks. Until now, not many works have addressed the corresponding topics from a theoretic perspective. Thus, developing mathematical frameworks for studying malware diffusion for cyber-physical systems in general and cloud systems in particular seems a fruitful path, and similarly for IoT systems, which also exhibit various degrees of topological complexity. However, their market penetration and anticipated proliferation call for more research. The presented frameworks can have many applications in various problems emerging in the cloud-IoT worlds and can potentially simplify solutions obtained. The cloud-IoT world is expected to be one of the prominent fields of applications of the state-of-the-art methodologies presented in this book.
- **Centrality considerations.** In Network Science, the notion of centrality has been identified as very important for discovering key nodes that control various

operations. For this reason, multiple and diverse definitions of centrality exist as well. This feature can be exploited in the presented frameworks for identifying important nodes that govern malware diffusion dynamics. Centrality features can be used in the analysis of malware diffusion from both attack and defense perspectives. It can be used for correlating nodes with propagation dynamics and then exploit results for improving attacks/countermeasures in any of the frameworks presented.

- **Malware-dependent recovery.** The above point motivates the need for studying malware-dependent recoveries. In many cases of malware, the recovery depends on the specific type of malware, in the sense that a more sophisticated and harmful malware is expected to take more time to dispose of on average. Under this regime, the overall system of recoveries loses the lack of memory property assumed in Chapter 4 and more general queues have to be eventually analyzed in order to obtain the results provided already. This could be very tough and involve rather advanced elements from the theory of queues and stochastic processes, but useful results nevertheless.

- **Bigdata and malware.** The field of bigdata analytics has exploded in the last decade, with numerous and diverse applications. Analytics is about identifying correct patterns within vast amounts of data. Security concerns have been raised from the conception of this field, and various types of malware are expected to emerge in the future. The proposed approaches can be applied in various aspects of bigdata analytics, where information diffusion is involved and aid in the more efficient and accurate operation of the corresponding systems. Looking for applications of the malware modeling frameworks in the bigdata world is expected to be a fruitful endeavor.

CHAPTER

Conclusions

11

11.1 LESSONS LEARNED

In the previous chapters, this book presented a multitude of malware diffusion modeling frameworks and methodologies, some older and some state-of-the-art. With respect to the employed mathematical tools, objectives and obtained results, all these approaches exhibit characteristic diversity. However, within this variety of techniques and results, several emerging trends may be identified, which can be of significant aid in exploiting these approaches in the future, and more generally for rethinking the current state-of-the-art.

In this chapter, we provide a summary of the most general and most important principles that emerged from the presented methodologies in a codified manner, hoping that they could be exploited both for extending the same frameworks themselves, and applied in other disciplines or similar processes as well.

As mentioned before, the main application domain of the presented framework was networking and more specifically wireless complex and multihop networks. However, all the provided approaches can be extended in other types of wireless and wired networks, and in fact some of them already have, e.g. those presented in Chapters 4, 5 and 9. Additionally, the demonstrated methodologies and techniques can be applied in other domains of Network Science (Chapter 9). The following lessons learned/emerging trends can aid toward this direction and toward tackling some of the open problems provided in the previous chapter as well.

Lesson 11.1 *(Queuing and Malware). The queuing framework presented in Chapter 4 is a holistic approach for modeling malware diffusion in spreading/propagative networks, fixed networks, and networks with churn and even formulate optimization problems for studying attack strategies. Currently, available analytic results cover wireless multihop networks and for fixed topologies complex networks as well.*

Lesson 11.2 *(MRFs and Malware). The MRF framework is a very powerful and generic framework that is capable to describe malware dynamics and information dissemination in broader cases, even when correlations emerge between information/malware and recoveries or between nodes. The framework has been demonstrated for all types of complex networks. It is a simple approach, even though of suboptimal nature.*

Lesson 11.3 *(Optimal Control and Malware). The optimal control framework pre-sented in Chapter 6 is also a very powerful framework, which can be used almost identically for malware diffusion and information dissemination modeling. It can span many different scenarios and types of objective functions and operational scenarios. Furthermore, stochastic optimal control can be further exploited on the specific basis for more generic modeling approaches. We showed that tradeoffs in reaching an objective do not necessarily imply an intermediate value of control, but rather, when the degree of freedom of having a time-dependent decision is introduced, the tradeoffs imply different decisions at different times as the state of the system is evolved. We showed that by exploring the necessary conditions that an optimal control has to satisfy, significant insights about the structure of the solution can be obtained even before the closed-form or numerical solution is obtained.*

Qualitatively, we showed that static behavior of a new malware should not be interpreted as its natural trait, as there is a possibility that this behavior is part of a dynamic strategy of the worm, and will be changed at a later optimal time.

Lesson 11.4 *(Game Theory and Malware). In Chapter 7, we showed how to use differential game theory to model a situation when the worm as well as the network defender can dynamically manipulate the state of the nodes with respect to spread of the malware. We introduce the notion of saddle-point strategies as a reasonable expectation of how the game will be played as mutually optimal responses over the course of the diffusion. As in the optimal control case, we showed through an example how investigation of the necessary conditions that saddle-point strategies need to satisfy can be used to extract key structural properties of the solution, helping both in providing insight and in the computation of such strategies.*

Lesson 11.5 *(Malware Spreading and Propagation). Malware has been segregated in two categories: spreading and propagative. The presented frameworks cover either both or one type, but extensions to both cases are possible. Thus, when studying specific malware, it is important to first identify the correct type and then choose the most appropriate framework, while also taking into account the behavior required to model, e.g. churn, mobility, energy, and control.*

Lesson 11.6 *(Malware Diffusion and Network Churn). Network churn is a process observed rather frequently in practical scenarios. The queuing framework presented in Chapter 4 can be effectively used to model malware diffusion in networks with churn. On the other hand, the MRF-based approach faces several convergence issues when the nodes of the network vary. The same holds for the game theory based frame-work, while the optimal control framework can be potentially extended to cover such cases. Thus, for networks with churn, the queuing approach seems to be currently the most suitable. However, one needs to carefully examine the churn type, since, e.g. the MRF approach may accommodate edge churn easily, and investigate whether a specific framework can cope with the specific features of churn needed to model.*

Lesson 11.7 *(Complexity and Resources). Computational and resource requirements can be decisive for the type of modeling approach to use. Among the four state-of-the-art frameworks presented, the MRF one is the less demanding of all from all perspectives, while the optimal control and game theory based can be rather demanding. In general, such aspects should be considered early when developing a malware modeling approach, since typically the scale of operation of malware in practice is very rapid and the framework one develops will need to take such scales into account.*

Lesson 11.8 *(Mobility and Energy). Mobility and energy features are essential elements of wireless mobile networks. However, they are rather problematic in terms of modeling and analysis of their impact on network operations and malware diffusion. All four frameworks presented exhibit various difficulties and complications taking into account mobility or energy constraints, and thus more dedicated research is required toward this direction.*

Lesson 11.9 *(Malware and Control). Malware diffusion and control is a topic extensively covered in this book. The optimal control, game theory, and queuing theory based approaches all are capable of addressing one form or another of control on the malware they model. The MRF framework can be also extended to do so. Thus, a very broad spectrum of control techniques is available. Consequently, researchers should take these into account, carefully analyze the objectives they want to attain, and select the control technique appropriately. Furthermore, the guidelines obtained in various parts of the book can be extrapolated and used for potentially developing more intelligent frameworks and studying new malware dynamics.*

Lesson 11.10 *(Malware and Network Robustness). The presented frameworks allow comparing the robustness of the analyzed networks against malware by assessing the expected damage a specific parameterization of malware can cause. This was feasible for the queuing and MRF-based frameworks, but extending the two to different types of complex networks should also be viable. From the obtained results, it is thus possible to characterize the robustness of complex networks and take it into account in future studies. Thus, random networks have emerged as the most robust, followed by scale-free, which in turn are tightly followed by small-world. The random geometric (multihop) seems to have the worse performance, while regular is between small-world and random geometric.*

Lesson 11.11 *(Malware and Connectivity). Among all the frameworks and models presented, a common emerging trend is that the average connectivity of a complex network is very critical for malware diffusion dynamics and their eventual outcome. It can essentially determine whether an epidemic will become pandemic, endemic, or die out completely, and thus should be always one of the employed assessment factors.*

Lesson 11.12 *(Malware Modeling Frameworks and Flexibility). For each of the four frameworks presented in Part 2, various settings were employed and results were obtained. At the same time, several directions for further extending these frameworks, analytically or in terms of applications, have been identified. Among all, the MRF framework appears as the most flexible, requiring less effort to extend it mathematically and in applications. The rest require various simple or more complex modifications, or they can even face fundamental difficulties, e.g. ergodicity in the queuing-based approach. In terms of extending each framework, careful analysis of the malware features (spreading-propagation, homogeneity of mixing, infection-recovery processes, etc.) and network structure (topology type, churn, etc.) is needed to evaluate properly the modification that will be required to the corresponding framework, and its feasibility.*

11.2 FINAL CONCLUSIONS

Considering cumulatively the content of the book, it becomes evident that malware diffusion theory can be rather useful for network designers, administrators, and professionals separately. Depending on the network type and employed applications, one of the four frameworks presented in Part 2 can be utilized for properly predicting the behavior of a network under attack and design proper countermeasures. More importantly, these approaches enable designing dynamic response mechanisms, which are able to intelligently adapt to the fundamental nature and features of the threats, thus more effectively securing the underlying infrastructure. This was not possible in the past at the magnitude attained by the approaches presented in the book.

At the same time, the analyzed models and frameworks have been shown to be generic enough, so that their analytic properties cover broader application areas. Similar phenomena to malware diffusion emerge in information flow applications and the proposed frameworks could be easily extended and adapted to cover more general problems of information diffusion over complex communication networks. Other similar application domains can be identified within the areas of future wireless Internet [86] and other complex networks, while properly extrapolating the techniques presented for obtaining faster the desired outcomes.

The techniques and models presented in this book may be considered as the first steps of a broader vision to develop holistic frameworks describing the flow of information in communication networks. Starting with the diffusion of malware, similar attempts for other problems and application areas of content dissemination can be inspired. This would signify the successful potential of the content of this book and provide even more efficient mechanisms for designing infrastructures and information management mechanisms of the future.

Appendices

Systems of ordinary differential equations

A.1 INITIAL DEFINITIONS

A differential equation is an equation that contains one or more derivatives of an unknown function. Differential equations are extensively used in mathematical modeling of many systems. When the independent variable is time, then the differential equation describes the temporal dynamics of a monolithic quantity, e.g. population, capital investment, chemical compound, decayed radioactive material, and so on. Other independent random variables such as the spatial coordinates can give more detailed description of a spatiotemporal quantity with spatiotemporal dynamics, e.g. diffusion of heat and fluids. In the rest of this appendix, we assume that the independent variable is time and denote it by t. Any other interpretation of t is valid too.

A differential equation that only contains ordinary derivatives is called an ODE. If, on the other hand, one or more partial derivatives are involved as well, it is referred to as a *partial differential equations* (PDEs) [81]. Each of these two broad categories is further classified based on the highest order of derivative that appears in them. In this appendix, we only consider ODEs.

Depending on the number of unknown functions that are involved, a classification to single ODE and system of ODEs is obtained. If there is a single function to be determined, then one equation is sufficient. However, if there are two or more unknown functions, then a system of equations is required.

Let x in a space $X \subset \mathbb{R}^m$ (for some proper dimension m of the space) be a function of a single-dimensional real-valued variable t. Furthermore, for any positive integer i, let $x^{(i)}(\tau)$ represent the ith order derivative of x with respect to t evaluated at τ, i.e. $x^{(i)}(\tau) := d^i(x(t))/dt^i$ evaluated at $t = \tau$. Then, we have the following.

Definition A.1. *A differential equation that can be put in the form of*

$$F\left(t, x(t), x^{(1)}(t), \dots, x^{(n-1)}(t), x^{(n)}(t)\right) = 0 \tag{A.1}$$

for a given $F : D \subseteq \mathbb{R} \times X^{n+1} \to X$ constitutes an ODE of order n.

In the rest of this appendix, whenever not ambiguous, we will suppress the explicit dependence on t for simplicity of notation. Note that the first and second order derivatives are also often denoted by x' and x'', and specially when the independent variable is time, also by $\dot{x}$ and $\ddot{x}$ as well.

ODE (A.1) is said to be linear if F is a linear function of the variables $x(t), x^{(1)}(t), \ldots, x^{(n-1)}(t), x^{(n)}(t)$. A similar definition applies to PDEs. An equation not in linear form is called nonlinear ODE or PDE, respectively.

An important class of first-order equations are those in which the independent variable does not appear explicitly. Such equations are called *autonomous* and have the form $x' = f(x)$.

Definition A.2. *A solution of ODE*(A.1) *over an interval* $t \in I \subseteq \mathbb{R}$ *is a function* $\varphi(t)$ *that identically satisfies* (A.1) *over the interval of I. That is* (1) *the first n-derivatives of x exist for any point in the interval of I;* (2) *for every* $t \in I$, F *is defined, i.e.* $(t, \varphi, \varphi', \ldots, \varphi^{(n)})$ *is in the domain of F; and* (3) $F(t, \varphi, \varphi', \ldots, \varphi^n) = 0, \forall t \in I$.

This definition describes an exact solution of the ODE, as opposed to an approximate one. When a solution of the ODE needs to satisfy the initial condition of $\varphi(t_0) = x_0$ for a given $(t_0, x_0) \in D$, we refer to the problem as an *initial value problem* (IVP). Some very important questions about a solution of an ODE (PDE) or an IVP is the existence, uniqueness, and stability of solution or solutions.

A.2 FIRST-ORDER DIFFERENTIAL EQUATIONS

By Eq. (A.1) and when F is invertible with respect to $x^{(n)}(t)$, a first-order ODE may also be expressed in the following form:

$$x' = f(t, x), \tag{A.2}$$

for a function $f : D \subseteq \mathbb{R} \times X \to X$. For linear first-order ODEs, the *integrating factor* solution approach can be used to obtain the desired result. Details on the application of such techniques can be found in [40, Section 2.1].

In cases where x is a single-valued function of t, this is just a single ODE. If, on the other hand, x is a multivalued function of t, i.e. $X = X_1 \times X_2 \times \ldots X_m$ for a positive integer m, then the ODE can be decomposed and expressed in terms of each element of the vector function x. For such a case, the f function is also multivalued, i.e. $f = (f_1, \ldots, f_m)$ such that each $f_i : D \subseteq \mathbb{R} \times X \to X_i$. This leads to the following *system* of m differential equations:

$$x_i' = f_i(t, x_1, \ldots, x_m), \quad i = 1, \ldots, m \tag{A.3}$$

that need to be simultaneously satisfied over the interval of solution. The converse direction yields another interpretation of the above system of first-order ODEs. According to this if there are m unknown functions $x_1, \ldots x_m$ that need to simultaneously satisfy m first-order differential equations, then by introducing an m-dimensional vector-valued function one may obtain a single vectorized differential equation.

The form of first-order ODE presented in (A.2) can also model higher order ODEs through introduction of auxiliary dependent variables. Specifically, consider an n-th order ODE of the form $x^{(n)} = F_0(t, x, x', \ldots, x^{(n-1)})$. Now consider the auxiliary functions $y_1, \ldots, y_n$ such that $y_1 = x$, $y_2 = x'$, $y_3 = x''$, and so on till $y_n = x^{(n-1)}$. Then, we have $y_1' = x' = y_2$, $y_2' = x'' = y_3, \ldots, y_{n-1}' = x^{(n-1)} = y_n$. Moreover, $y_n' = x^{(n)}$, which must be equal to $F_0(t, x, x', \ldots, x^{(n-1)})$, hence $y_n' = F_0(t, x, x', \ldots, x^{(n-1)}) = F_0(t, y_1, y_2, \ldots, y_n)$. Hence, the nth order ODE $x^{(n)} = F_0(t, x, x', \ldots, x^{(n-1)})$ is equivalent to the following system of ODEs:

$$y_i' = y_{i+1} \quad (i = 1, \ldots, (n-1)),$$
$$y_n' = F_0(t, y_1, y_2, \ldots, y_n).$$

In (A.3), if f_i functions are linear functions in x, specifically,

$$x_i' = \sum_{j=1}^{m} a_{ij}(t)x_j + b_i(t) \quad i = 1, \ldots, m, \tag{A.4}$$

where a_{ij}, b_i are real-valued continuous functions of t over the interval I, then we have a *first-order system of linear ODEs*. Using the technique of introducing auxiliary functions for higher order derivatives of the unknown function x, we can convert the *n-th order linear ODE* of $x^{(n)} = a_1(t)x^{(n-1)} + a_2(t)x^{(n-2)} + \ldots + a_n(t)x + b(t)$ into a first-order system of linear ODEs too [40]. The analysis and properties of (systems of) linear ODEs are well-developed, and consequently, whenever meaningful, nonlinear ODE systems are analyzed by approximating them through "linearization" (around an operation point) [40].

Some first-order ODEs of the form $\frac{dx}{dn} = f(x, n)$ can be transformed as $M(x, n) + N(x, n)\frac{dx}{dn} = 0$ and cast in the form $M(n)dn + N(x)dx = 0$ when M is a function of n and N a function of x. Such equations are called *separable*. The solution to such equation can be easily obtained. More on separable ODEs and their solution methodology can be found in [40].

A.3 EXISTENCE AND UNIQUENESS OF A SOLUTION

Existence of a solution to any ODE of the form $x' = f(x, t)$ with initial condition $x(0) = x_0$ is guaranteed for any continuous finite dimensional function f. More formally, we have the following.

Theorem A.1 (*Solution Existence [73, 2.4.4]*). *Let X (the $\mathbb{R}^m$ space for the values of the unknown function x) be finite dimensional, let D be an open set in $\mathbb{R} \times X$, let $(t_0, x_0) \in D$ and let $f : D \to X$ be continuous (for $t \geq t_0$). Then there exists at least one solution φ of the ODE $x' = f(t, x)$ satisfying the initial condition of $\varphi(t_0) = x_0$ and reaching the boundary of D on the right.*

In (A.2), we say function $f : D \to X$ satisfies a *K-Lipschitz* condition in x over D, where $D \subseteq \mathbb{R} \times X$, if there exists a $K > 0$, such that

$$\text{for all } (t, x), (t, y) \in D, \|f(t, x) - f(t, y)\| \leq K\|x - y\|.$$

Note that in the above Lipschitz condition, the right hand side does not have any explicit appearance of t, and the t in the two terms of the left hand side are the same. For ODEs $x' = f(t,x)$, where f satisfies the Lipschitz condition, there is a result stronger than Theorem A.1, which guarantees not only existence but also uniqueness of the solution.

Theorem A.2 (*Solution Existence and Uniqueness [73, 2.7.4]*). *Let I be an interval in $\mathbb{R}$ and assume $(t_0, x_0) \in I \times X$. Consider $f : I \times X \to X$ to be a continuous function that is K-Lipschitz in x over $J \times X$, where J is any compact subinterval of I (K may depend on J). Then, the IVP of $x' = f(t,x)$ and $x(0) = x_0$ has one and only one solution over I.*

The continuity requirement of this theorem for f can be relaxed as long as discontinuities constitute a null-measure set [72]. A *measure* is a function that assigns a non-negative real number or ∞ to (certain) subsets of a set X [210, 226]. It must assign zero to the empty set and be (countably) additive: the measure of a "large" subset that can be decomposed into a finite (or countable) number of "smaller" disjoint subsets is the sum of the measures of the "smaller" subsets. Thus, a measure on a set is a systematic way to assign a number to each suitable subset of that set, intuitively interpreted as its size [210, 226]. Any set of measure zero is called a *null-set* (or simply a null-measure set).

A.4 LINEAR ORDINARY DIFFERENTIAL EQUATIONS

In this section, we review some elements of the theory of systems of linear ODEs. These can be helpful in the understanding of the content of Chapters 3, 6 and 7, where systems of ODEs emerge.

Consider the first-order system of linear ODEs as in (A.4). Using a matrix representation, we can rewrite (A.4) in the form

$$\mathbf{x}' = \mathbf{A}(t)\mathbf{x} + \mathbf{b}(t),$$

where $\mathbf{x}$ and $\mathbf{b}$ are m-dimensional vectors of continuous functions from $I \subseteq \mathbb{R} \to X$, and for finite dimensional X, $\mathbf{A}(t)$ is the $m \times m$ matrix $[a_{ij}(t)]$ where a_{ij} are all real-valued and continuous. Consider the initial condition of $x(0) = (x_{10}, \ldots, x_{m0})$. The existence and uniqueness of the solution to this IVP follows from Theorem A.2. Since the functions are linear, they automatically satisfy the Lipschitz condition, with K associated with the supremum/maximum of $\mathbf{A}(t)$'s norm over the compact interval mentioned in the statement of Theorem A.2.

When $\mathbf{b}(t)$ is identically zero, the resulting linear ODE, $\mathbf{x}' = \mathbf{A}(t)\mathbf{x}$, is called *homogeneous* linear ODE. Then, $\mathbf{x}(t) = \vec{0}$ is the unique solution of the homogeneous linear ODE with the initial state of $\mathbf{x_0} = \vec{0}$. Clearly, addition of a solution of a linear ODE with any solution of its associated homogeneous linear ODE leads to a new solution of that linear ODE. Moreover, subtraction of two solutions of a linear

ODE results in a solution of its associated homogeneous counterpart. The solution space of the homogeneous linear ODE forms a vector space. Let $\varphi_1, \ldots, \varphi_p$ be p solutions of the homogeneous linear ODE $\mathbf{x}' = \mathbf{A}(t)\mathbf{x}$. Then $\varphi_1(t), \ldots, \varphi_p(t)$ are linearly independent at each $t \in I$ if and only if they are linearly independent at some point $t_0 \in I$. The proof of this key result can be found in [73, 2.8.2]. Therefore, if X has finite dimension m and $\varphi_1, \ldots, \varphi_m$ are m linearly independent solutions of the homogeneous linear ODE, then every other solution of it can be written as a unique linear combination of $\varphi_1, \ldots, \varphi_m$, which is known as the *general solution of the homogeneous linear ODE*. Hence, if v is a solution of the nonhomogeneous linear ODE (called a *particular* solution), then every solution of the nonhomogeneous problem ψ is the sum of the general solution of the associated homogeneous problem and a particular solution, i.e. $\psi = \sum_{j=1}^{m} \alpha_j \varphi_j + v$ for a unique set of scalars $\alpha_1, \ldots, \alpha_m$.

The solution of the IVP $\mathbf{x}' = \mathbf{A}(t)\mathbf{x} + \mathbf{b}(t)$, $\mathbf{x}(0) = \mathbf{x_0}$, can be expressed as the following:

$$\mathbf{x} = \Lambda(t)(\Lambda(t_0)^{-1}\mathbf{x_0}) + \Lambda(t) \int_{t_0}^{t} \Lambda(\tau)^{-1}\mathbf{b}(\tau)\, d\tau,$$

where $\Lambda(t)$ is a fundamental kernel[1] of the linear transformation associated with $\mathbf{A}(t)$. When X has a finite dimension equal to m, $\Lambda(t)$ is a matrix whose columns are the coordinates of m linearly independent solutions of the homogeneous linear ODE $\mathbf{x}' = \mathbf{A}(t)\mathbf{x}$. Additionally, when the coefficients of the linear ODE are constant, i.e. we have an *autonomous* linear ODE, the basis-forming solutions of the homogeneous ODE $\mathbf{x}' = \mathbf{A}\mathbf{x}$ are of the form $e^{\lambda_l t}\phi_{lp}(t)$ for $l = 1, \ldots, L$ and $p = 1, \ldots, P_l$ for polynomials ϕ_{lp} of degree of at most $P_l - 1$, where $\lambda_1, \ldots, \lambda_L$ are the distinct eigenvalues of matrix $\mathbf{A}$, and $P_1, \ldots, P_L$ are their respective multiplicity.

A.5 STABILITY

Consider the autonomous ODE of $\mathbf{x}' = f(\mathbf{x})$ (where there is no explicit dependence of f on t) with the solution of $\varphi(\mathbf{x_0}, t)$ for the initial value of $\mathbf{x}(0) = \mathbf{x_0}$. Then a point $\hat{\mathbf{x}} \in X$ is an *equilibrium* (also known as a *steady state*, *fixed point*, or *critical state*) if the constant function $\phi(t) = \hat{\mathbf{x}}, \forall t$, is a solution of the ODE with the initial condition $\mathbf{x}(0) = \hat{\mathbf{x}}$, i.e. the solution remains at $\hat{\mathbf{x}}$ if it starts there as the initial state. We say an equilibrium $\hat{\mathbf{x}}$ is stable if any solution of the IVP $\mathbf{x}' = f(\mathbf{x})$, $\mathbf{x}(0) = \mathbf{x_0}$ with $\mathbf{x_0}$ close to $\hat{\mathbf{x}}$ remains in a small neighborhood of $\hat{\mathbf{x}}$ for all $t \geq 0$. It is further called *asymptotically stable* if it is stable and there is a small neighborhood of $\hat{\mathbf{x}}$ such that if the initial value $\mathbf{x_0}$ is in that neighborhood we have $\lim_{t \to \infty} \varphi(\mathbf{x_0}, t) = \hat{\mathbf{x}}$.

---

[1] In linear algebra and functional analysis, the kernel (null-space) of a linear map $L : V \to W$ between two vector spaces V and W, is the set of all elements $\mathbf{v}$ of V for which $L(\mathbf{v}) = \mathbf{0}$, where $\mathbf{0}$ denotes the zero vector in W. That is, $\ker(L) = \{\mathbf{v} \in V | L(\mathbf{v}) = \mathbf{0}\}$.

The following result governs the stability of equilibrium points derived from $f(\mathbf{x}) = 0$:

Theorem A.3 *([73]). Let $\hat{\mathbf{x}}$ be an equilibrium of the ODE $\mathbf{x}' = f(\mathbf{x})$. If for every eigenvalue ξ of the Jacobian matrix of f at $\hat{\mathbf{x}}$ we have $Re(\xi) < 0$, then $\hat{\mathbf{x}}$ is asymptotic stable. If there exists an ξ with $Re(\xi) > 0$, then $\hat{\mathbf{x}}$ is unstable.*

It should be noted that a finite dimensional space is assumed. In vector calculus, the *Jacobian matrix* is the matrix of all first-order partial derivatives of a vector-valued function. Specifically, suppose $\mathbf{f} : \mathbb{R}^n \to \mathbb{R}^m$ is a function which takes as input the vector $\mathbf{x} \to \mathbb{R}^n$ and produces as output the vector $\mathbf{f(x)} \to \mathbb{R}^m$. Then, the Jacobian matrix $\mathbf{J}$ of $\mathbf{f}$ is an $m \times n$ matrix, usually defined and arranged as follows:

$$\mathbf{J} = \frac{d\mathbf{f}}{d\mathbf{x}} = \begin{bmatrix} \dfrac{\partial \mathbf{f}}{\partial x_1} & \cdots & \dfrac{\partial \mathbf{f}}{\partial x_n} \end{bmatrix} = \begin{bmatrix} \dfrac{\partial f_1}{\partial x_1} & \cdots & \dfrac{\partial f_1}{\partial x_n} \\ \vdots & \ddots & \vdots \\ \dfrac{\partial f_m}{\partial x_1} & \cdots & \dfrac{\partial f_m}{\partial x_n} \end{bmatrix}. \tag{A.5}$$

The Jacobian matrix is important because if the function $\mathbf{f}$ is differentiable at a point $\mathbf{x}$ (this is a slightly stronger condition than merely requiring that all partial derivatives exist there), then the Jacobian matrix defines a linear map $\mathbb{R}^n \to \mathbb{R}^m$, which is the best linear approximation of the function $\mathbf{f}$ near the point $\mathbf{x}$. This linear map is thus the generalization of the usual notion of derivative and is called the derivative or the differential of $\mathbf{f}$ at $\mathbf{x}$.

Elements of queuing theory and queuing networks

B

B.1 INTRODUCTION

The term *queuing* is defined as *the time delay experienced by various entities*, e.g. customers, network packets, and cars, *in the different facets of life and operations they participate in.* Queuing emerges frequently in natural and artificial systems in diverse application settings, e.g. waiting lines, toll stations, and packet routers in computer networks. Queuing theory is the branch of stochastic processes [66, 71, 88, 174, 187] that provides formal methodologies and techniques for analyzing queuing phenomena, as the ones described above.

In this book, elements from queuing theory have been used extensively in Chapter 4 to analyze malware diffusion by quantifying the time spent by legitimate nodes in various states, e.g. susceptible and infected. The purpose of this primer is to provide the fundamental knowledge required to understand the contents of Chapter 4. In addition, the primer aspires to aid in the understanding of the methodologies presented in other parts of the book that are relevant to Markov processes, such as techniques in Chapter 5. Appendix B focuses only on the analysis of simple Markovian systems and simple open/closed queuing networks. More advanced treatments may be found in dedicated monographs and books available in the literature, such as [25, 33, 36, 90, 142, 209, 211, 229].

The rest of Appendix B is organized as follows. Initially, it explains the basic components of queues, notation, elementary arrival-service processes, and Little's law. Then, it reviews simple queuing models, namely, birth-death Markov systems in equilibrium, followed by some elementary analysis of queues-in-tandem. The latter leads to the notion of reversibility, which together with Burke's theorem pave the way for a brief overview of open and closed Jackson networks. The latter consists of the most essential knowledge required to understand the methodologies presented in Chapter 4.

B.2 BASIC QUEUING SYSTEMS, NOTATION, AND LITTLE'S LAW

In this section, we start by introducing the components of a queuing system, and then explain the employed notation. Then, fundamental stochastic processes and simple but useful results, e.g. Little's law, are presented.

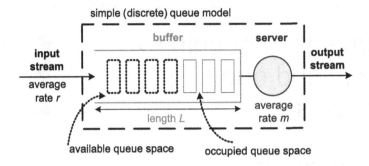

FIGURE B.1

A generic independent queuing system.

B.2.1 ELEMENTS OF A QUEUING SYSTEM

The simplest queuing system is shown in Fig. B.1. It consists of two distinct components, namely, a generic *queue* (buffer space) and a generic *server*. The queue/buffer provides storage space for the customers that enter the system and cannot be served immediately (delayed in the system), while the server corresponds to the processing (service) required by each customer. For example, if customers correspond to network packets, the processing can be the time required to determine routing and transmit packets, etc. In the instance shown in Fig. B.1, the queue is considered discrete, e.g. it accommodates humans in a bank queue, packets in a communications network, as opposed to a continuous queue. An example of the latter is a communications network, where one would treat the information content of the packets as a continuous quantity, e.g. liquid and draw an analogy between a router and a water dam, where water corresponds to information content and the dam valve to the router processor. In that sense, the continuous queue is an idealization of the actual queue, since in nature all quantities emerge discrete, which is used for mathematical tractability.

Consequently, each queuing system consists of a waiting component and a serving component, as well as a generic input (*arrival*) and generic output (*service*) streams. Multiple input/output streams might be implemented in real systems, both of which are typically of stochastic nature, i.e. proper stochastic processes [187].

The service discipline of a system determines the output. Together with the buffering discipline, which partially determines the time a customer spends in the system, the arrival and service processes are critical for the evolution and behavior of the queuing system. For the simple but generic system shown in Fig. B.1, the queuing time and the service time of each customer determine the total time spent by the customer in the system. The arrival process is characterized by the times between two successive arrivals (*interarrival times*) and the service process by the times between two successive services (*service times*). In simple queuing systems, successive services are independent of each other and independent of the sequence of interarrival times, while in more complex ones this is not the case [25, 142, 229].

The service and queuing disciplines can vary considerably. For instance, with reference to Fig. B.1, the simplest queuing discipline is the first-in-first-out (FIFO) [25, 33, 211], where arriving customers always enter the queue before being served, while a customer is removed from the queue and assigned to the server when the latter becomes free, according to a specific policy. In FIFO, customers enter the queue sequentially (queuing policy) and served (service policy) at the same order. Other similar disciplines like last-in-first-out (LIFO or commonly known as stack), heap implementing queuing with priorities (analyzed in detail in Section 3.5.3 of [25]), etc., might be applied and considered, and for more details the interested reader may find more details in [25, 33, 211]. The above disciplines are *work-conserving,* meaning that if one customer is served, the next in queue will take its place. Hence, the serving policy will prevent the system from becoming idle, if customers (work) still remain in the system (in queue). Queuing behavior can also depend on different queuing/serving policies determined by priorities, fairness factors, etc. More details are out of scope of this primer and can be found in relevant references [25, 33, 36, 90, 142, 209, 211, 229].

B.2.2 FUNDAMENTAL NOTATION AND QUANTITIES OF INTEREST

In order to characterize queuing systems, a notation system has been proposed by Kendall in 1951 and it was adopted by the research and industrial communities. A three or four part symbol is used in the form of $X/Y/z/w$, where X specifies the stochastic input process (interarrival distribution), Y specifies the service policy (service time distribution defined by the corresponding stochastic process), z denotes the number of servers, and w specifies the total holding capacity of the system including any served customers, if any. Table B.1 summarizes the most frequently used symbol used to explicitly denote the input/service processes (time interval distribution between successive arrivals or services) in Kendall notation. Symbol "M" stands for Poisson or exponential process (memoryless/Markovian), symbol "D" for deterministic arrival/service disciplines, "E_n" represents Erlangian distribution policies, "G" for an arbitrary distribution function, and "GI" for general but independent distribution policies. Thus, for example, a $M/M/2$ notation defines a queuing system with two servers with common queue of infinite capacity, Poisson arrivals, and exponential service times, while $M/M/2/4$ defines a similar system, where the buffer holds only two incoming customers, discarding any excess arrivals. This is a system of maximum capacity four, two customers at service and two waiting in the best case. A fifth customer will be discarded as opposed to an $M/M/2$ system.

In general, it should be pointed that interarrival and service times in systems typically described by Kendall notation are assumed independent of each other and in addition identically distributed (interarrival times and service times separately). This

Table B.1 Arrival-Service Discipline Characterization in Kendall Notation

Symbol	Meaning
M	Poisson or exponential (Markovian/memoryless)
D	Deterministic
E_n	Erlangian distribution
$Geom$	Geometric distribution
G	Arbitrary distribution function
GI	Arbitrary independent distribution function

is usually denoted as "i.i.d." Of course, the distribution of interarrival times may be different than that of service times, as explained above.

In many cases, it is possible to analytically study queuing systems of interest. When analyzing such queuing systems, the most important quantities of interest for identifying the key factors and assessing the performance of the corresponding system are summarized in the following list:

- The average number of customers in the buffer (queue) space,
- the average total number of customers in the system (queue and server),
- the average waiting time for a customer (expected time spent in the queue only),
- the average system time (expected total time spent in the system),
- the average busy period time (expected time the server works before becoming idle),
- the average system throughput (number of successfully served customers in the time unit).

In real complex systems, it is typically tough to obtain analytic expressions for quantities as the ones mentioned above. However, in some simple yet useful systems, it is possible to obtain such expressions and study explicitly the behavior of their evolution, as shown for some of them in Appendix B.3. Building suitable approximations allows the analysis and study of even more these complex queuing systems, as can be seen in more dedicated treatments of the topic [25, 36, 209, 229].

B.2.3 RELATION BETWEEN ARRIVAL-DEPARTURE PROCESSES AND LITTLE'S LAW

Both arrival and departure processes of a queuing system are associated with two corresponding counting processes, the first counting the number of arriving customers and the second counting the departed customers (those that finished waiting and service and leaving the system completely). The difference between the arrival and departing counting processes yields the number of customers currently in the system (waiting and served cumulatively). If $a(t)$ is the arrival and $b(t)$ the departure counting process, respectively, then $N(t) = a(t) - b(t)$ is the number

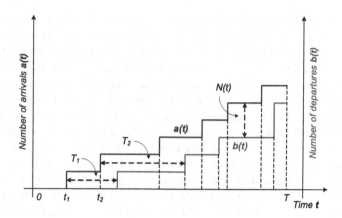

FIGURE B.2

Graphical presentation of the relation between the arrival-departure counting processes and visual explanation of Little's law.

of customers in the system, assuming the system starts empty. Fig. B.2 shows an instance of such counting processes for the evolution of a queuing system with a FIFO discipline over time in the time interval $[0,t]$. If T_i represents the time spent in the system by customer i, the area between the two curves is given by $\int_0^t N(\tau)d\tau = \sum_{i=1}^{\alpha(t)} T_i$. Dividing this by t and taking the limit as t tends to ∞,

$$\frac{1}{t}\int_0^t N(\tau)d\tau = \frac{1}{t}\sum_{i=1}^{\alpha(t)} T_i = \frac{\alpha(t)}{t}\frac{\sum\limits_{i=1}^{\alpha(t)} T_i}{\alpha(t)}$$

yields a very important result, most widely known as *Little's law*.

Theorem B.1 (*Little's Law*). *If N is the average number of customers in the system, λ the mean arrival rate, and T the mean system time, these are related with the simple formula*

$$N = \lambda T. \tag{B.1}$$

It should be highlighted that the terms "averages" in Little's law are long-term time averages, which for a stationary system these averages are also equal to the expected values of the corresponding random variables.

Even though a proof can be found in [25], Little's law does not qualify for a formal mathematical theorem, since it essentially expresses a conservation principle with respect to the customers of a queuing system. Little's law expresses the intuitive idea that the number of customers in the system depends on the average delay and vice versa. This is intuitively justified if one considers that in each queuing system that does not absorb customers, i.e. all customers entering the system will eventually depart, the more customers enter in the system, the more the average delay experienced by a random customer will be and vice versa, i.e. the more the average delay experienced by an individual customer, the more customers are expected to be found in the system. The relation between the expected number of customers and the average delay of a customer is given by Little's law. Little's law is used extensively in Chapter 4 for relating the expected number of infected nodes with the expected time these nodes remain in the infected state under the SIS infection model. Essentially, the expected infection duration can be obtained in each scenario via Little's law.

Although it looks intuitively reasonable, it is quite a remarkable result, as the corresponding relationship it expresses is not influenced by the arrival process distribution, the service distribution, the service order, or practically anything else. In addition, Little's law has more general value and can be applied recursively to arbitrary components of a queuing system, i.e. in systems within systems. For instance, if only the queue is considered, $N_Q = \lambda W$ gives the number of customers in the queue as a function of the arrival rate (in the queue) and the mean waiting time W. The only requirements are that the system is stable and non-preemptive, thus ruling out transition states such as initial startup or shutdown transient phenomena. If the system has n different inputs, each with rate λ_i, then Little's theorem yields $N = \sum_{i=1}^{n} \lambda_i T$.

B.3 MARKOVIAN SYSTEMS IN EQUILIBRIUM

B.3.1 DISCRETE-TIME MARKOV CHAINS

A Markov chain discrete-time Markov chain - DTMC is a random process that undergoes transitions from one state to another on a state space. It must possess the "memorylessness" property, namely, that the probability distribution of the next state depends only on the current state and not on the sequence of events that preceded it. The material presented in this subsection will be useful for the study of queuing systems treated in the later subsections of Appendix B.3 and Chapters 4 and 5.

More formally, a stochastic process $\{X_n, n = 0, 1, 2, ...\}$ that takes a finite or countable number of possible values is said to be in state i at time n if $X_n = i$. We assume that there is a fixed probability P_{ij} that if the process is in state i it will next be in state j,

$$\Pr\{X_{n+1} = j | X_n = i, X_{n-1} = i_{n-1}, ..., X_1 = i_1, X_0 = i_0\}$$
$$= \Pr\{X_{n+1} = j | X_n = i\} = P_{ij} \tag{B.2}$$

for all states $i_0, i_1, ..., i_{n-1}, i, j$ and all $n \geq 0$. Such a stochastic process is known as a *(discrete) Markov chain*. The *Markovian* property, expressed by Eq. (B.2), is that the conditional distribution of any future state X_{n+1} given the past states $X_0, X_1, ..., X_{n-1}$ and the present state X_n is independent of the past states and depends only on the present state. Thus, a Markov chain is a sequence of random variables $X_1, X_2, X_3, \ldots$ with the Markov property. Since the process must make a transition into some state, $\sum_{j=0}^{\infty} P_{ij} = 1, i = 0, 1, \ldots, P_{ij} \geq 0, i, j \geq 0$.

A Markov chain is said to be *homogeneous* in time if the transition probability between two distinct states depends only on the time step difference, i.e. $\Pr\{X_{m+n} = j | X_m = i\} = P_{ij}^n$. Here, we focus on homogeneous Markov processes.

Eq. (B.2) defines the so-called one-step transition probabilities, namely probabilities that describe the transition of the chain in one time step. These transitions can be depicted via a *state diagram*, where a state diagram is a directed graph used to picture the state transitions. The states that the chain transitions to will be called neighboring states. Similarly, the n-step probabilities that the process will be in state j starting at i after n successive transitions (n successive time steps) are

$$P_{ij}^n = \Pr\{X_{n+m} = j | X_m = i\}, \quad n, i, j \geq 0. \tag{B.3}$$

In order to compute the n-step transition probabilities, the *Chapman-Kolmogorov* equations shown in the following can be employed:

$$P_{ij}^{n+m} = \sum_{k=0}^{\infty} P_{ik}^n P_{kj}^m, \quad \text{for all } n, m \geq 0, \text{ all } i, j. \tag{B.4}$$

If f_{ij}^n is the probability that starting in state i, the first transition to j occurs at time n, then state j is said to be *recurrent* if $f_{jj} = 1$, which means that starting at state j the process will return to state j with probability 1 for some time n ($f_{jj} = \sum_{n=1}^{\infty} f_{ij}^n$). A state is transient if $f_{jj} < 1$. If state j is recurrent and $\mu_j = \sum_{n=1}^{\infty} n f_{jj}^n$ denotes the expected number of transitions needed to return to start j (mean recurrence time - expected return time in state j), then state j is positive recurrent if $\mu_j < \infty$ and null recurrent if $\mu_j = \infty$. A state j is said to be accessible from a state i (written $i \to j$) if a system that started in state i has a nonzero probability of transitioning into state j at some point ($\Pr\{X_n = j | X_0 = i\} = P_{ij}^n > 0$ for some $n > 0$). A state i has period k if any return to state i must occur in multiples of k time steps. If $k = 1$, then the state is said to be aperiodic, namely, returns to state i can occur at irregular times. A Markov chain is aperiodic if every state is aperiodic and an irreducible Markov chain only needs one aperiodic state to imply all states are aperiodic. A state i is called absorbing if it is impossible to leave this state.

A state i is said to be *ergodic* if it is aperiodic and positive recurrent, namely, if it is recurrent, has a period of 1, and it has finite mean recurrence time. If all states in an irreducible Markov chain are ergodic, then the chain is said to be *ergodic*. It can be shown that a finite state irreducible Markov chain is ergodic if it has an aperiodic state.

If an irreducible aperiodic Markov chain is positive recurrent, then we can define as $\pi_j = \lim_{n \to \infty} P_{ij}^n$, where P_{ij}^n is obtained from (B.3) for $n = 1$. Let S denote the set of possible states that the process can make a transition to. If there is a probability distribution over states π such that

$$\pi_j = \sum_{i \in S} \pi_i \Pr\{X_{n+1} = j | X_n = i\} \tag{B.5}$$

for every state j and every time n then π is a stationary distribution of the Markov chain. An irreducible chain has a stationary distribution if and only if all of its states are positive recurrent. In that case, π is unique and related to the expected return time M_j by $\pi_j = \frac{C}{M_j}$, where C is a normalizing constant independent of the state. The chain converges to the stationary distribution regardless of the initial state distribution. Such π is called the *equilibrium distribution* of the chain. For a finite state space, a stationary distribution π is a (row) vector, whose entries are non-negative and sum to 1, it is unchanged by the operation of transition matrix $\mathbf{P}$ and it is defined by $\pi\mathbf{P} = \pi$. In addition, $0 \leq \pi_j \leq 1$ and $\sum_{j \in S} \pi_j = 1$. Then the transition probability distribution can be represented by the *transition matrix* $\mathbf{P}$, with the (i, j)th element of $\mathbf{P}$ equal to $\pi_j = \Pr\{X_{n+1} = j | X_n = i\}$.

B.3.2 CONTINUOUS-TIME MARKOV PROCESSES

A *continuous-time Markov process* may be in one of the finite or infinite many (however countably infinite many) states at a time. The probability that the process transitions from state i to j at time $t_0 + t$ is defined as $\Pr\{X(t_0 + t) = j | X(t_0) = j\}$ and it is independent of the behavior of $X(t)$ prior to t_0. If $X(t)$ is a *homogeneous Markov process*, the above probability depends only on the elapsed time required for the transition and not the initial time t_0. Thus in this case, $P_{ij}(t) = \Pr\{X(t_0 + t) = j | X(t_0) = i\}$ or $P_{ij}(t) = \Pr\{X(t) = j | X(0) = i\}$, denoted by *state transition rates* as opposed to the corresponding state transition probabilities in the discrete case (P_{ij}^n). The state transition rates are obtained as derivatives of the state transition probabilities at $t = 0$. Markov processes, in general, are characterized by the basic property that the past has no influence on the future if the present is specified and can be expressed in the continuous form as

$$\Pr\{X(n) = x_n | X(n-1), ...X(1)\} = \Pr\{X(n) = x_n | X(n-1)\}. \qquad \text{(B.6)}$$

All Markov processes share the interesting property that the time it takes for a change of state, called *sojourn time* is an *exponentially* distributed random variable [174].

The initial probability distribution of state i is given by $P_i(0) = P\{X(0)\} = i$, defined analogously for the rest of states, and the unconditional probability that the process is in state j at time t is defined as $P_j(t) = \sum_i P_i(0)P_{ij}(t)$. For arbitrary t and s, the continuous version of the Chapman-Kolmogorov equations is given by

$$P_{ij}(t + s) = P\{X(t + s) = j | X(0) = i\} = \sum_k P_{ik}(t)P_{kj}(s). \qquad \text{(B.7)}$$

B.3.3 BIRTH-AND-DEATH PROCESSES

The *birth-death process* is a special case of continuous-time Markov chain (CTMC), where the state transitions are of two types only, namely, "births," which increase the state variable by one and "deaths," which decrease the state by one [174]. A typical example is that of some banks which have a security door that allows only

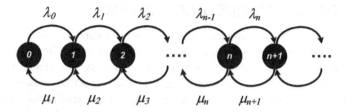

FIGURE B.3

State diagram for the birth-death process.

one person at a time to enter/leave the bank. In this case, irrespective of the system state, at each time instant the next event is either the arrival of a customer in the bank, or the departure of a customer that completed the required service and leaves the building. In these and similar cases, when a birth occurs, the process goes from state n to $n + 1$. When a death occurs, the process goes from state n to state $n - 1$. The state diagram evolution for such a process is depicted in Fig. B.3. Such processes are specified by their birth rates $\{\lambda_i\}_{i=0...\infty}$ and death rates $\{\mu_i\}_{i=1...\infty}$. A pure birth process is a birth-death process where $\mu_i = 0$ for all $i \geq 0$, while a pure death process is a birth-death process where $\lambda_i = 0$ for all $i \geq 0$. A (homogeneous) Poisson process is a pure birth process where $\lambda_i = \lambda$ for all $i \geq 0$.

The global balance equations (also known as full balance equations) are a set of equations that in principle can be solved to give the equilibrium distribution of a Markov process (when such a distribution exists). For a continuous-time Markov process with state space S, transition rate from state i to j given by q_{ij} and equilibrium distribution given by π, the global balance equations are given for every state i in S by

$$\sum_{j \in S\{i\}} \pi_i q_{ij} = \sum_{j \in S\{i\}} \pi_j q_{ji}, \qquad (B.8)$$

where $\pi_i q_{ij}$ represents the probability flux for state i to state j.

For a CTMC with transition rate matrix $\mathbf{Q} = [q_{ij}]$, if π_i can be found such that for every pair of states i and j, $\pi_i q_{ij} = \pi_j q_{ji}$ holds, then the global balance equations are satisfied and π is the stationary distribution of the process. If such a solution can be found, the resulting equations are usually much easier than directly solving the global balance equations. Furthermore, a CTMC is *reversible* if and only if the detailed balance conditions are satisfied for every pair of states i and j.

In queuing theory, the birth-death process is the most fundamental example of a queuing model. In the following, we provide a brief overview of simple and characteristic birth-death queuing systems that have particular interest for the analysis of queuing networks and some of the methodologies presented in this book, especially in Chapter 4. The rest of the systems presented in this subsection with an infinite queue are all characteristic examples of birth-death systems defined in this subsection.

B.3.4 THE *M/M/*1 QUEUING SYSTEM

The $M/M/1$ queue is a system having a single server, where arrivals are determined by a Poisson process and job service times have an exponential distribution. Arrivals are assumed to occur at rate λ and service times have an exponential distribution with mean $1/\mu$. The single server operates according to a FIFO discipline. When the service of a customer is complete, the customer at the head of the queue leaves the storage space entering the server, the customer that completed service leaves the systems, and the number of customers in the system reduces by one. The buffer (queue) of the system is of infinite size.

The model is considered stable only if $\lambda < \mu$. If, on average, arrivals happen faster than service completions the queue will grow indefinitely long and the system will not have a stationary distribution. Assuming $\rho = \lambda/\mu$ for the utilization of the buffer and require $\rho < 1$ for the queue to be stable, ρ represents the average proportion of time for which the server is occupied. In order to see this more intuitively, consider the version of Little's law with respect to the server, which gives the average number of customers in the server N_w with respect to the arrival rate λ and the mean service time $1/\mu$, $Nw = \frac{\lambda}{\mu} = \rho$. For the $M/M/1$ system to be stable, at most one customer at a time should be in service, as expected by the definition of its operation.

The corresponding state diagram is of birth-death type, similar to Fig. B.3. Useful closed-form expressions may be obtained for various quantities of interest. More details on the analysis methodology of the $M/M/1$ queue can be found in [25, 142], etc. The most useful results obtained include the following:

- The steady-state probability distribution is given by $p_n = \rho^n(1 - \rho), n = 0, 1, \dots$
- The average number of customers in the system in the steady state is $N = \frac{\lambda}{\mu-\lambda}$.
- The average system delay per customer (waiting time plus service time) is $T = \frac{1}{\mu-\lambda}$.
- The average waiting time in queue is $W = \frac{\rho}{\mu-\lambda}$.
- The average number of customers in the queue is $N_Q = \frac{\rho^2}{1-\rho}$.

Additional quantities of interest can be obtained via analysis as shown in [25] and other references targeted explicitly to queuing theory analysis [33, 36, 90, 142, 209, 211, 229].

B.3.5 THE *M/M/m* SYSTEM AND OTHER MULTISERVER QUEUING SYSTEMS

The $M/M/1$ queue is a very simple system that emerges frequently in various occasions. However, oftentimes, multiserver systems are employed. In this subsection, we overview the most fundamental of them. As with the $M/M/1$ system, for all these multiserver systems, it is possible to derive a set of equations from the corresponding state transition diagram and solve for the steady-state (equilibrium) probabilities. Using Little's law allows computing the average delay per customer and other quantities of interest.

The *M/M/m* queuing system

The *M/M/m* system is similar to the *M/M/1* with respect to the arrival process (Poisson), the independence of service times, and the exponentially distributed and independent interarrival times. However, this system has *m* parallel servers, meaning that a customer at the head of the queue (ready to receive service) can be routed to any of the server currently available, once a customer completes service in one of them.

Applying the analysis methodology of balance equations, the corresponding outcomes for the quantities of interest obtained in the *M/M/1* case are as follows:

- The probability that an arriving customer will find all servers buys and will be forced to queue is $P_Q = \frac{p_0(m\rho)^m}{m!(1-\rho)}$, where p_0 is given by the expression

$$p_0 = \left[1 + \sum_{n=1}^{m-1} \frac{(m\rho)^n}{n!} + \sum_{n=m}^{\infty} \frac{(m\rho)^n}{m!} \frac{1}{m^{n-m}} \right]^{-1}. \tag{B.9}$$

- The average number of customers in the system in the steady state is $N = m\rho + \frac{\rho P_Q}{1-\rho}$.
- The average system delay per customer (waiting time plus service time) is $T = \frac{1}{\mu} + \frac{P_Q}{m\mu-\lambda}$.
- The average waiting time in queue is $W = \frac{\rho P_Q}{\lambda(1-\rho)}$.
- The average number of customers in the queue is $N_Q = P_Q \frac{\rho}{1-\rho}$.

Now the utilization of the system is defined as $\rho = \frac{\lambda}{m\mu} < 1$. In order to understand this stability condition for the system, note that now *m* servers operate in parallel each with an exponentially distributed service rate of μ. This means that the combined server will be operating as a single server with service rate which will be the minimum of *m* exponential distributions. This has been proven to be again exponentially distributed with rate the sum of partial rates $m\mu$. Thus, for the mean service time expression in the similar application of Little's law that we performed for the *M/M/1* system, now for the case for the *M/M/m* would yield $\frac{\lambda}{m\mu}$.

The *M/M/∞* queuing system

In several cases, the available service has such a capacity that practically, any customer entering the system receives immediate service. Typical examples include web platforms receiving requests and serving them immediately (up to a certain point of course). In practice, the number of available servers is always finite. However, whenever the number of available servers is such that no queuing at all is observed, the system might well be considered as of an *M/M/∞* type.

The *M/M/∞* system can be considered as the limit of the *M/M/m* system when $m \rightarrow \infty$. Consequently, in the limiting case where the number of servers is high (theoretically infinite number of servers), $m = \infty$, one obtains a $M/M/\infty$ system for which the quantities of interest become the following:

- The average number of customers in the system in the steady state $N = \frac{\lambda}{\mu}$.
- The average system delay per customer (waiting time plus service time) $T = \frac{1}{\mu}$.

Clearly, for such a system where a customer always finds an available empty server, there is no queuing and thus $N_Q = 0, W = 0$. The response time for each arriving job is a single exponential distribution with parameter μ as already explained and it thus coincides with the average system delay per customer given above. One should also notice that the stationary distribution of a more general queue with practical applications, the $M/G/\infty$ queue where the service discipline can be arbitrary, but arrivals are Poisson and there exists an infinite number of servers, is the same as that of the $M/M/\infty$ queue. This implies that an overprovisioned system, i.e. one with practically infinite number of servers, behaves the same irrespective of the service policy and the behavior is rather simple, since each customer is essentially served immediately. Thus, the mean number of customers in the system will be $N = \frac{\lambda}{\mu}$ in any case.

The steady-state distribution of the system is given by the expression

$$p_n = \left(\frac{\lambda}{\mu}\right)^n \frac{e^{-\lambda/\mu}}{n!}, \quad n = 0, 1, \ldots \tag{B.10}$$

with $p_0 = e^{-\lambda/\mu}$. From the above expression, it can be seen that the system is always stable, as no special constraint on the latter distribution is required. This is also intuitively verified by the fact that no matter how heavy the input load of customers, there are always available servers to satisfy it (infinite number). Thus, the number of customers waiting in the systems never grows arbitrarily and the system will be stable. In fact, it can be shown that the number of customers in the system is Poisson distributed even if the service time distribution is not exponential, e.g. in $M/G/\infty$ [25]. The $M/M/\infty$ system is also known as the Erlang-C model, as opposed to the Erlang-B that will be described in the following.

The M/M/m/m queuing system

This system is identical to an $M/M/m$ system with the difference that there is no queuing buffer space. The system consists of a facility where arrivals form a single queue and are governed by a Poisson process, there are m servers and job service times are exponentially distributed. Thus, the total capacity of the system is defined by the number of customers the system can serve instantaneously, which also means that when an arriving customer finds all m servers of the system busy, it is blocked from entering the system and thus discarded. For this reason, the model is called a loss system. The lost customers are said to be blocked. Practical examples of such systems are telephone exchanges, cellular phones, and other more general flow-service based systems.

In general, it is tough to obtain analytic expressions for the quantities of interest mentioned above for the other systems presented. However, it is possible to obtain the probability of i customers in the system and the blocking probability p_m, namely, the probability that an arriving customer will find all servers busy and will be discarded from the system

$$p_m = \frac{(\lambda/\mu)^m/m!}{\sum_{n=0}^{m}(\lambda/\mu)^n/n!}.$$ (B.11)

The latter is known as the Erlang B formula. This is also called the Erlang-B loss function and tabular solutions are usually provided in the related references. More details can be obtained in the corresponding references [25, 33, 36, 90, 142, 209, 211, 229]. It should be noted that as $m \to \infty$, the above distribution approaches the Poisson distribution of the $M/M/\infty$ system, described above.

B.4 REVERSIBILITY

A discrete Markov chain is said to be reversible if there is a probability distribution over states π, such that

$$\pi_i \Pr\{X_{n+1} = j \mid X_n = i\} = \pi_j \Pr\{X_{n+1} = i \mid X_n = j\}$$ (B.12)

for all times n and all states i and j. With a time-homogeneous Markov chain, $\Pr\{X_{n+1} = j|X_n = i\}$ does not change with time n and it can be written more simply as p_{ij}. In this case, the above detailed balance equation can be written more compactly as

$$\pi_i p_{ij} = \pi_j p_{ji}.$$ (B.13)

For reversible Markov chains, π is always a steady-state distribution of $\Pr(X_{n+1} = j|X_n = i)$ for every n. If one considers the sequence of states going backward in time, i.e. starting at some state n and following $X_n, X_{n-1}, ...$, this sequence is itself a Markov chain and it is called the *reversed* chain. Then the chain will be time reversible if the transition probabilities of the forward and reversed chains are identical.

Time reversibility generally occurs when a stochastic process can be broken up into subprocesses that undo the effects of each other. Most of the previously presented Markovian systems turn out to be reversible.

Consequently, the notion of time reversibility extends naturally to CTMCs. More specifically, the reversed chain is a CTMC with the same stationary distribution as the forward chain and with transition rates $q_{ij}^* = \frac{p_j q_{ij}}{p_i}, i, j \geq 0$, where $p_j, j \geq 0$ is the stationary (equilibrium) distribution of the forward chain. It should be also noted that the transition rates q_{ij} are obtained from the corresponding transition probabilities p_{ij} in the limit of time tending to zero. The forward chain is time reversible if and only if its stationary distribution and transition rates satisfy the detailed balance equations $\pi_i p_{ij} = \pi_j p_{ji}$, for all $i, j \geq 0$.

In general, a reversible process is stationary [127]. A very useful result associated with the queuing systems we presented above and the latter is the following.

Theorem B.2 *(Burke's Theorem).* *Consider an M/M/1, M/M/m, or M/M/∞ system with arrival rate* λ*. Assume the system starts in the steady-state. Then the following hold:*

- *The departure process is Poisson with rate* λ*.*
- *At each time t, the number of customers in the system is independent of the sequence of departure times prior to t.*
- *If customers are served in the order they arrive, then, given that a customer departs at time t, the arrival time of that customer is independent of the departure process prior to t.*

The first two assertions of Burke's theorem are counterintuitive, since one would expect that a recent stream of closely spaced departures implies a rather busy system with a very large number of customers in the queue. However, Burke's theorem asserts this is not the case. Such theorem will be extremely useful in the analysis of queuing networks in the sequel, which is also of significant importance in the analysis of the corresponding methodologies for malware diffusion modeling in Chapters 4 and 5.

The theorem can be generalized for "only a few cases," but remains valid for $M/M/m$ queues and Geom/Geom/1 queues. It is thought that Burke's theorem does not extend to queues fed by a Markovian arrival process (MAP) and is conjectured that the output process of a $MAP/M/1$ queue is a MAP only if the queue is an $M/M/1$ queue. It is noted that a MAP process is a mathematical model for the time between job arrivals to a system, modeling the case where the Markov property holds for the corresponding distribution, e.g. Poisson. The above described issues are still open and of great interest and importance for the mathematics and engineering communities with numerous practical applications.

B.5 QUEUES IN TANDEM

Realistic queuing systems are more complex than the ones presented above in Appendix B.3. However, this means that the involved analysis becomes more complex as well. This can be depicted very easily, even with the very simple queuing system of two $M/M/1$ queues connected in tandem, as will be shown in the sequel. Such a system is shown in Fig. B.4. The service times at the two queues are exponentially distributed and mutually independent. The queue of each system is infinite. This system can be analyzed using the methods presented previously.

Using Burke's theorem provided above, it can be shown that the number of customers in queues 1 and 2 is independent at a given time and the stationary probability is given by

$$\Pr\{n \text{ at queue } 1, m \text{ at queue } 2\} = \rho_1^n(1-\rho_1)\rho_2^n(1-\rho_2), \qquad (B.14)$$

where $\rho_i, i = 1, 2$ is the traffic intensity of queues 1 and 2, respectively. The above result implies that the two queues behave as if they are independent $M/M/1$ queues.

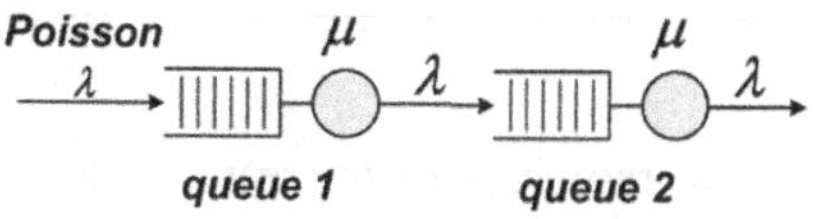

FIGURE B.4

Two queues in tandem.

This fact comes from the assumption that the service times of a customer at the first and second queue are mutually independent as well as independent of the arrival process (Kleinrock independence approximation [25]). The proof of this result can be found in [25, 142].

The above example is essentially a simple case of an acyclic (open) queuing network, where the output is not connected to the input. The routing of customers is straightforward; however, even in this case many complications might emerge. These could have significant impact on the analysis and control of such systems.

For instance, consider the degenerate case of two transmission lines of equal transmission rate in tandem (essentially the system of Fig. B.4, where there is no buffering at each queue and the service rate of each server is identical). Furthermore, assume that Poisson arrivals at rate λ enter the first queue and all packets have the same length (in general customers require the same service time). This means that the first queue is essentially an M/D/1 queue and the average interarrival time is given by $W = \frac{\rho}{2(1-\rho)}$, where $\rho = \lambda \bar{X}$ and $\bar{X}$ is the average service time [25] (the latter is a result obtained by application of the more general expression for an M/G/1 queue, but showing how to obtain this is out of the scope of this primer—the interested reader may consult [25] or other similar references to obtain the associated details). However, the interarrival times at the second queue will be greater than or equal to $1/\mu$, as determined by the output of the first queue. Also, since the service times are equal at the two queues, each packet arriving at the second queue will complete service at or before the time the next packet arrives, which means there will be no queuing in the second line. Therefore, a Poisson model for the second line seems to be inappropriate even in this simple case.

Even in the case where an $M/M/1$ queue was in the first stage, the second cannot be modeled as an $M/M/1$ as well because the interarrival times at the second queue will be strongly correlated with the packet lengths. To see this intuitively, consider a busy period of the first queue. In this case, the interarrival time of two such packets in the second queue equals the transmission time of the second packet transmitted in the first queue. This means that long packets will typically wait less at the second queue than short packets, since the second queue will have more time to empty out. Up to date there exists no analytic expression for the simple case of tandem queues shown involving Poisson arrivals and exponentially distributed services shown in Fig. B.4 [25].

B.6 QUEUING NETWORKS

Networks of queues emerge oftentimes when the resources of interest are shared by a set of customers, or whenever multiple or hierarchical processing stages exist. Typically in these cases, customers enter the system, join the queue of a service stage and once served, they proceed with another stage in the system, until they complete service, and leave the system. Characteristic examples include public services, where a citizen might need to move from one queue to another to complete service, traffic highways, communications networks, etc. In these examples, the various service stages correspond to different queues-servers that the customers need to go through potentially in arbitrary order in the most general case. For example, in a chain of queues, an incoming item has to go through all queues and service stages before it exits the system.

In general, whenever there exists an interconnection of queues, the system may be characterized as a *queuing network*. In reality queuing networks are the norm, e.g. traffic highways and streets. In the following, we will focus on simple queuing systems emerging frequently, and will present techniques for analyzing such systems. The behavior of networks of queues is characterized by their output distributions, which describe the expected number of customers in each individual queue of the queuing network, and the service time distributions of the servers.

The two queues in tandem presented above are the simplest possible queuing network. As already mentioned, this was in fact an *open queuing network*, since there existed at least one output of a queue driving the served customers completely out of the system. On the contrary, when customers cannot leave the system and always reside in one of the servers/queues therein, the network will be characterized as a *closed queuing network*. In turn, the simplest closed queuing network is obtained from the two queues in tandem system where now the output of the second becomes the input of the first. This system is shown in Fig. B.5. We note that such system could start at a transient initial state before reaching eventually its stationary steady-state (if one exists), or it could be directly started at the equilibrium state, where on average, a subset of the customers will lie in the one queue and the rest in the other queue.

As it was explained before, queuing networks, even the simple open two queues in tandem, can be very complicated to analyze and obtain expressions for their equilibrium distributions. This is due to the fact that after the first queue, the interarrival times of customers become correlated with their service times. However, if several assumptions are met, it is possible to show that their steady-state distribution probability is of product-form type (as in Eq. (B.14)), in which case they are called Jackson networks. More specifically, it has been shown, see [25] and references therein, that if this correlation was bypassed and randomization was used to divide traffic among different routes, the system can be solved as if there was no correlation between queues. Thus, such approximation is feasible. These necessary assumptions constitute the requirements for the application of the so-called *Kleinrock's Independence Approximation* [25, 142]. The assumptions are shortlisted as follows:

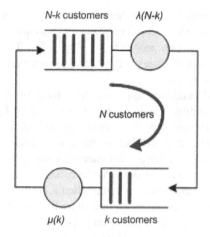

N-k customers   λ(N-k)

N customers

μ(k)   k customers

FIGURE B.5

A simple two-queue closed network.

- The arrival statistics are Poisson.
- The service times at each queue server are exponentially distributed and independently distributed from queue to queue.
- The route from one queue to another is selected randomly, with fixed routing probabilities

and whenever they can be ensured, the corresponding Kleinrock's independence approximation can be applied for analyzing the corresponding queuing networks.

In the following, we assume that these assumptions are valid for the considered queuing networks and Jackson's theorem, which will be provided in the following, can be applied for analyzing them. More specifically, the system is considered as a CTMC with states $n_1, n_2, ..., n_K$, where n_i denotes the number of customers at queue i and K queues in total are assumed in the network. If $P(n_1, n_2, ..., n_K)$ is the stationary distribution of the chain, where n_i is the number of customers in queue i of the queuing network, then we have the following.

Theorem B.3 *(Jackson's Theorem). Assuming $\rho_i < 1, i = 1, ..., K$ we have for all $n_1, ..., n_K \geq 0$ that $P(n_1, n_2, ..., n_K) = P(n_1)P(n_2)....P(n_K)$, where $P(n_i) = \rho_i^n(1 - \rho_i), n \geq 0$.*

Parameter $\rho_i = \lambda_i/\mu_i$ is the utilization factor of each queue, assuming exponentially distributed service times with mean $1/\mu_i$, and that the input of each queue is determined as the solution to the system of linear equations

$$\lambda_j = r_j + \sum_{i=1}^{K} \lambda_i P_{ij}, \quad j = 1, ..., K, \tag{B.15}$$

where external customers arrive at each queue i in accordance with independent Poisson streams of rate r_i, and once each customer is served in queue i it joins queue j with probability P_{ij} or abandons the network with probability $1 - \sum_{j=1}^{K} P_{ij}$. The unique solution of the above system is guaranteed under very general assumptions [25] and it is valid for many different types of systems.

Jackson's theorem and Kleinrock's independence approximation have many implications for the analysis of the corresponding queuing networks in Chapter 4. First of all, Jackson's theorem implies that the number of customers in each queue at a given time is independent of the corresponding number in the rest of queues. Second, it implies that the distribution for the number of customers at each queue i is identical to that of an $M/M/1$ queue, even though the arrival process at each queue need not be a Poisson process (a case frequently emerging when a customer follows a feedback route in the queuing network and visits a queue more than once). Thus, by applying them in a queuing network for which the assumptions mentioned above hold, one may consider every link as an independent $M/M/1$ queue, define the corresponding input/service rates, and perform analysis with sufficient accuracy for many practical purposes.

B.6.1 ANALYTICAL SOLUTION OF TWO-QUEUE CLOSED QUEUING NETWORK

In this subsection, as an example analysis of the framework presented previously, we provide the analytical solution for a typical two-queue closed queuing network, shown in Fig. B.5, using standard Markov chain analysis methods. This solution approach is exploited and adapted in Chapter 4 to obtain closed-form expressions for the corresponding malware diffusion models that are based on principles and modeling tools inspired by queuing theory.

Thus, let us consider the simple two-queue closed queuing network of Fig. B.5 in the general case, where N customers in total oscillate between the two queuing units. The service rate of the lower queue is assumed to be $\mu(k)$, if k customers are in queue, while then $N - k$ customers will be in the upper queue and its service rate is $\lambda(N - k)$. Thus, in the general case, we consider state-dependent queues. We will see also partial cases with fixed service rates for the two queues. A critical observation is that the output of one queue becomes the input of the other and vice versa. Without loss of generality, we focus on the lower queue. If k customers are in the lower queue (from now on denoted as "queue 1") at a time, then $N - k$ are in the upper queue (from now on denoted as "queue 2"). The state diagram shown in Fig. B.6 describes the possible state transitions of the associated discrete (embedded) Markov chain obtained from the corresponding continuous Markov chain. Using balance equations between state $i - 1$ and state i, we can obtain

$$\lambda(i - 1)\pi(i - 1) = \mu(i)\pi(i), \tag{B.16}$$

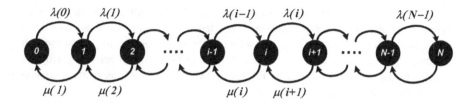

FIGURE B.6

State diagram of a two-queue closed queuing network with state-dependent service rates.

yielding $\pi(i) = \frac{\lambda(i-1)}{\mu(i)}\pi(i)$ and in general

$$\pi(i) = \pi(0) \prod_{j=1}^{i} \frac{\lambda(j-1)}{\mu(j)}. \tag{B.17}$$

The normalization condition $\sum_{i=0}^{N} \pi(i) = 1$ imposes a constraint on the previous expression yielding

$$\pi(0) = \left(\sum_{i=0}^{N} \prod_{j=1}^{i} \frac{\lambda(j-1)}{\mu(j)} \right)^{-1}. \tag{B.18}$$

The above expression is general. But more specific results can be obtained, depending on the specific operational assumptions of each queuing stage. For instance, in the event that both service rates are fixed for both queues, i.e. $\lambda(k) = \lambda$, $\mu(k) = \mu$ for state k and λ, μ fixed, then

$$\pi(0) = \frac{1}{\sum\limits_{i=0}^{N} \rho^i}, \tag{B.19}$$

where $\rho = \lambda/\mu$. Thus,

- if $\rho = 1$ then $\pi(0) = \frac{1}{N}$ and $\pi(i) = \frac{\rho^i}{N}$;
- if $\rho \neq 1$ then $\pi(0) = \frac{\rho-1}{\rho^{N+1}-1}$ and $\pi(i) = \frac{\rho-1}{\rho^{N+1}-1}\rho^i$,

which yield the following results for the expected number of customers in the lower queue denoted as $E[N_1]$ correspondingly:

- if $\rho = 1$ then $E[N_1] = \frac{N+1}{2}$;
- if $\rho \neq 1$ then $E[N_1] = \frac{\rho}{\rho-1} \cdot \frac{N\rho^{N+1}-(N+1)\rho+1}{\rho^{N+1}-1}$.

Of course, the corresponding expected number of customers in the upper queue will be $N - E[N_1]$.

Now assume the state-dependent service rates of the two queues are of the following form $\lambda(j) = \alpha \cdot \lambda \cdot (N - i + j + 1)$ and $\mu(j) = \beta \cdot \mu \cdot j$, where α, β are just scaling factors for the fixed service rates λ and μ, depending on system service factors. Both arrival and service rates are state-dependent, and in the above form they are expressed for the state with j customers in the upper queue, or j customers in the lower queue. In this case, the steady-state probability of the zero state (for the instance of the system described by Eq. (B.18) with $N - j$ customers in one queue and j in the other) becomes

$$
\pi(0) = \frac{1}{\sum\limits_{i=0}^{N} \prod\limits_{j=1}^{i} \frac{\alpha \cdot \lambda \cdot (N-i+j)}{\beta \cdot \mu \cdot j}} = \frac{1}{\sum\limits_{i=0}^{N} \left(\frac{\alpha}{\beta}\rho\right)^i \frac{N!}{i!(N-i)!}} = \frac{1}{\sum\limits_{i=0}^{N} \left(\frac{\alpha}{\beta}\rho\right)^i \binom{N}{i}} = \frac{1}{\left(\frac{\alpha}{\beta}\rho + 1\right)^N},
$$

(B.20)

where the binomial identity $(x + y)^n = \sum_{k=0}^{n} \binom{n}{k} x^k y^{n-k}$ has been used in the final step. Thus, the closed-form expression of the stationary probabilities is

$$
\pi(i) = \frac{\left(\frac{\alpha}{\beta}\rho\right)^i \binom{N}{i}}{\left(\frac{\alpha}{\beta}\rho + 1\right)^N}.
$$

(B.21)

Now, it is possible to obtain the closed-form expression for the expected number of customers in the lower queue straightforwardly, since distribution (B.21) expresses a binomial $B(N, p = \frac{\frac{\alpha}{\beta}\rho}{\frac{\alpha}{\beta}\rho+1})$ distribution, with an expected value of $Np = \frac{N\left(\frac{\alpha}{\beta}\rho\right)}{\left(\frac{\alpha}{\beta}\rho+1\right)}$. Alternatively, one may proceed analytically to compute the value in detail,

$$
E[N_1] = \sum_{i=0}^{N} i \cdot \pi(i) = \frac{1}{\left(\frac{\alpha}{\beta}\rho + 1\right)^N} \sum_{i=1}^{N} i \binom{N}{i} \left(\frac{\alpha}{\beta}\rho\right)^i
$$

$$
= \frac{1}{\left(\frac{\alpha}{\beta}\rho + 1\right)^N} \sum_{i=1}^{N} N \binom{N-1}{i-1} \left(\frac{\alpha}{\beta}\rho\right)^i.
$$

(B.22)

In the last step, the binomial factor identity,

$$
k\binom{n}{k} = n\binom{n-1}{k-1},
$$

(B.23)

was employed, thus allowing further computation of the expected number of customers in the lower queue,

$$
E[N_1] = \frac{N}{\left(\frac{\alpha}{\beta}\rho + 1\right)^N} \sum_{i=1}^{N} \binom{N-1}{i-1} \left(\frac{\alpha}{\beta}\rho\right)^i = \frac{N\left(\frac{\alpha}{\beta}\rho\right)}{\left(\frac{\alpha}{\beta}\rho + 1\right)^N} \sum_{k=0}^{N-1} \binom{N-1}{k} \left(\frac{\alpha}{\beta}\rho\right)^k
$$

$$
= \frac{N\left(\frac{\alpha}{\beta}\rho\right)}{\left(\frac{\alpha}{\beta}\rho + 1\right)},
$$

(B.24)

verifying the result obtained above via the binomial distribution argument. Obviously, the expected number of customers in the upper queue will be $N - E[N_1]$.

Optimal control theory and Hamiltonians

C.1 BASIC DEFINITIONS, STATE EQUATION REPRESENTATIONS, AND BASIC TYPES OF OPTIMAL CONTROL PROBLEMS

In this section, we will introduce the basic terminology for optimal control problems, set the basic problems, and present the main problems emerging. In general, for each optimal control problem, there exist three necessary components. The first (mathematical mode) determines the behavior, the second (physical constraints) is dictated by the environment, and the third (performance criterion) reflects the objectives of the system.

In optimal control problems, the considered systems are characterized by their input signals, the state variables denoted by $x_1(t), x_2(t), ..., x_n(t)$ that describe the internal state of the system, and the control inputs denoted as $u_1(t), u_2(t), ..., u_m(t)$, which can be cast in vector form as $\mathbf{x}(t) = [x_1(t)\ x_2(t)....x_n(t)]^T$, $\mathbf{u}(t) = [u_1(t)\ u_2(t)....u_m(t)]^T$, with dimensions n, m, respectively, with $\mathbf{x}(t) \subset \mathfrak{R}^n$ and $\mathbf{u}(t) \subset \mathfrak{R}^m$. The state vector $\mathbf{x}(t)$ is also called trajectory, since in motion problems it usually describes the evolution of the system's location. Within the time observation interval $[t_0, t_f]$, the system is described by a set of state equations representing ODEs, typically in the form

$$\dot{x}_1(t) = a_1(x_1(t), x_2(t), ..., x_n(t), u_1(t), u_2(t), ..., u_m(t), t),$$
$$\dot{x}_2(t) = a_2(x_1(t), x_2(t), ..., x_n(t), u_1(t), u_2(t), ..., u_m(t), t), \quad \text{(C.1)}$$

$$....$$

$$\dot{x}_n(t) = a_n(x_1(t), x_2(t), ..., x_n(t), u_1(t), u_2(t), ..., u_m(t), t),$$

which can be also cast in the mode compact vector state equation form

$$\dot{\mathbf{x}}(t) = \mathbf{a}(\mathbf{x}(t), \mathbf{u}(t), t). \quad \text{(C.2)}$$

With respect to the above equation, an *admissible control* is the control history $\mathbf{u}(t)$ that satisfies the imposed control constraints in $[t_0, t_f]$. The set of admissible controls is denoted by $\mathcal{U}$. Similarly, *admissible trajectory* is the state trajectory $\mathbf{x}(t)$ that satisfies the imposed state variable constraints in the entire $[t_0, t_f]$. The set of admissible trajectories is denoted by X. The target set S is the set of states where the final state is required to belong to. For a tangible example, consider a rocket, where the control describes the telemetry signals, the state trajectory $\mathbf{x}(t)$ describes the coordinates history of the rocket, the admissible control the set of allowed telemetry signals, the admissible trajectory the set of coordinates the rocket is allowed to follow, and the target set, the set of coordinates where the rocket is desired to land.

Given that the *optimal control* is defined as an admissible control that minimizes (maximizes) the employed performance measure, denoted by J, the performance measure is a scalar functional that depends on the final state $x(t_f)$, and a function of the state trajectory for the entire duration of $[t_0, t_f]$,

$$J = h(x(t_f), t_f) + \int_{t_0}^{t_f} g(\mathbf{x}(t), \mathbf{u}(t), t)dt, \tag{C.3}$$

where h, g are scalar functions.

Based on the previous definitions, the broader definition of the *optimal control problem* can now be formally stated as follows:

Definition C.1. *An optimal control problem is the problem of finding an admissible* $\mathbf{u}^*$ *that causes the system*

$$\dot{\mathbf{x}}(t) = \mathbf{a}(\mathbf{x}(t), \mathbf{u}(t), t) \tag{C.4}$$

to follow an admissible trajectory $\mathbf{x}^*$ *that minimizes (maximizes) the performance metric*

$$J = h(x(t_f), t_f) + \int_{t_0}^{t_f} g(\mathbf{x}(t), \mathbf{u}(t), t)dt, \tag{C.5}$$

where $\mathbf{u}^*$ *is the optimal control and* $\mathbf{x}^*$ *the optimal trajectory.*

A number of emerging issues regarding optimal control problems are summarized in the following list. Techniques to address them can be found in the literature, e.g. [137].

- **Existence of the optimal control**—A key question regarding the optimal control problems is always whether an optimal control actually exists. This is not always guaranteed and there exist several methodologies to answer this question.
- **Uniqueness of the optimal control**—Another important question following the existence of an optimal control is whether this is the sole solution or many more (and how many actually) exist. This is important and sometimes even critical, e.g. for rocket trajectory planning.
- **Absolute (global) optimality of the performance measure** J—If more than one optimal control exists, it is necessary to find which is the global optimal. This is not always easy and several tools may be employed for this goal.
- **Feasibility of optimal control**—Finally, one critical choice is whether the obtained optimal control can be actually implemented in practice. This might be very crucial depending on each problems' setting and criticality.

If a function (functional)[1] $\mathbf{u}^* = \mathbf{f}(\mathbf{x}(t),t)$ can be found for the optimal control at time t, function (functional) $\mathbf{f}$ is the optimal control law (or optimal policy). Function (functional) $\mathbf{f}$ determines $\mathbf{u}$ for any time t, for any admissible $\mathbf{x}(t)$.

Based on the type of the vector-form state equations, above system (C.2) can be classified as

- *nonlinear* and *time-varying*: $\dot{\mathbf{x}}(t) = \mathbf{a}(\mathbf{x}(t),\mathbf{u}(t),t)$,
- *nonlinear* and *time-invariant*: $\dot{\mathbf{x}}(t) = \mathbf{a}(\mathbf{x}(t),\mathbf{u}(t))$,
- *linear* and *time-varying*: $\dot{\mathbf{x}}(t) = \mathbf{A}(t)\mathbf{x}(t) + \mathbf{B}(t)\mathbf{u}(t)$,
- *linear* and *time-invariant*: $\dot{\mathbf{x}}(t) = \mathbf{A}\mathbf{x}(t) + \mathbf{B}\mathbf{u}(t)$,

where $\mathbf{A}(t)$, $\mathbf{B}(t)$ are matrices with time-varying elements and $\mathbf{A}$, $\mathbf{B}$ are considered constant matrices. Similarly, referring to physical quantities that can be measured in the system, the outputs can be in the following forms:

- *nonlinear* and *time-varying*: $\dot{\mathbf{y}}(t) = \mathbf{c}(\mathbf{x}(t),\mathbf{u}(t),t)$,
- *linear* and *time-invariant*: $\dot{\mathbf{y}}(t) = \mathbf{C}\mathbf{x}(t) + \mathbf{D}\mathbf{u}(t)$,

where $\mathbf{C}$, $\mathbf{D}$ are again considered constant matrices. A simplifying assumption that does not harm the generality is that the system states are all available for measurement, i.e. $\mathbf{y}(t) = \mathbf{x}(t)$.

According to the type of the performance measure, the optimal control problem can be classified in one of the following categories:

- *Minimum-time problems*: In these problems, it is required to take the system from $x(t_0) = x_0$ to a specified target set S in minimum time, i.e. considering a performance measure in the form $J = t_f - t_0 = \int_{t_0}^{t_f} dt$.
- *Terminal control problems*: In these problems, the objective is to minimize the deviation of the final state of a system from its desired value $\mathbf{r}(t_f)$. In the general case, this is expressed as a performance measure: $J = [\mathbf{x}(t_f) - \mathbf{r}(t_f)]^T \mathbf{H}[\mathbf{x}(t_f) - \mathbf{r}(t_f)]$, or $J = \|\mathbf{x}(t_f) - \mathbf{r}(t_f)\|_{\mathbf{H}}^2$, where $\mathbf{H}$ is real, symmetric, positive semi-definite $n \times n$ matrix, and $\|\cdot\|$ denotes the 2-norm of matrix $\mathbf{H}$.
- *Minimum control-effort problems*: The objective in these types of problems is to transfer a system from an arbitrary initial state $\mathbf{x}(t_0) = x_0$ to a specified target set of states S, with minimum expenditure of control effort, expressed as $J = \int_{t_0}^{t_f} [\mathbf{u}^T(t)\mathbf{R}\mathbf{u}(t)]dt$, where $\mathbf{R}$ is real, symmetric, positive definite weighting matrix.
- *Tracking problems*: The objective is to maintain the system state $\mathbf{x}(t)$, as close as possible to the desired state $\mathbf{r}(t)$ in the interval $[t_0,t_f]$, expressed in the form of the functional $J = \int_{t_0}^{t_f} \left[\|\mathbf{x}(t) - \mathbf{r}(t)\|_{\mathbf{Q}(t)}^2 + \|\mathbf{u}(t)\|_{\mathbf{R}(t)}^2\right] dt$, where $\mathbf{Q}(t)$ is real, symmetric, positive semi-definite for $t \in [t_0,t_f]$, $n \times n$ matrix whose elements weight the relative importance of different components of the state vector and $\mathbf{R}(t)$ is real, symmetric, positive definite for $t \in [t_0,t_f]$, $m \times m$ matrix.

---

[1] The definition of a functional and the difference from ordinary functions is explained in Definition C.2 in Appendix C.2.

- *Regulator problems*: This is a special case of the above class of tracking problems, when the desired states are zero, i.e. $\mathbf{r}(t) = \mathbf{0}, t \in [t_0, t_f]$.

Based on the framework presented in Chapter 6, the above problems can be used to study various attack strategies and their objectives.

In linear systems, it is possible to obtain the solution of state equations with respect to the state transition matrix $\phi(t,t_0)$, which is defined as

$$\phi(t,t_0) = e^{\mathbf{A}(t-t_0)} = \mathcal{L}^{-1}\{\Phi(s)\} = \mathcal{L}^{-1}\left\{[s\mathbf{I} - \mathbf{A}]^{-1}\right\}, \tag{C.6}$$

where $\mathcal{L}^{-1}(\cdot)$ is the inverse Laplace transform [43]. Thus, it can be also obtained as

$$e^{\mathbf{A}t} = \mathbf{I} + \mathbf{A}t + \frac{1}{2}\mathbf{A}^2t^2 + \frac{1}{3!}\mathbf{A}^3t^3 + \dots + \frac{1}{k!}\mathbf{A}^kt^k + \dots. \tag{C.7}$$

Then, the solution of the state equations for a linear system is obtained as

$$\mathbf{x}(t) = \phi(t,t_0)\mathbf{x}(t_0) \int_{t_0}^{t_f} \phi(t,\tau)\mathbf{B}(\tau)d\tau. \tag{C.8}$$

Analytic solutions, in addition to more specific details regarding the above optimal control problems in linear systems can be found in various books in the literature, e.g. [114, 188] and references therein. Based on the above system models and solutions, two very important notions can be defined, namely, controllability and observability. The first denotes the ability to control a system in the general case and the second the ability to observe its output.

State $\mathbf{x}_0$ at t_0 is controllable if there exists $t_1 \geq t_0$ finite and control $\mathbf{u}(t), t \in [t_0, t_1]$ that transfers $\mathbf{x}_0$ to the origin (zero vector) at t_1. A system is controllable if all values of $\mathbf{x}_0$ are controllable for all t_0. A linear time-invariant system is controllable if the $n \times mn$ matrix

$$\mathbf{E} = [\mathbf{B}|\mathbf{AB}|\mathbf{A}^2\mathbf{B}|\dots|\mathbf{A}^{n-1}\mathbf{B}]$$

has rank[2] n, i.e. $\mathbf{E}$ is right-invertible[3] [188].

Similarly, state $\mathbf{x}(t_0) = \mathbf{x}_0$ at t_0 is observable if by observing output $\mathbf{y}(t)$ during the finite interval $t \in [t_0, t_1]$, $\mathbf{x}_0$ can be determined. A system is observable if all states $\mathbf{x}_0$ are observable for every t_0. A linear time-invariant system is observable if the $n \times qn$ matrix

$$\mathbf{G} = [\mathbf{C}^T|\mathbf{A}^T\mathbf{C}^T|(\mathbf{A}^T)^2\mathbf{C}^T|\dots|(\mathbf{A}^T)^{n-1}\mathbf{C}^T]$$

has rank n, i.e. $\mathbf{G}$ is again right-invertible [188].

---

[2]The rank of a matrix $\mathbf{A}$ is the size of the largest collection of linearly independent columns of $\mathbf{A}$ (the column rank), or equivalently the size of the largest collection of linearly independent rows of $\mathbf{A}$ (the row rank). Thus, the row rank equals the column rank and they can be used interchangeably.
[3]If non-square matrix $\mathbf{M}$ is $m \times n$ and the rank of $\mathbf{M}$ is equal to n, then $\mathbf{M}$ has a left inverse: an $n \times m$ $\mathbf{N}$ such that $\mathbf{NM} = \mathbf{I}$. If $\mathbf{M}$ has rank m, then it has a right inverse: an $n \times m$ $\mathbf{N}$ such that $\mathbf{MN} = \mathbf{I}$.

C.2 CALCULUS OF VARIATIONS

In the above definition of the optimal control problem, the notion of a functional was employed. In this section, we will provide basic definitions on functionals and present several properties of interest. A functional is a more general notion of a function, which can be used to treat similar types of functions cumulatively. More specifically, the following definition describes formally a functional:

Definition C.2. *A functional J is a rule of correspondence (function) that assigns to each function* $\mathbf{x}$ *in* Ω *a unique real number. Class* Ω *is called the* domain *of J and the set of real numbers associated with the functions in* Ω *is called the* range *of J.*

A functional may be considered as a function of a function. A functional J is linear, if is satisfies the homogeneity and additivity properties:

1. $J(a\mathbf{x}) = aJ(\mathbf{x})$, for every $\mathbf{x} \in \Omega, a \in \mathcal{R}$, such that $a\mathbf{x} \in \Omega$ (homogeneity).
2. $J(\mathbf{x}^{(1)} + \mathbf{x}^{(2)}) = J(\mathbf{x}^{(1)}) + J(\mathbf{x}^{(2)})$, for every $\mathbf{x}^{(1)}, \mathbf{x}^{(2)}, \mathbf{x}^{(1)} + \mathbf{x}^{(2)} \in \Omega$ (additivity).

The norm of a function is a rule of correspondence (functional) that assigns a real number to each function $\mathbf{x} \in \Omega$ defined for $t \in [t_0, t_f]$. The norm of $\mathbf{x}$, denoted by $||\mathbf{x}||$, satisfies the following properties:

- $||\mathbf{x}|| \geq 0$ and $||\mathbf{x}|| = 0$ if and only if $\mathbf{x}(t) = 0$ for every $t \in [t_0, t_f]$.
- $||a\mathbf{x}|| = |a| \cdot ||\mathbf{x}||$ for every $a \in \mathcal{R}$.
- $||\mathbf{x}^{(1)} + \mathbf{x}^{(2)}|| \leq ||\mathbf{x}^{(1)}|| + ||\mathbf{x}^{(2)}||$ (triangle inequality).

If $\mathbf{q}$ and $\mathbf{q} + \Delta\mathbf{q}$ are elements in the domain of function f, then the increment of f is defined as

$$\Delta f = \Delta f(\mathbf{q}, \Delta\mathbf{q}) = f(\mathbf{q} + \Delta\mathbf{q}) - f(\mathbf{q}). \tag{C.9}$$

Similarly, if $\mathbf{x}$ and $\mathbf{x} + \delta\mathbf{x}$ are functions in the domain of functional J, then the increment of J is defined as

$$\Delta J = \Delta J(\mathbf{x}, \delta\mathbf{x}) = J(\mathbf{x} + \delta\mathbf{x}) - J(\mathbf{x}). \tag{C.10}$$

Quantity $\delta\mathbf{x}$ is called the variation of function $\mathbf{x}$ and it may be considered analogous to the differential for a function of a single variable. The increment of a function of n variables, which will be of more interest in optimal control problems, can be written as

$$\Delta f(\mathbf{q}, \Delta\mathbf{q}) = df(\mathbf{q}, \Delta\mathbf{q}) + g(\mathbf{q}, \Delta\mathbf{q}) \cdot ||\Delta\mathbf{q}||, \tag{C.11}$$

where df is a linear function of $\Delta\mathbf{q}$. If $\lim_{||\Delta\mathbf{q}|| \to 0} g(\mathbf{q}, \Delta\mathbf{q})) = 0$, f is said to be differentiable at $\mathbf{q}$. The differential is given by

$$df = \frac{\partial f}{\partial q_1}\Delta q_1 + \frac{\partial f}{\partial q_2}\Delta q_2 + \dots + \frac{\partial f}{\partial q_n}\Delta q_n \tag{C.12}$$

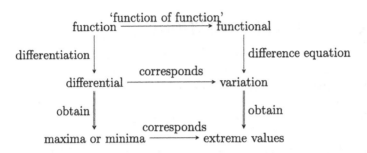

FIGURE C.1

Analogy between functions, functionals, and extreme values.

when f differentiable. Thus, the increment of a functional can be eventually written as

$$\Delta J(\mathbf{x}, \delta\mathbf{x}) = \delta J(\mathbf{x}, \delta\mathbf{x}) + g(\mathbf{x}, \delta\mathbf{x}) \cdot ||\delta\mathbf{x}||, \tag{C.13}$$

where δJ is linear in $\delta\mathbf{x}$. If $\lim_{||\delta\mathbf{x}|| \to 0} g(\mathbf{x}, \delta\mathbf{x})) = 0$, then J is said to be differentiable in $\mathbf{x}$ and δJ is the variation of J evaluated for the function $\mathbf{x}$.

A functional J with domain Ω has a relative extremum at $\mathbf{x}^*$ if there is an $\epsilon > 0$ such that for all functions $\mathbf{x} \in \Omega$ which satisfy $||\mathbf{x} - \mathbf{x}^*|| < \epsilon$ the increment of J has the same sign:

- If $\Delta J = J(\mathbf{x}) - J(\mathbf{x}^*) \geq 0$, $J(\mathbf{x}^*)$ is a relative minimum.
- If $\Delta J = J(\mathbf{x}) - J(\mathbf{x}^*) \leq 0$, $J(\mathbf{x}^*)$ is a relative maximum.

If the above inequalities are satisfied for arbitrarily large ϵ, then $J(\mathbf{x}^*)$ is a global (absolute) minimum/maximum and $\mathbf{x}^*$ is called an extremal, while $J(\mathbf{x}^*)$ is referred to as an extremum.

Theorem C.1 *(Fundamental Theorem of Calculus of Variations). Let $\mathbf{x}$ be a vector function of t in Ω, and $J(\mathbf{x})$ be a differentiable functional of $\mathbf{x}$. Assume that the functions in Ω are not constrained by any boundaries. Then if $\mathbf{x}^*$ is an extremal, the variation of J must vanish at $\mathbf{x}^*$, i.e.*

$$\delta J(\mathbf{x}^*, \delta\mathbf{x}^*) = 0$$

for all admissible $\delta\mathbf{x}$.

Another useful and important result associated with the Fundamental Theorem of Calculus of Variations is the following:

Lemma C.1 *(Fundamental Lemma of Calculus of Variations). If a function h is continuous and $\int_{t_0}^{t_f} h(t)\delta x(t)dt = 0$ for every function δx that is continuous in the interval $[t_0, t_f]$, then h must be zero everywhere in $[t_0, t_f]$.*

Both the Fundamental Theorem of Calculus of Variations, as well as the Fundamental Lemma of Calculus of Variations are frequently employed to ensure optimality of solutions in various optimal control problems. Their proofs can be found in [137] and references therein. Fig. C.1 highlights the correspondence between functions and functionals as "functions of functions" and the correspondence between extrema of functions-extreme values of functionals.

C.3 FINDING TRAJECTORIES THAT MINIMIZE PERFORMANCE MEASURES

The problem classes presented in this section provide characteristic methodologies that can be exploited in broader optimal control problems, some of which have been exploited in this book. The rest can be used in various extensions to the study of intelligent attack strategies. In general, the Fundamental Theorem of Calculus of Variations can be exploited in optimal control theory for finding trajectories that minimize performance measures expressed as functionals. This is also the case in the corresponding modeling of malware diffusion in Chapter 6 and potential extensions to attack strategies.

C.3.1 FUNCTIONALS OF A SINGLE FUNCTION

When the performance metric is a function of a single scalar function x in the class of continuous first derivatives, the functional can be written in the form

$$J(x) = \int_{t_0}^{t_f} g(x(t), \dot{x}(t), t) dt. \tag{C.14}$$

The objective is to find x^* for which $J(x)$ has a relative extremum. Functions (curves) in Ω that satisfy the end conditions are called admissible.

When $x(t_0)$, $x(t_f)$ are free and t_f specified, the functional increment can be written as

$$\Delta J(x, \delta x) = \int_{t_0}^{t_f} [g(x(t) + \delta x(t), \dot{x}(t) + \delta \dot{x}(t), t) - g(x(t), \dot{x}(t), t)] dt, \tag{C.15}$$

which using a Taylor series expansion about $(x(t), \dot{x}(t))$ can be expressed as

$$\delta J(x, \delta x) = \int_{t_0}^{t_f} \left\{ \left[\frac{\partial g}{\partial x}(x(t), \dot{x}(t), t) \right] \delta x(t) + \left[\frac{\partial g}{\partial \dot{x}}(x(t), \dot{x}(t), t) \right] \delta \dot{x}(t) \right\} dt. \tag{C.16}$$

By the fundamental theorem of calculus of variations, a necessary condition for x^* to be an extremal is

$$\frac{\partial g}{\partial x}(x^*(t), \dot{x}^*(t), t) - \frac{d}{dt} \left[\frac{\partial g}{\partial \dot{x}}(x^*(t), \dot{x}^*(t), t) \right] = 0, \quad \forall t \in [t_0, t_f]. \tag{C.17}$$

The latter is an Euler differential equation. More specifically, it is a second order nonlinear ordinary and time-varying differential equation, which is hard to solve since it yields a two-point boundary-value problem with split boundary conditions.

When an extremal for the metric given in (C.15) with $t_0, t_f, x(t_0)$ specified and $x(t_f)$ free (which corresponds to the case where one boundary condition is given and the second is obtained from Euler equation-differential boundary condition) is required, the same approach as before applies as well, where now the metric functional is integrated by parts. It turns out that the necessary condition for x^* to be an extremal is Eq. (C.17), and we may also obtain a natural boundary condition

$$\frac{\partial g}{\partial \dot{x}}(x^*(t_f), \dot{x}^*(t_f), t_f) = 0. \tag{C.18}$$

Thus, an extremal for a free endpoint problem is also an extremal for the fixed endpoint problem with the same endpoints and the same functional.

If one is given metric (C.15) and $t_0, x(t_0) = x_0, x(t_f) = x_f$ are specified while t_f is free, Euler condition (C.17) is satisfied, along with the boundary condition

$$g(x^*(t_f), \dot{x}^*(t_f), t_f) - \left[\frac{\partial g}{\partial \dot{x}}(x^*(t_f), \dot{x}^*(t_f))\right] \dot{x}^*(t_f) = 0.$$

Finally, when an extremal is sought for metric (C.15) with $t_0, x(t_0) = x_0$ specified and $t_f, x(t_f) = x_f$ is free, again Euler condition (C.17) is satisfied, and when t_f and $x(t_f)$ are unrelated, $\delta x_f, \delta t_f$ are independent and arbitrary yielding $g(x^*(t_f), \dot{x}^*(t_f), t_f) = 0$, while if $t_f - x(t_f)$ are related, e.g. in the form $x(t_f) = \theta(t_f)$, then $\delta x_f = \frac{d\theta}{dt}(t_f)\delta t_f$ and the transversality condition

$$\left[\frac{\partial g}{\partial \dot{x}}(x^*(t_f), \dot{x}^*(t_f), t_f)\right]\left[\frac{d\theta}{dt}(t_f) - \dot{x}^*(t_f)\right] + g(x^*(t_f), \dot{x}^*(t_f), t_f) = 0 \tag{C.19}$$

is obtained for solving the problem.

C.3.2 FUNCTIONALS OF SEVERAL INDEPENDENT FUNCTIONS

When the problem involves several independent functions, the performance metric can be written in the form

$$J(\mathbf{x}) = \int_{t_0}^{t_f} g(\mathbf{x}(t), \dot{\mathbf{x}}(t), t)dt, \tag{C.20}$$

with $t_0, t_f, \mathbf{x}(t_0) = \mathbf{x_0}, \mathbf{x}(t_f) = \mathbf{x_f}$ all specified (fixed endpoints). This leads to a more convenient matrix form of the Euler equation

$$\frac{\partial g}{\partial \mathbf{x}}(\mathbf{x}^*(t), \dot{\mathbf{x}}^*(t), t) - \frac{d}{dt}\left[\frac{\partial g}{\partial \dot{\mathbf{x}}}(\mathbf{x}^*(t), \dot{\mathbf{x}}^*(t), t)\right] = 0, \quad \forall t \in [t_0, t_f]. \tag{C.21}$$

When the metric is given by (C.20) and $t_0, \mathbf{x}(t_0) = \mathbf{x}_0$ are specified, but $t_f, \mathbf{x}(t_f) = \mathbf{x}_f$ are free, then Euler condition (C.21) is satisfied along with boundary conditions at the final time of the form

$$
\left[\frac{\partial g}{\partial \mathbf{x}} (\mathbf{x}^*(t_f), \dot{\mathbf{x}}^*(t_f), t_f) \right]^T \delta \mathbf{x}_f
$$

$$
+ \left[g(\mathbf{x}^*(t_f), \dot{\mathbf{x}}^*(t_f), t_f) - \left[\frac{\partial g}{\partial \dot{\mathbf{x}}} (\mathbf{x}^*(t_f), \dot{\mathbf{x}}^*(t_f), t) \right]^T \dot{\mathbf{x}}^*(t_f) \right] \delta t_f = 0. \quad \text{(C.22)}
$$

C.3.3 PIECEWISE-SMOOTH EXTREMALS

When a function has piecewise continuous first derivatives except for a finite number of times (corner points), where $\dot{\mathbf{x}}$ is discontinuous, then the functional of the performance metric can be defined in intervals that do not include the corner points. In this case, necessary conditions for $\mathbf{x}^*$ to be an extremal for J in each of the corresponding time intervals excluding the corner points of the function are the Weierstrass-Erdmann corner conditions, which for functions of many variables, can be written as

$$
\frac{\partial g}{\partial \dot{\mathbf{x}}} (\mathbf{x}^*(t_1), \dot{\mathbf{x}}^*(t_1^-), t_1) = \frac{\partial g}{\partial \dot{\mathbf{x}}} (\mathbf{x}^*(t_1), \dot{\mathbf{x}}^*(t_1^+), t_1) \quad \text{and}
$$

$$
g(\mathbf{x}^*(t_1), \dot{\mathbf{x}}^*(t_1^-), t_1) - \left[\frac{\partial g}{\partial \dot{\mathbf{x}}} (\mathbf{x}^*(t_1), \dot{\mathbf{x}}^*(t_1^-), t_1) \right] \dot{\mathbf{x}}^*(t_1^-)
$$

$$
= g(\mathbf{x}^*(t_1), \dot{\mathbf{x}}^*(t_1^+), t_1) - \left[\frac{\partial g}{\partial \dot{\mathbf{x}}} (\mathbf{x}^*(t_1), \dot{\mathbf{x}}^*(t_1^+), t_1) \right] \dot{\mathbf{x}}^*(t_1^+),
$$

where t_1 is corner point of g.

C.3.4 CONSTRAINED EXTREMA

State-control equations involve finding extrema of functionals of $n + m$ functions, namely, the state $\mathbf{x}$ and controls $\mathbf{u}$, where only m of the functions are independent, i.e. the controls $\mathbf{u}$. For the constrained minimization of functions, two alternative methodologies can be used, namely, the elimination method and the Lagrange multipliers method. The first can be employed if the dependence of variables/functions can be solved with respect to one of them and substituted in the differential $df(\mathbf{y}^*) = 0$.

More specifically, assume the problem of finding the extreme values for a function of $n + m$ variables $y_1, ..., y_{n+m}$ $f(y_1, ..., y_{n+m})$, given n constraints of the form

$$
a_i(y_1, ..., y_{n+m}) = 0, \quad i = 1, 2, \cdots, n,
$$

where only m variables out of the $n + m$ are independent.

Under the elimination method and the above constraints,

$$
y_i = e_i(y_1, ..., y_{n+m}) = 0, \quad i = 1, 2, \cdots, n,
$$

and then by substituting into f, a function $f(y_{n+1}, ..., y_{n+m})$ of m independent variables is obtained. Then, this satisfies the following system of equations:

$$\frac{\partial f}{\partial y_{n+1}}(y_{n+1}^*, ..., y_{n+m}^*) = 0,$$

$$....$$

$$\frac{\partial f}{\partial y_{n+m}}(y_{n+1}^*, ..., y_{n+m}^*) = 0,$$

which can be solved for $y_{n+1}^*, ..., y_{n+m}^*$ and substituted back in $y_1, ..., y_n$ to obtain $y_1^*, ..., y_n^*$. Then $f(y_1^*, ..., y_{n+m}^*)$ can be obtained.

In the Lagrange multipliers method, the augmented function

$$f_a(y_1, ..., y_{n+m}, p_1, ..., p_n) = f(y_1, ..., y_{n+m}) + p_1 [a_1(y_1, ..., y_{n+m})]$$
$$+ ... + p_n [a_n(y_1, ..., y_{n+m})] \qquad \text{(C.23)}$$

is defined and then the differential of the augmented function is obtained

$$df_a = \frac{\partial f_a}{\partial y_1} \Delta y_1 + ... \frac{\partial f_a}{\partial y_{n+m}} \Delta y_{n+m} + \frac{\partial f_a}{\partial p_1} \Delta p_1 + ... + \frac{\partial f_a}{\partial p_1} \Delta p_n$$

$$= \frac{\partial f_a}{\partial y_1} \Delta y_1 + ... \frac{\partial f_a}{\partial y_{n+m}} \Delta y_{n+m} + a_1 \Delta p_1 + ... a_n \Delta p_n,$$

from which $2n + m$ equations from the KKT conditions ([27, 42])

$$a_i(y_1^*, ..., y_{n+m}^*) = 0, \quad i = 1, 2, ..., n,$$

$$\frac{\partial f_a}{\partial y_j}(y_1^*, ..., y_{n+m}^*, p_1^*, ...p_n^*) = 0, \quad j = 1, 2, ..., n + m$$

can be obtained, leading to $(y_1^*, ..., y_{n+m}^*)$.

The above are valid for functions of single or multiple variables. In case of minimization of functionals, however, additional constraints in the form of differential equation or point constraints are required, both of which lead to the necessary condition

$$\frac{\partial g_a}{\partial \mathbf{w}}(\mathbf{w}^*(t), \dot{\mathbf{w}}^*(t), \mathbf{p}^*(t), t) - \frac{d}{dt}\left[\frac{\partial g_a}{\partial \dot{\mathbf{w}}}(\mathbf{w}^*(t), \dot{\mathbf{w}}^*(t), \mathbf{p}^*(t), t)\right] = 0, \qquad \text{(C.24)}$$

where $g_a(\mathbf{w}(t), \dot{\mathbf{w}}(t), \mathbf{p}(t), t) = g(\mathbf{w}(t), \dot{\mathbf{w}}(t), t) + \mathbf{p}^T(t)[\mathbf{f}(\mathbf{w}(t), \dot{\mathbf{w}}(t), t)]$.

Then the general form of the problem is to minimize the functional $J(\mathbf{w}) = \int_{t_0}^{t_f} g(\mathbf{w}(t), \dot{\mathbf{w}}(t), t)dt$ and the problem can be eventually solved using either the elimination or the Lagrange multipliers (augmented functional) method.

For the case of point constraints, the optimal $\mathbf{w}^*$ is required to satisfy constraints of the form $f_i(\mathbf{w}(t), t) = 0, i = 1, 2, ..., n$, where only m out of $n + m$ components of $\mathbf{w}$

are independent. The augmented functional is

$$J_a(\mathbf{w},\mathbf{p}) = \int_{t_0}^{t_f} \{g(\mathbf{w}(t),\dot{\mathbf{w}}(t),t) + p_1(t)[f_1(\mathbf{w}(t),t)] + \dots + p_n(t)[f_n(\mathbf{w}(t),t)]\}\,dt$$

$$= \int_{t_0}^{t_f} \left[g(\mathbf{w},\dot{\mathbf{w}}(t),t) + \mathbf{p}^T(t)[\mathbf{f}(\mathbf{w}(t),t)] \right] dt.$$

On an extremal, the point constraints must be satisfied $\mathbf{f}(\mathbf{w}(t),t) = \mathbf{0}$, as well as $\delta J_a(\mathbf{w}^*,\mathbf{p}) = 0$, where

$$\delta J_a(\mathbf{w},\delta\mathbf{w},\mathbf{p},\delta\mathbf{p}) = \int_{t_0}^{t_f} \left\{ \left[\frac{\partial g^T}{\partial \mathbf{w}}(\mathbf{w}(t),\dot{\mathbf{w}}(t),t) + \mathbf{p}^T(t)\left[\frac{\partial \mathbf{f}}{\partial \mathbf{w}}(\mathbf{w}(t),t) \right] \right] \delta\mathbf{w}(t) \right.$$
$$\left. + \left[\frac{\partial g^T}{\partial \mathbf{w}}(\mathbf{w}(t),\dot{\mathbf{w}}(t),t) \right] \delta\dot{\mathbf{w}}(t) + [\mathbf{f}^T(\mathbf{w}(t),t)]\delta\mathbf{p}(t) \right\} dt,$$

leading to aforementioned condition (C.24).

In the cases when the constraints are given in the form of differential equations $f_i(\mathbf{w}(t),\dot{\mathbf{w}}(t),t) = \mathbf{0}, i = 1,2,\dots,n$, where again only m out of $n + m$ components of $\mathbf{w}$ are independent. In this case, the elimination approach is almost impossible, and Lagrange multipliers have to be adopted. The augmented functional is as before and the conditions on the extremals $\mathbf{f}(\mathbf{w}^*(t),\dot{\mathbf{w}}^*(t),t) = \mathbf{0}$ and $\delta J_a(\mathbf{w}^*,\mathbf{p}) = 0$ again yield condition (C.24), where

$$\delta J_a(\mathbf{w},\delta\mathbf{w},\mathbf{p},\delta\mathbf{p}) = \int_{t_0}^{t_f} \left\{ \left[\frac{\partial g^T}{\partial \mathbf{w}}(\mathbf{w}(t),\dot{\mathbf{w}}(t),t) + \mathbf{p}^T(t)\left[\frac{\partial \mathbf{f}}{\partial \mathbf{w}}(\mathbf{w}(t),\dot{\mathbf{w}}(t),t) \right] \right] \delta\mathbf{w}(t) \right.$$
$$+ \left[\frac{\partial g^T}{\partial \dot{\mathbf{w}}}(\mathbf{w}(t),\dot{\mathbf{w}}(t),t) + \mathbf{p}^T(t)\left[\frac{\partial \mathbf{f}}{\partial \dot{\mathbf{w}}}(\mathbf{w}(t),\dot{\mathbf{w}}(t),t) \right] \right] \delta\dot{\mathbf{w}}(t)$$
$$\left. + [\mathbf{f}^T(\mathbf{w}(t),\dot{\mathbf{w}}(t),t)]\delta\mathbf{p}(t) \right\} dt.$$

Finally, the extremal $\mathbf{w}^*$ for the functional could be subject to isoperimetric[4] constraints of the form $\int_{t_0}^{t_f} e_i(\mathbf{w}(t),\dot{\mathbf{w}}(t),t)dt = c_i, i = 1,2,\dots,n$. In this case, the auxiliary variable $z_i = \int_{t_0}^{t_f} e_i(\mathbf{w}(t),\dot{\mathbf{w}}(t),t)dt = c_i$ is defined, so that in vector form $\mathbf{z}(t) = \mathbf{e}(\mathbf{w}(t),\dot{\mathbf{w}}(t),t)$ with boundary values $z_i(t_0) = 0$ and $z_i(t_f) = c_i$. In this case, the augmented functional becomes

$$g_a(\mathbf{w}(t),\dot{\mathbf{w}}(t),\mathbf{p}(t),t) = g(\mathbf{w}(t),\dot{\mathbf{w}}(t),t) + \mathbf{p}^T(t)\left[\mathbf{e}(\mathbf{w}(t),\dot{\mathbf{w}}(t),t) - \dot{\mathbf{z}}(t)\right].$$

Then, the overall system can be solved by using the $n+m$ available equations obtained from

$$\frac{\partial g_a}{\partial \mathbf{w}}(\mathbf{w}^*(t),\dot{\mathbf{w}}^*(t),\mathbf{p}^*(t),\dot{\mathbf{z}}^*(t),t) - \frac{d}{dt}\left[\frac{\partial g_a}{\partial \dot{\mathbf{w}}}(\mathbf{w}^*(t),\dot{\mathbf{w}}^*(t),\mathbf{p}^*(t),\dot{\mathbf{z}}^*(t),t) \right] = \mathbf{0},$$

---

[4]The isoperimetric problem is to determine a plane figure of the largest possible area whose boundary has a specified length.

the r equations available from the following:

$$\frac{\partial g_a}{\partial \mathbf{z}}(\mathbf{w}^*(t),\dot{\mathbf{w}}^*(t),\mathbf{p}^*(t),\dot{\mathbf{z}}^*(t),t) - \frac{d}{dt}\left[\frac{\partial g_a}{\partial \dot{\mathbf{z}}}(\mathbf{w}^*(t),\dot{\mathbf{w}}^*(t),\mathbf{p}^*(t),\dot{\mathbf{z}}^*(t),t)\right] = \mathbf{0}$$

and the r equations available from

$$\mathbf{z}(t) = \mathbf{e}(\mathbf{w}(t),\dot{\mathbf{w}}(t),t),$$

while in addition it holds that $\dot{\mathbf{p}}(t) = \mathbf{0}$.

C.4 VARIATIONAL APPROACH FOR OPTIMAL CONTROL PROBLEMS

The variational approach can be used to find the admissible $\mathbf{u}^*$ causing the system $\dot{\mathbf{x}}(t) = \mathbf{a}(\mathbf{x}(t),\mathbf{u}(t),t)$ to follow the admissible $\mathbf{x}^*$ that minimizes the performance metric $J(\mathbf{u}) = h(\mathbf{x}(t_f),t_f) + \int_{t_0}^{t_f} g(\mathbf{x}(t),\mathbf{u}(t),t)dt$. This is the optimal control problem form that will be mostly needed in Chapter 6.

C.4.1 NECESSARY CONDITIONS FOR OPTIMAL CONTROL

The state-control regions are initially considered unbounded, while $x(t_0) = x_0$ and t_0 are explicitly specified. The problem has $n + m$ functions satisfying n differential equations constraints, while the m control inputs are independent. By applying the Lagrange multipliers approach, the augmented function becomes

$$g_a(\mathbf{x}(t),\dot{\mathbf{x}}(t),\mathbf{u}(t),\mathbf{p}(t),t) = g(\mathbf{x}(t),\mathbf{u}(t),t) + \mathbf{p}^T(t)\left[\mathbf{a}(\mathbf{x}(t),\mathbf{u}(t),t) - \dot{\mathbf{x}}(t)\right]$$

$$+ \left[\frac{\partial h}{\partial \mathbf{x}}(\mathbf{x}(t),t)\right]^T + \frac{\partial h}{\partial t}(\dot{\mathbf{x}}(t),t),$$

where $\mathbf{p}(t)$ are the Lagrange multipliers. The augmented functional in this case becomes

$$J_a(\mathbf{u}) = \int_{t_0}^{t_f} g_a(\mathbf{x}(t),\dot{\mathbf{x}}(t),\mathbf{u}(t),\mathbf{p}(t),t)dt.$$

The condition for an extremal is that $\delta J_a(\mathbf{u}^*) = 0$ regardless of the boundary conditions. Following some algebraic manipulation of the terms of the integral involving only function h and the chain rule, the definition of the Hamiltonian function, denoted by

$$\mathcal{H}(\mathbf{x}(t),\mathbf{u}(t),\mathbf{p}(t),t) = g(\mathbf{x}(t),\mathbf{u}(t),t) + \mathbf{p}^T(t)\left[\mathbf{a}(\mathbf{x}(t),\mathbf{u}(t),t)\right] \qquad (C.25)$$

and by the requirement that the integral must vanish on an extremal, the following necessary conditions can be obtained:

$$\dot{\mathbf{x}}^*(t) = \frac{\partial \mathcal{H}}{\partial \mathbf{p}}(\mathbf{x}^*(t),\mathbf{u}^*(t),\mathbf{p}^*(t),t),$$

$$\dot{\mathbf{p}}^*(t) = -\frac{\partial \mathcal{H}}{\partial \mathbf{p}}(\mathbf{x}^*(t), \mathbf{u}^*(t), \mathbf{p}^*(t), t), \tag{C.26}$$

$$0 = \frac{\partial \mathcal{H}}{\partial \mathbf{p}}(\mathbf{x}^*(t), \mathbf{u}^*(t), \mathbf{p}^*(t), t),$$

for all $t \in [t_0, t_f]$, and

$$\left[\frac{\partial h}{\partial \mathbf{x}}(\mathbf{x}^*(t_f), t_f) - \mathbf{p}^*(t_f) \right]^T \delta \mathbf{x}_f$$

$$+ \left[\mathcal{H}(\mathbf{x}^*(t_f), \mathbf{u}^*(t_f), \mathbf{p}^*(t_f), t_f) + \frac{\partial h}{\partial t}(\mathbf{x}^*(t_f), t_f) \right] \delta t_f = \mathbf{0}. \tag{C.27}$$

According to the different objectives and types of problems, various boundary conditions may be available. The latter can vary from problems with fixed final time to problems with free final time and various subcases. More details on these types of problems and the corresponding boundary conditions that may be available can be found in [137].

C.4.2 PONTRYAGIN'S MINIMUM PRINCIPLE

When the controls and states are bounded, then $\delta J(\mathbf{u}^*, \delta \mathbf{u}) \geq \mathbf{0}$ if $\mathbf{u}^*$ lies on the boundary during any portion of the time interval $[t_0, t_f]$. On the other hand, $\delta J(\mathbf{u}^*, \delta \mathbf{u}) = \mathbf{0}$ if $\mathbf{u}^*$ lies within the boundary for the entire time interval $[t_0, t_f]$. By considering the form of $\Delta J(\mathbf{u}^*, \delta \mathbf{u})$ and segregating higher-order terms, in order for $\mathbf{u}^*$ to be a minimum control, it is necessary that

$$\int_{t_0}^{t_f} [\mathcal{H}(\mathbf{x}^*(t), \mathbf{u}^*(t) + \delta \mathbf{u}(t), \mathbf{p}^*(t), t) - \mathcal{H}(\mathbf{x}^*(t), \mathbf{u}^*(t), \mathbf{p}^*(t), t)] \, dt \geq 0,$$

for all admissible $\delta \mathbf{u}(t)$ and for all $t \in [t_0, t_f]$.

In terms of the Hamiltonian, the necessary conditions for $\mathbf{u}^*$ to be an optimal control are given by (C.26) and (C.27), and $\mathbf{u}^*$ is a control causing the Hamiltonian to assume a global minimum value. These conditions are necessary, but not in general sufficient.

In the unbounded case, a necessary but not sufficient condition for $\mathbf{u}^*$ to minimize the Hamiltonian is

$$\frac{\partial \mathcal{H}}{\partial \mathbf{u}}(\mathbf{x}^*(t), \mathbf{u}^*(t), \mathbf{p}^*(t), t) = 0. \tag{C.28}$$

If $\frac{\partial^2 \mathcal{H}}{\partial \mathbf{u}^2}(\mathbf{x}^*(t), \mathbf{u}^*(t), \mathbf{p}^*(t), t)$ is positive definite, then $\mathbf{u}^*(t)$ causes $\mathcal{H}$ to be a local minimum. If $\mathcal{H}(\mathbf{x}(t), \mathbf{u}(t), \mathbf{p}(t), t) = f(\mathbf{x}(t), \mathbf{u}(t), \mathbf{p}(t), t) + [\mathbf{c}(\mathbf{x}(t), \mathbf{p}(t), t)]^T \mathbf{u}(t) + \frac{1}{2}\mathbf{u}^T(t)\mathbf{R}(t)\mathbf{u}(t)$, where $\mathbf{c}$ is an m-long vector not depending on $\mathbf{u}(t)$, then the above are necessary and sufficient conditions for $\mathcal{H}(\mathbf{x}^*(t), \mathbf{u}^*(t), \mathbf{p}^*(t), t)$ to be a global minimum.

Additional necessary conditions may be obtained if the final time is fixed and $\mathcal{H}(.)$ does not depend explicitly on time. Then $\mathcal{H}(.)$ must be a constant when evaluated on an extremal trajectory $\mathcal{H}(\mathbf{x}^*(t), \mathbf{u}^*(t), \mathbf{p}^*(t)) = c_1, t \in [t_0, t_f]$. If the final time is free and $\mathcal{H}(.)$ does not explicitly depend on time, then $\mathcal{H}(.)$ must be identically zero when evaluated on an extremal trajectory $\mathcal{H}(\mathbf{x}^*(t), \mathbf{u}^*(t), \mathbf{p}^*(t)) = 0$, for $t \in [t_0, t_f]$.

When state variable inequality constraints of the form $\mathbf{f}(\mathbf{x}(t), t) \geq \mathbf{0}$, where $\mathbf{f}$ is a ℓ-vector function, are given, then the additional state

$$\dot{x}_{n+1}(t) = [f_1(\mathbf{x}(t), t)]^2 \mathbb{1}(-f_1) + [f_2(\mathbf{x}(t), t)]^2 \mathbb{1}(-f_2) + \ldots + [f_\ell(\mathbf{x}(t), t)]^2 \mathbb{1}(-f_\ell) \quad \text{(C.29)}$$

can be defined, where

$$\mathbb{1}(-f_i) = \begin{cases} 0 & \text{for } f_i(\mathbf{x}(t), t) \geq 0 \\ 1 & \text{for } f_i(\mathbf{x}(t), t) < 0 \end{cases} \quad \text{(C.30)}$$

and require $x_{n+1}(t) = \int_{t_0}^{t_f} \dot{z}_{n+1}(t) dt + x_{n+1}(t_0)$, satisfying $x_{n+1}(t_0) = 0$ and $x_{n+1}(t_f) = 0$. $\dot{x}_{n+1}(t)$ must be zero throughout the interval $[t_0, t_f]$, but this occurs only if the constraints are satisfied for all $t \in [t_0, t_f]$.

The Hamiltonian can be cast in the form $\mathcal{H}(\mathbf{x}(t), \mathbf{u}(t), \mathbf{p}(t), t) = g(\mathbf{x}(t), \mathbf{u}(t), t) + \mathbf{p}^T(t) \mathbf{a}(\mathbf{x}(t), \mathbf{u}(t), t)$, where $a_{n+1}(\mathbf{x}(t), t) = [f_1(\mathbf{x}(t), t)]^2 \mathbb{1}(-f_1) + \ldots + [f_\ell(\mathbf{x}(t), t)]^2 \mathbb{1}(-f_\ell)$, and $\mathbf{x}(t), \mathbf{p}(t)$ are $(n+1)$-vectors in this case. The necessary conditions for optimality become

$$\dot{x}_1^*(t) = a_1(\mathbf{x}^*(t), \mathbf{u}^*(t), t), \quad \text{(C.31)}$$

$$\ldots$$

$$\dot{x}_{n+1}^*(t) = a_{n+1}(\mathbf{x}^*(t), \mathbf{u}^*(t), t), \quad \text{(C.32)}$$

$$\dot{p}_1^*(t) = -\frac{\partial \mathcal{H}}{\partial x_1}(\mathbf{x}^*(t), \mathbf{u}^*(t) \mathbf{p}^*(t), t), \quad \text{(C.33)}$$

$$\ldots$$

$$\dot{p}_{n+1}^*(t) = -\frac{\partial \mathcal{H}}{\partial x_{n+1}}(\mathbf{x}^*(t), \mathbf{u}^*(t), \mathbf{p}^*(t), t), \quad \text{(C.34)}$$

$$\mathcal{H}(\mathbf{x}^*(t), \mathbf{u}^*(t), \mathbf{p}^*(t), t) \leq \mathcal{H}(\mathbf{x}^*(t), \mathbf{u}(t), \mathbf{p}^*(t), t), \quad \text{(C.35)}$$

for all admissible $\mathbf{u}(t)$, for all $t \in [t_0, t_f]$ with boundary conditions $x_{n+1}^*(t_0) = 0$, $x_{n+1}^*(t_f) = 0$.

C.4.3 MINIMUM-TIME PROBLEMS

In this class of problems, called minimum-time problems, the objective is to transfer the system $\dot{\mathbf{x}}(t) = \mathbf{a}(\mathbf{x}(t), \mathbf{u}(t), t)$ from arbitrary initial state x_0 to a specified target set $S(t)$ in minimum time t^*. This corresponds to minimizing the performance functional $J(\mathbf{u}) = \int_{t_0}^{t_f} dt = t_f - t_0$. Typically, $|u_i(t)| \leq 1, i = 1, 2, \ldots, m, t \in [t_0, t^*]$. The optimal control policy can be obtained by using the minimum principle.

If a system with initial state $\mathbf{x}(t_0) = \mathbf{x}_0$ is subject to all admissible control histories for a time interval $[t_0, t]$, the collection of state values $\mathbf{x}(t)$ is called the set of states that are *reachable* (from $\mathbf{x}_0$ at time t), or simply the set of reachable states.

Assuming the system $\dot{\mathbf{x}}(t) = \mathbf{a}(\mathbf{x}(t),t) + \mathbf{B}(\mathbf{x}(t),t)\mathbf{u}(t)$, where $\mathbf{B}$ is an $n \times m$ matrix, and $M_{-i} \leq u_i(t) \leq M_{+i}, i = 1,2,...,m, t \in [t_0,t^*]$, the form of the Hamiltonian becomes $\mathcal{H}(\mathbf{x}(t),\mathbf{u}(t),\mathbf{p}(t),t) = 1 + \mathbf{p}^T(t)[\mathbf{a}(\mathbf{x}(t),t) + \mathbf{B}(\mathbf{x}(t),t)\mathbf{u}(t)]$, yielding a necessary condition for an extremal

$$\mathbf{p}^{*^T}(t)\mathbf{B}(\mathbf{x}^*(t),t)\mathbf{u}^*(t) \leq \mathbf{p}^{*^T}(t)\mathbf{B}(\mathbf{x}^*(t),t)\mathbf{u}(t), \tag{C.36}$$

where $\mathbf{u}^*(t)$ is the control that causes $\mathbf{p}^{*^T}(t)\mathbf{B}(\mathbf{x}^*(t),t)\mathbf{u}(t)$ to assume its minimum value.

If $\mathbf{B}(\mathbf{x}^*(t),t) = [\mathbf{b}_1(\mathbf{x}^*(t),t)|...|\mathbf{b}_m(\mathbf{x}^*(t),t)]$, then in this case,

$$\mathbf{p}^{*^T}(t)\mathbf{B}(\mathbf{x}^*(t),t)\mathbf{u}(t) = \sum_{i=1}^{m} \mathbf{p}^{*^T}(t)[\mathbf{b}_i(\mathbf{x}^*(t),t)\mathbf{u}_i(t). \tag{C.37}$$

If the control components are independent of one another, $\mathbf{p}^{*^T}(t)[\mathbf{b}_i(\mathbf{x}^*(t),t)\mathbf{u}_i(t)$ must be minimum with respect to $u_i(t)$ for $i = 1,2,...,m$. If the $u_i(t)$ coefficient is positive, then $u_i^*(t)$ must be the smallest possible, while if the $u_i(t)$ coefficient is negative, then $u_i^*(t)$ must be the largest possible. The latter indicates a characteristic optimal control policy, which is widely known as "bang-bang" type of policy, where the control exercised is either maximum or minimum control, in order to attain optimality as fast as possible,

$$u_i^*(t) = \begin{cases} M_{i+}, \text{ if } \mathbf{p}^{*^T}(t)\mathbf{B}(\mathbf{x}^*(t),t) < 0 \\ M_{i-}, \text{ if } \mathbf{p}^{*^T}(t)\mathbf{B}(\mathbf{x}^*(t),t) > 0 \\ \text{undetermined, if } \mathbf{p}^{*^T}(t)\mathbf{B}(\mathbf{x}^*(t),t) = 0 \end{cases} \cdot \tag{C.38}$$

If $\mathbf{p}^{*^T}(t)\mathbf{B}(\mathbf{x}^*(t),t) = 0$ is zero for some finite time interval, no information can be obtained for selecting $u_i^*(t)$.

C.4.4 MINIMUM CONTROL-EFFORT PROBLEMS

In the problems entitled "minimum control-effort" problems the objective is to drive the system $\dot{\mathbf{x}}(t) = \mathbf{a}(\mathbf{x}(t),\mathbf{u}(t),t)$ from an arbitrary initial state x_0 to a specified target set $S(t)$ with minimum expenditure of control effort, $M_{-i} \leq u_i(t) \leq M_{+i}, i = 1,2,...,m$, expressed in the form of minimum "fuel" problems with $J_1(\mathbf{u}) = \int_{t_0}^{t_f} \left[\sum_{i=1}^{m} \beta_i |u_i(t)| \right] dt$, or minimum "energy" problems with performance functional $J_2(\mathbf{u}) = \int_{t_0}^{t_f} \left[\sum_{i=1}^{m} r_i |u_i^2(t)| \right] dt$, where β_i, r_i are non-negative weights. The "fuel" terminology has been derived from traditional, airplane, and rocket control problems, where the objective is to minimize the amount of actual fuel or energy supplies required for the transferring of the mass from one location to another.

The characteristics of minimum energy problems are similar to those of minimum fuel problems and in the following we focus on the second class of problems. For minimum fuel problems, the solution is given by the intersection of the target set $S(t)$ with a set of reachable states $R(t)$, which requires the smallest amount of consumed "fuel." Assuming the system $\dot{\mathbf{x}}(t) = \mathbf{a}(\mathbf{x}(t),t) + \mathbf{B}(\mathbf{x}(t),t)\mathbf{u}(t)$, with $\mathbf{B}^{n\times m}$, $-1 \leq u_i(t) \leq 1, i = 1,2,...,m, t \in [t_0,t_f]$, and $J(\mathbf{u}) = \int_{t_0}^{t_f} \left[\sum_{i=1}^{m} |u_i(t)|\right] dt$, the Hamiltonian will have the form

$$\mathcal{H}(\mathbf{x}(t),\mathbf{u}(t),\mathbf{p}(t),t) = \sum_{i=1}^{m} |u_i(t)| + \mathbf{p}^T(t)\mathbf{a}(\mathbf{x}(t),t) + \mathbf{p}^T(t)\mathbf{B}(\mathbf{x}(t),t)\mathbf{u}(t), \quad (C.39)$$

according to the aforementioned minimum principle, it is required for all admissible $\mathbf{u}(t)$ and for all $t \in [t_0,t_f]$ that

$$\sum_{i=1}^{m} |u_i^*(t)| + \mathbf{p}^{*^T}(t)\mathbf{a}(\mathbf{x}^*(t),t) + \mathbf{p}^{*^T}(t)\mathbf{B}(\mathbf{x}^*(t),t)\mathbf{u}^*(t)$$

$$\leq \sum_{i=1}^{m} |u_i(t)| + \mathbf{p}^{*^T}(t)\mathbf{a}(\mathbf{x}^*(t),t) + \mathbf{p}^{*^T}(t)\mathbf{B}(\mathbf{x}^*(t),t)\mathbf{u}(t). \quad (C.40)$$

Assuming $\mathbf{B}(\mathbf{x}^*(t),t) = [\mathbf{b}_1(\mathbf{x}^*(t),t)|...|\mathbf{b}_m(\mathbf{x}^*(t),t)]$, and if the components of $\mathbf{u}$ are independent of one another, allows obtaining

$$|u_i^*(t)| + \mathbf{p}^{*^T}(t)\mathbf{b}_i(\mathbf{x}^*(t),t)\mathbf{u}^*(t) \leq |u_i(t)| + \mathbf{p}^{*^T}(t)\mathbf{b}_i(\mathbf{x}^*(t),t)\mathbf{u}(t),$$

for $i = 1,2,...,m$, which yields

$$|u_i(t)| = \begin{cases} u_i(t) & u_i(t) \geq 0 \\ -u_i(t) & u_i(t) \leq 0 \end{cases} \quad (C.41)$$

and therefore,

$$|u_i(t)| + \mathbf{p}^{*^T}(t)\mathbf{b}_i(\mathbf{x}^*(t),t)u_i(t) = \begin{cases} \left[1 + \mathbf{p}^{*^T}(t)\mathbf{b}_i(\mathbf{x}^*(t),t)\right]u_i(t) & \text{for } u_i(t) \geq 0 \\ \left[-1 + \mathbf{p}^{*^T}(t)\mathbf{b}_i(\mathbf{x}^*(t),t)\right]u_i(t) & \text{for } u_i(t) \leq 0 \end{cases}. \quad (C.42)$$

The optimal control is eventually obtained in the form of a bang-off-bang policy (a bang-off-bang policy means that the control policy alternates between bang-bang type controls—namely, on/off optimal controls—and other free forms of controls; see [137] and references therein, as well as some types of controls in Chapter 6), such as the following:

$$u_i^*(t) = \begin{cases} 1, & \text{for } \mathbf{p}^{*^T}(t)\mathbf{b}_i(\mathbf{x}^*(t),t) < -1 \\ 0, & \text{for } -1 < \mathbf{p}^{*^T}(t)\mathbf{b}_i(\mathbf{x}^*(t),t) < 1 \\ -1, & \text{for } 1 < \mathbf{p}^{*^T}(t)\mathbf{b}_i(\mathbf{x}^*(t),t). \\ \text{undetermined non-negative value, if } \mathbf{p}^{*^T}(t)\mathbf{b}_i(\mathbf{x}^*(t),t) = -1 \\ \text{undetermined non-positive value, if } \mathbf{p}^{*^T}(t)\mathbf{b}_i(\mathbf{x}^*(t),t) = 1 \end{cases}$$

$$(C.43)$$

C.4.5 SINGULAR INTERVALS IN OPTIMAL CONTROL PROBLEMS

In many problems, the case where there exists an interval $[t_1, t_2]$ of finite duration during which the Pontryagin's necessary condition, given by

$$\mathcal{H}(\mathbf{x}^*(t), \mathbf{u}^*(t), \mathbf{p}^*(t), t) \le \mathcal{H}(\mathbf{x}^*(t), \mathbf{u}(t), \mathbf{p}^*(t), t), \tag{C.44}$$

for all $t \in [t_0, t_f]$ and all admissible $\mathbf{u}(t)$, does not provide information about $\mathbf{x}^*(t), \mathbf{u}^*(t), \mathbf{p}^*(t)$. Then the problem is singular, and $[t_1, t_2]$ is a singular interval.

Assume the following singular interval minimum-time optimal problem that requires driving the system $\dot{\mathbf{x}}(t) = \mathbf{A}\mathbf{x}(t) + \mathbf{b}u(t)$ from $\mathbf{x}_0$ at $t = 0$ to $S(t)$ in minimum time, $|u(t)| \le 1$. Then, the Hamiltonian of the system becomes

$$\mathcal{H}(\mathbf{x}(t), u(t), \mathbf{p}(t)) = 1 + \mathbf{p}^T(t)\mathbf{A}\mathbf{x}(t) + \mathbf{p}^T(t)\mathbf{b}u(t), \tag{C.45}$$

which according to the necessary condition and since the final time is free yields

$$\mathcal{H}(\mathbf{x}^*(t), \mathbf{u}^*(t), \mathbf{p}^*(t)) = 1 + \mathbf{p}^{*T}(t)\mathbf{A}\mathbf{x}^*(t) + \mathbf{p}^{*T}(t)\mathbf{b}u^*(t) = 0. \tag{C.46}$$

Now interval $[t_1, t_2]$ is singular if $\mathbf{p}^{*T}(t)\mathbf{b} = 0$ for all $t \in [t_1, t_2]$. However, $\mathbf{p}^{*T}(t) \ne \mathbf{0}$ for all $t \in [0, t_f]$, and $\mathbf{b} \ne \mathbf{0}$. Thus, $\mathbf{p}^{*T}(t)\mathbf{b} = 0$ and so $\frac{d^k}{dt^k}\left[\mathbf{p}^{*T}(t)\mathbf{b}\right] = 0$, for $k = 1, 2, ...$, while $\frac{d^k}{dt^k}\left[\mathbf{p}^{*T}(t)\mathbf{b}\right] = \frac{d^k}{dt^k}\left[\mathbf{p}^{*T}(t)\right]\mathbf{b} = \mathbf{p}^{*T(k)}\mathbf{b}$.

From the Hamiltonian, it is obtained that $\dot{\mathbf{p}}^*(t) = -\mathbf{A}^T\mathbf{p}^*(t)$, yielding $\mathbf{p}^*(t) = e^{-\mathbf{A}^T t}\mathbf{c}$, where $\mathbf{c}$ is the vector of initial costate variables. Differentiating this result, one obtains $\dot{\mathbf{p}}^{*T}(t)\mathbf{b} = -\left[\mathbf{A}^T e^{-\mathbf{A}^T t}\mathbf{c}\right]^T \mathbf{b} = 0$, which means that $\left[e^{-\mathbf{A}^T t}\mathbf{c}\right]^T \mathbf{A}\mathbf{b} = \mathbf{0}$. Similarly, $\ddot{\mathbf{p}}^{*T}(t) = -\mathbf{A}^T \dot{\mathbf{p}}^*(t)$, which yields $\left[e^{-\mathbf{A}^T t}\mathbf{c}\right]^T \mathbf{A}^2\mathbf{b} = \mathbf{0}$, and in general, for $k = 0, 1, 2, ...$, one obtains $\left[e^{-\mathbf{A}^T t}\mathbf{c}\right]^T \mathbf{A}^{n-1}\mathbf{b} = \mathbf{0}$. By putting all of the above together yields that the following matrix $\mathbf{E} = [\mathbf{b}|\mathbf{A}\mathbf{b}|\mathbf{A}^2\mathbf{b}|...|\mathbf{A}^{n-1}\mathbf{b}]$ must be singular.

In summary, in linear, stationary, and minimum-time problems, for a singular interval to exist, it is necessary that the system is uncontrollable. On the contrary, if a system is completely controllable, a singular interval cannot exist. Finally, if $\mathbf{E}$ is singular, a singular interval must exist.

In case of existence of singular intervals in linear fuel-optimal problems, i.e. problems of the form $\dot{\mathbf{x}}(t) = \mathbf{A}\mathbf{x}(t) + \mathbf{b}u(t)$, $|u(t)| \le 1$ to be driven from $\mathbf{x}_0$ to $S(t)$, with a performance metric $J(u) = \int_0^{t_f} |u(t)|dt$, where t_f is free, the Hamiltonian is given by $\mathcal{H}(\mathbf{x}(t), u(t), \mathbf{p}(t)) = |u(t)| + \mathbf{p}^T(t)\mathbf{A}\mathbf{x}(t) + \mathbf{p}^T(t)\mathbf{b}u(t)$. According to the necessary condition $\mathbf{p}^{*T}(t)\mathbf{b} = 1$ that holds for all $t \in [t_1, t_2]$ or the condition $\mathbf{p}^{*T}(t)\mathbf{b} = -1$ that holds for all $t \in [t_1, t_2]$. Then $\mathbf{p}^{*T}(t)\mathbf{A}\mathbf{x}^*(t) = 0$, which implies $\frac{d^k}{dt^k}\left[\mathbf{p}^{*T}(t)\mathbf{b}\right] = 0$, for $k = 1, 2, ...$ and $t \in [t_1, t_2]$. Using the necessary condition $\mathcal{H}(\mathbf{x}^*(t), \mathbf{u}^*(t), \mathbf{p}^*(t)) = 0$, it is obtained that $[\mathbf{b}|\mathbf{A}\mathbf{b}|....|\mathbf{A}^{n-1}\mathbf{b}]^T \mathbf{A}^T$ is singular. Consequently, in order for a singular interval to exist, either the studied system must not be completely controllable or matrix $\mathbf{A}$ is singular.

For general systems, one should examine the Hamiltonian $\mathcal{H}(\mathbf{x}(t), \mathbf{u}(t), \mathbf{p}(t))$ to determine if there are situations in which the minimum principle does not yield sufficient information to determine the relationship between $\mathbf{x}^*(t), \mathbf{u}^*(t), \mathbf{p}^*(t)$. If this occurs, then setting $\mathcal{H}(\mathbf{x}(t), \mathbf{u}(t), \mathbf{p}(t)) = 0$ and requiring that $\dot{\mathcal{H}}(\mathbf{x}(t), \mathbf{u}(t), \mathbf{p}(t)), \ddot{\mathcal{H}}(\mathbf{x}(t), \mathbf{u}(t), \mathbf{p}(t))$ equal zero, the necessary conditions for the existence of singular intervals are determined. The existence of singular intervals, although complicating the solution of optimal control problems, may turn out to be helpful, e.g. indicate that an optimal control in nonunique, implying various outcomes for system behavior.

C.5 NUMERICAL DETERMINATION OF OPTIMAL TRAJECTORIES

In the previously presented approach (variational approach to optimal control problems), the problems usually emerge as nonlinear, two-point boundary-value problems, which also typically cannot be solved analytically to obtain an optimal control law, or an optimal open-loop control. In this case, iterative numerical techniques may be used to determine open-loop optimal controls. In this subsection, we briefly present some candidate numerical approaches that can be used in order to compute optimal control laws, or optimal open-loop controls whenever this is not analytically possible. We present the general concepts of these numerical approaches and point toward suitable references for detailed presentations for the more interested reader.

The general problem addressed is given by the following system of equations:

$$\dot{\mathbf{x}}^*(t) = \frac{\partial \mathcal{H}}{\partial \mathbf{p}} = \mathbf{a}(\mathbf{x}^*(t), u^*(t), t),$$

$$\dot{\mathbf{p}}^*(t) = -\frac{\partial \mathcal{H}}{\partial \mathbf{x}} = -\left[\frac{\partial \mathbf{a}}{\partial \mathbf{x}}(\mathbf{x}^*(t), \mathbf{u}^*(t), t)\right]^T \mathbf{p}^*(t) - \frac{\partial g}{\partial \mathbf{x}}(\mathbf{x}^*(t), \mathbf{u}^*(t), t),$$

$$\mathbf{0} = \frac{\partial \mathcal{H}}{\partial \mathbf{u}} = \left[\frac{\partial \mathbf{a}}{\partial \mathbf{u}}(\mathbf{x}^*(t), \mathbf{u}^*(t), t)\right]^T \mathbf{p}^*(t) + \frac{\partial g}{\partial \mathbf{u}}(\mathbf{x}^*(t), \mathbf{u}^*(t), t), \qquad \text{(C.47)}$$

$$\mathbf{x}(t_0) = \mathbf{x}_0,$$

$$\mathbf{p}^*(t_f) = \frac{\partial h}{\partial \mathbf{x}}(\mathbf{x}^*(t_f)).$$

It considers unbounded states/controls with t_f, $x(t_f)$ free. From the system of state/costate equations and boundary conditions, it is possible to obtain an explicit relationship between $\mathbf{x}^*(t)$ and $\mathbf{u}^*(t)$, $t \in [t_0, t_f]$ dependent on $\mathbf{x}_0$. Assuming $\mathbf{u}^*(t) = f(\mathbf{x}^*(t), \mathbf{p}^*(t), t)$ and substituted in the state/costate equations, a set of $2n$ first-order ODEs (actually reduced ODEs can be obtained and considered with the boundary conditions $\mathbf{x}(t_0) = \mathbf{x}_0$ and $\mathbf{p}^*(t_f) = \frac{\partial h}{\partial \mathbf{x}}(\mathbf{x}^*(t_f))$. If the boundary conditions are all known at t_0 or t_f, then the optimal control law $\mathbf{x}^*(t), \mathbf{p}^*(t), t \in [t_0, t_f]$ can be obtained by numerical integration and the optimal control history can be found by substitution in the previous relation.

The basic strategy in all of following three numerical approaches considered (steepest descent, variation of extremals, and quasilinearization) is to initially guess a solution so that one of the three set of equations (state/costate/Hamiltonian) is not satisfied. Then a solution is used to adjust the initial guess, i.e. make the next solution closer to satisfying all necessary conditions. Finally, the steps are repeated until the iterative procedure converges.

When the boundary values are split[5] (for nonlinear equations), the above cannot be applied and the gradient projection is a more suitable approach.

C.5.1 STEEPEST DESCENT

For a function of multiple variables to have a relative minimum at a point, it is necessary that the gradient of the function is zero at this point, which in general yields a system of nonlinear differential equations. Suppose $\mathbf{u}^{(i)}(t), t \in [t_0, t_f]$ is known and used to solve

$$\dot{\mathbf{x}}^{(i)}(t) = \mathbf{a}(\mathbf{x}^{(i)}(t), \mathbf{u}^{(i)}(t), t), \quad \mathbf{x}^{(i)}(t_0) = \mathbf{x}_0, \tag{C.48}$$

$$\dot{\mathbf{p}}^{(i)}(t) = -\frac{\partial \mathcal{H}}{\partial \mathbf{x}}(\mathbf{x}^{(i)}(t), \mathbf{u}^{(i)}(t), \mathbf{p}^{(i)}(t), t), \quad \mathbf{p}^{(i)}(t_f) = \frac{\partial h}{\partial \mathbf{x}}(\mathbf{x}^{(i)}(t_f)). \tag{C.49}$$

If $\mathbf{u}^{(i)}(t)$ satisfies $\frac{\partial h}{\partial \mathbf{x}}(\mathbf{x}^{(i)}(t_f)) = \mathbf{0}$, for $t \in [t_0, t_f]$ as well, then $\mathbf{x}^{(i)}(t), \mathbf{u}^{(i)}(t), \mathbf{p}^{(i)}(t)$ are extremal.

If the above are not satisfied, then

$$\delta J_a = \left[\frac{\partial h}{\partial \mathbf{x}}(\mathbf{x}^{(i)}(t_f)) - \mathbf{p}^{(i)}(t_f)\right]^T \delta \mathbf{x}(t_f)$$

$$+ \int_{t_0}^{t_f} \left\{\left[\dot{\mathbf{p}}^{(i)}(t) + \frac{\partial \mathcal{H}}{\partial \mathbf{x}}(\mathbf{x}^{(i)}(t)), \mathbf{u}^{(i)}(t)\mathbf{p}^{(i)}(t), t)\right]^T \delta \mathbf{x}(t)\right.$$

$$+ \left[\frac{\partial \mathcal{H}}{\partial \mathbf{u}}(\mathbf{x}^{(i)}(t)), \mathbf{u}^{(i)}(t)\mathbf{p}^{(i)}(t), t)\right]^T \delta \mathbf{u}(t)$$

$$+ \left.\left[\mathbf{a}(\mathbf{x}^{(i)}(t), \mathbf{u}^{(i)}(t), t) - \dot{\mathbf{x}}^{(i)}(t)\right]^T \delta \mathbf{p}(t)\right\} dt,$$

where $\delta \mathbf{x}(t) = \mathbf{x}^{(i+1)}(t) - \mathbf{x}^{(i)}(t)$, $\delta \mathbf{u}(t) = \mathbf{u}^{(i+1)}(t) - \mathbf{u}^{(i)}(t)$ and $\delta \mathbf{p}(t) = \mathbf{p}^{(i+1)}(t) - \mathbf{p}^{(i)}(t)$.
If the necessary conditions are satisfied however, then

$$\delta J_a = \int_{t_0}^{t_f} \left[\frac{\partial \mathcal{H}}{\partial \mathbf{u}}(\mathbf{x}^{(i)}(t)), \mathbf{u}^{(i)}(t)\mathbf{p}^{(i)}(t), t)\right]^T \delta \mathbf{u}(t) dt.$$

[5] A boundary-value problem (BVP) is a system of ODEs with solution and derivative values specified at more than one point. Most commonly, the solution and derivatives are specified at just two points (the boundaries) defining a two-point boundary-value problem. Most practically arising two-point BVPs have separated boundary conditions where the boundary function may be split into two parts (one for each endpoint)—split boundary-value problem.

If the norm of $\delta\mathbf{u}$, $\|\mathbf{u}^{(i+1)} - \mathbf{u}^{(i)}\|$ is small, the sign of $\Delta J_a = J_a(\mathbf{u}^{(i+1)}) - J_a(\mathbf{u}^{(i)})$ will be determined by the sign of δJ_a. Then, if $\delta\mathbf{u}(t) = \mathbf{u}^{(i+1)} - \mathbf{u}^{(i)} = -\tau\frac{\partial\mathcal{H}^{(i)}}{\partial\mathbf{u}}(t), t \in [t_0, t_f]$ with $\tau > 0$,

$$\delta J_a = -\tau\int_{t_0}^{t_f}\left[\frac{\partial\mathcal{H}^{(i)}}{\partial\mathbf{u}}(t)\right]^T\left[\frac{\partial\mathcal{H}^{(i)}}{\partial\mathbf{u}}(t)\right] dt \leq 0,$$

and in this case the equality holds if and only if $\frac{\partial\mathcal{H}^{(i)}}{\partial\mathbf{u}}(t) = 0$, for all $t \in [t_0, t_f]$.

The value of τ is usually selected *ad hoc*, possibly as a value that effects a certain value of ΔJ_a or using a single variable search. The steepest descent algorithm can be found in detail in [137] and references therein.

C.5.2 VARIATION OF EXTREMALS

In the optimal control problem defined in the previous sections, the condition $\frac{\partial\mathcal{H}(\mathbf{x}^*(t),\mathbf{u}^*(t),\mathbf{p}^*(t))}{\partial\mathbf{u}} = \mathbf{0}$ was used to obtain a set of differential equations of the form $\dot{x}(t) = a(x(t),p(t),t)$ and $\dot{p}(t) = d(x(t),p(t),t)$ (for a first-order system), where the admissible state/control values were not bounded. In order to proceed with a numerical computation of the optimal trajectory, the value of the initial $p^{(0)}(t_0)$ needs to be guessed and used for integrating and further obtaining every succeeding trajectory. For this purpose, Newton's method [11] may be employed, which for the case of $2n$ differential equations would yield the matrix equation

$$\mathbf{p}^{(i+1)}(t_0) = \mathbf{p}^{(i)}(t_0) - \left[\mathbf{P}_p(\mathbf{p}^{(i)}(t_0),t_f)\right]^{-1}\mathbf{p}^{(i)}(t_f), \tag{C.50}$$

where the costate influence function matrix $\mathbf{P}_p$ is given by

$$\mathbf{P}_p(\mathbf{p}^{(i)}(t_0),t) = \begin{bmatrix} \frac{\partial p_1(t)}{\partial p_1(t_0)} & \frac{\partial p_1(t)}{\partial p_2(t_0)} & \cdots & \frac{\partial p_1(t)}{\partial p_n(t_0)} \\ \cdots & \cdots & \cdots & \cdots \\ \frac{\partial p_n(t)}{\partial p_1(t_0)} & \frac{\partial p_n(t)}{\partial p_2(t_0)} & \cdots & \frac{\partial p_n(t)}{\partial p_n(t_0)} \end{bmatrix}_{\mathbf{p}^{(i)}(t_0)}, \tag{C.51}$$

which is appropriate only if the desired value of the final costate is zero, which in addition occurs if the term $h(\mathbf{x}(t_f))$ is missing from performance measure (C.3). If however, $h(\mathbf{x}(t_f))$ exists, then

$$\mathbf{p}^{(i+1)}(t_0) = \mathbf{p}^{(i)}(t_0) - \left\{\left[\left[\frac{\partial^2 h}{\partial\mathbf{x}^2}(\mathbf{x}(t_f))\right]\mathbf{P}_x(\mathbf{p}^{(i)}(t_0),t_f) - \mathbf{P}_p(\mathbf{p}^{(i)}(t_0),t_f)\right]_i\right\}^{-1}$$
$$\times\left[\mathbf{p}^{(i)}(t_f) - \frac{\partial h}{\partial\mathbf{x}}(\mathbf{x}(t_f))\right]_i, \tag{C.52}$$

where now the state influence function matrix is given by

$$
\mathbf{P}_x(\mathbf{p}^{(i)}(t_0),t) = \begin{bmatrix} \frac{\partial x_1(t)}{\partial p_1(t_0)} & \frac{\partial x_1(t)}{\partial p_2(t_0)} & \cdots & \frac{\partial x_1(t)}{\partial p_n(t_0)} \\ \cdots & \cdots & \cdots & \cdots \\ \frac{\partial x_n(t)}{\partial p_1(t_0)} & \frac{\partial x_n(t)}{\partial p_2(t_0)} & \cdots & \frac{\partial x_n(t)}{\partial p_n(t_0)} \end{bmatrix}_{\mathbf{p}^{(i)}(t_0)}, \tag{C.53}
$$

and the influence function matrices can be determined as in [137] (pp. 348–349). The appropriate initial conditions for influence matrices are

$$
\mathbf{P}_x(\mathbf{p}^{(i)}(t_0),t_0) = \frac{\partial \mathbf{x}(t_0)}{\partial \mathbf{p}(t_0)}\Big|_{\mathbf{p}^{(i)}(t_0)} = \mathbf{0},
$$

$$
\mathbf{P}_p(\mathbf{p}^{(i)}(t_0),t_0) = \frac{\partial \mathbf{P}(t_0)}{\partial \mathbf{p}(t_0)}\Big|_{\mathbf{p}^{(i)}(t_0)} = \mathbf{I}.
$$

A detailed description of the steps of the variation of extremals algorithm is provided also in [137] and references therein.

C.5.3 QUASILINEARIZATION

Consider a linear two-point boundary-value problem to be solved,

$$
\dot{x}(t) = a_{11}(t)x(t) + a_{12}(t)p(t) + e_1(t), \quad x(t_0) = x_0, \tag{C.54}
$$
$$
\dot{p}(t) = a_{21}(t)x(t) + a_{22}(t)p(t) + e_2(t), \quad p(t_f) = p_f, \tag{C.55}
$$

where $a_{11}, a_{12}, a_{21}, a_{22}, e_1, e_2$ are known functions and t_0, t_f, x_0, p_f are known constants.

For this system, a solution can be obtained by numerical integration and initial conditions $x''(t_0) = x_0, p''(t_0) = 0$, while a particular solution to the nonhomogeneous system can be again obtained with numerical integration, with initial conditions $x^P(t_0) = x_0, p^P(t_0) = 0$. Combining these two solutions, a third one may be obtained for the linear two-point boundary-value problem, in terms of the solution of a linear algebraic equation.

The quasilinearization approach is based on the linearization of the reduced state-costate equations, which in turn is based on the Taylor series expansion about the initial state-costate variables. To demonstrate this process, we assume one state and one costate variable and solve $\frac{\partial \mathcal{H}}{\partial u} = 0$ for $u(t)$. Then we substitute in the state-costate equations to obtain $\dot{x}(t) = a(x(t),p(t),t)$ and $\dot{p}(t) = d(x(t),p(t),t)$, where a,d are nonlinear functions, and $x^{(0)}(t), p^{(0)}(t), t \in [t_0,t_f]$ are known trajectories. The linearization using Taylor series expansion about $x^{(0)}(t), p^{(0)}(t)$ and some basic calculus yields

$$
\dot{x}^{(1)}(t) = a_{11}(t)x^{(t)}(t) + a_{12}(t)p^{(1)}(t) + e_1(t), \tag{C.56}
$$
$$
\dot{p}^{(1)}(t) = a_{21}(t)x^{(t)}(t) + a_{22}(t)p^{(2)}(t) + e_2(t), \tag{C.57}
$$

where

$$a_{11}(t) = \frac{\partial a}{\partial x}(x^{(0)}(t), p^{(0)}(t), t) \qquad a_{12}(t) = \frac{\partial a}{\partial p}(x^{(0)}(t), p^{(0)}(t), t),$$

$$a_{21}(t) = \frac{\partial d}{\partial x}(x^{(0)}(t), p^{(0)}(t), t) \qquad a_{22}(t) = \frac{\partial d}{\partial p}(x^{(0)}(t), p^{(0)}(t), t),$$

$$e_1(t) = a(x^{(0)}(t), p^{(0)}(t), t) - \left[\frac{\partial a}{\partial x}(x^{(0)}(t), p^{(0)}(t), t)\right]x^{(0)}(t)$$

$$- \frac{\partial a}{\partial p}(x^{(0)}(t), p^{(0)}(t), t)p^{(0)}(t),$$

$$e_2(t) = d(x^{(0)}(t), p^{(0)}(t), t) - \left[\frac{\partial d}{\partial x}(x^{(0)}(t), p^{(0)}(t), t)\right]x^{(0)}(t)$$

$$- \frac{\partial d}{\partial p}(x^{(0)}(t), p^{(0)}(t), t)p^{(0)}(t)$$

are all known functions.

Detailed implementation steps (in the form of pseudocode) of the quasilinearization approach can be found in [137]. Additional properties and aspects of its operation
and behavior can be also found in [137] and references therein.

The three previous iterative methods regard in general unconstrained, nonlinear,
two-point boundary-value problems. They solve problems (more accurately a sequence of partial problems that should converge to the analyzed problem), where
one or more necessary conditions is/are initially violated, but eventually satisfied
if the iterative process converges. Trying several different initial guesses, or if the
iterative procedure converges to the same control and trajectory for a variety of initial
guesses, there exists some assurance that a global minimum has been determined. A
table summary of the features of these three approaches, such as the initial guess,
storage requirements, required computations, is provided in [137].

C.5.4 GRADIENT PROJECTION

Gradient projection methods, on the other hand, are iterative numerical procedures
for finding extrema of function of several variables, which are required to satisfy
various constraining relations.

Assume a convex function $f(\mathbf{y})$ defined in the region R, which means that it
satisfies

$$(1 - \theta)f(\mathbf{y}^{(0)}) + \theta f(\mathbf{y}^{(1)}) \geq f((1 - \theta)\mathbf{y}^{(0)} + \theta\mathbf{y}^{(1)}),$$

for all $0 \leq \theta \leq 1$, $\mathbf{y}^{(0)}, \mathbf{y}^{(1)} \in R$. Variables $y_1, y_2, \dots y_k$ are constrained by L
linear inequalities of the form $\sum_{j=1}^{K} n_{ji} y_j - v_i \geq 0$ ($\mathbf{n}_i = [n_{1i} \ n_{2i} \ \dots \ n_{ki}]^T$). If
$\mathbf{N}_L = [\mathbf{n}_1 \ \mathbf{n}_2 \ \dots \ \mathbf{n}_L]$ are orthonormal vectors, defined as $\mathbf{v}_L = [v_1 \ v_2 \ \dots \ v_L]$ and
$\lambda(\mathbf{y}) = [\lambda_1(\mathbf{y}) \ \lambda_2(\mathbf{y}) \dots \lambda_L(\mathbf{y})]^T$, then the set of linear constraints can be expressed as
$\mathbf{n}_i^T \mathbf{y} - v_i \geq \lambda_i(\mathbf{y}) \geq 0$, for $i = 1, 2, \dots, L$ ($\mathbf{N}_L^T \mathbf{y} - \mathbf{v}_L = \lambda(\mathbf{y}) \geq \mathbf{0}$) and points of the
form $\lambda_i(\mathbf{y})$ will lie in a hyperplane H_i in the K-dimensional space.

Suppose $\mathbf{y}$ is a point $\mathbf{N}_q^T \mathbf{y} - \mathbf{v}_q = \mathbf{0}$ (intersection of q linearly independent hyperplanes) and points $\mathbf{w}$ are such that $\mathbf{N}_q^T \mathbf{w} = \mathbf{0}$ (intersection of q linearly independent hyperplanes, each of which contains the origin). Then Q (denoting the intersection defined above by $\mathbf{N}_q^T \mathbf{w} = \mathbf{0}$, which is a $(K - q)$-dimensional subspace of E^K, E^K being the K-dimensional Euclidean space) and Q' (the intersection of hyperplanes $H_1, H_2, ..., H_q$) are orthogonal differing only by vector $\mathbf{v}_q$. Then, the following two $K \times K$ symmetric matrices can be defined as

$$\tilde{\mathbf{P}}_q = \mathbf{N}_q [\mathbf{N}_q^T \mathbf{N}_q]^{-1} \mathbf{N}_q^T,$$
$$\mathbf{P}_q = \mathbf{I} - \mathbf{N}_q [\mathbf{N}_q^T \mathbf{N}_q]^{-1} \mathbf{N}_q^T = \mathbf{I} - \tilde{\mathbf{P}}_q,$$

where the first is the projection matrix that takes any vector in E^K into $\tilde{Q}$ and the second a projection matrix that takes any vector E^K into Q. Thus, $\tilde{Q}$ is a q-dimensional subspace of E^K spanned by $\mathbf{n}_1, ..., \mathbf{n}_L$ constituting the space $\mathbf{N}_q$.

Assuming that f is convex with continuous second partial derivatives in a closed and bounded convex region R of E^K, and letting $\mathbf{y}^*$ be a boundary point of R which lies on exactly q, $1 \leq q \leq K$ hyperplanes that are assumed to be linearly independent with Q' denoting the intersection of these hyperplanes, then point $\mathbf{y}^*$ is a constrained global minimum if and only if

$$\mathbf{P}_q \left[-\frac{\partial f}{\partial \mathbf{y}}(\mathbf{y}^*) \right] = \mathbf{0} \quad \text{and} \quad [\mathbf{N}_q^T \mathbf{N}_q]^{-1} \mathbf{N}_q^T \left[-\frac{\partial f}{\partial \mathbf{y}}(\mathbf{y}^*) \right] \leq \mathbf{0}. \qquad (C.58)$$

If $\mathbf{y}^*$ is an interior point in R, a necessary and sufficient condition is $\frac{\partial f}{\partial \mathbf{y}}(\mathbf{y}^*) = \mathbf{0}$.

Exploiting the above, the optimal trajectories of a constrained optimal control problem can be determined using gradient projection. More specifically, it is desired to find an admissible control history $\mathbf{u}^*$ causing $\dot{\mathbf{x}}(t) = \mathbf{a}(\mathbf{x}(t), \mathbf{u}(t))$, $\mathbf{x}(t_0) = \mathbf{x}_0$ to follow admissible $\mathbf{x}^*$ that minimizes the performance measure $J = h(x(t_f)) + \int_{t_0}^{t_f} g(\mathbf{x}(t), \mathbf{u}(t)) dt$, with $t_0 = 0$, t_f specified, and constraints

$$M_{i-} \leq u_i(t) \leq M_{i+}, t \in [0, t_f], \quad i = 1, 2, ..., m, \qquad (C.59)$$
$$S_{i-} \leq x_i(t) \leq S_{i+}, t \in [0, t_f], \quad i = 1, 2, ..., n, \qquad (C.60)$$
$$x_i(t_j) = T_{ij}, t_j \text{ specified}, \quad i = 1, 2, ..., n. \qquad (C.61)$$

Approximating the state differential equations by the difference equations (which can be nonlinear) we obtain

$$\mathbf{x}(t + \Delta t) \approx \mathbf{x}(t) + \mathbf{a}(\mathbf{x}(t), \mathbf{u}(t)) \times \Delta t, \quad t = 0, \Delta t, 2\Delta t, ..., N\Delta t.$$

The latter can be written together with the discrete performance metric in the simpler form

$$\mathbf{x}(k + 1) = \mathbf{x}(k) + \mathbf{a}(\mathbf{x}(k), \mathbf{u}(k)) \Delta t = \mathbf{a}_D(\mathbf{x}(k), \mathbf{u}(k)), \qquad (C.62)$$

$$J_D = h(\mathbf{x}(N)) + \Delta t \sum_{k=0}^{N-1} g(\mathbf{x}(k), \mathbf{u}(k)). \qquad (C.63)$$

Thus, it is required to find $(Nn + Nm)$ variables that minimize J_D and satisfy the approximate state equations and constraints. If the state equations are nonlinear, then, they can be linearized about a nominal state-control history, followed by solving a sequence of linearized problems.

The state-control history is considered known,

$$(\mathbf{x}^{(i)}(0), \mathbf{x}^{(i)}(1), ... \mathbf{x}^{(i)}(N); \mathbf{u}^{(i)}(0), \mathbf{u}^{(i)}(1), ..., \mathbf{u}^{(i)}(N - 1)),$$

and an initial state-control history can be guessed in the form

$$x^{(i+1)}(k + 1) = \mathbf{A}(k)x^{(i+1)}(k) + \mathbf{B}(k)\mathbf{u}^{(i+1)}(k) + \mathbf{c}(u), \qquad (C.64)$$

where $\mathbf{A}, \mathbf{B}, \mathbf{c}$ are known time-varying matrices. Since $\mathbf{x}(0) = \mathbf{x}_0$ is specified, $\mathbf{x}^{(i)}(0) = \mathbf{x}_0$ for all i, and thus in the general case

$$\mathbf{x}^{(i+1)}(k + 1) = \mathbf{A}(k)\mathbf{x}_H + \mathbf{c}(k) + \mathbf{A}(k)...\mathbf{A}(1)\mathbf{B}(0)\mathbf{u}^{(i+1)}(0) + ...$$

$$= \mathbf{x}_H(k + 1) + \mathbf{D}_0^{k+1}\mathbf{u}^{(i+1)}(0) + ... + \mathbf{D}_k^{k+1}\mathbf{u}^{(i+1)}(k)$$

$$= \mathbf{x}_H(k + 1) + \sum_{\ell=0}^{N-1} \left[\mathbf{D}_\ell^{k+1}\mathbf{u}^{(i+1)}(\ell) \right], \qquad (C.65)$$

where

$$\mathbf{x}_k(k + 1) = \mathbf{A}(k)\mathbf{x}_H(k) + \mathbf{c}(k), \qquad \mathbf{x}_H(0) = x_0$$

and

$$\mathbf{D}_{\ell k+1} = \begin{cases} \mathbf{A}(k)\mathbf{A}(k - 1)...\mathbf{A}(\ell + 1)\mathbf{B}(\ell), & k > \ell \\ \mathbf{B}(\ell), & k = \ell \\ \mathbf{0}, & k < \ell \end{cases}$$

In this fashion, the entire discrete state history can be written in terms of the discrete control history, as $\mathcal{X}^{(i+1)} = \mathbf{D}_0 \mathcal{U}^{(i+1)} + \mathcal{X}_0$. If there are inequality constraints

involving the states $\begin{bmatrix} S_{1-} \\ S_{2-} \\ ... \\ S_{n-} \end{bmatrix} \leq \mathbf{x}(k) \leq \begin{bmatrix} S_{1+} \\ S_{2+} \\ ... \\ S_{n+} \end{bmatrix}, k = 0, 1, ..., N$, these can be expressed

as linear constraints involving only control values

$$\begin{bmatrix} S_{1-} \\ S_{2-} \\ ... \\ S_{n-} \end{bmatrix} \leq \mathbf{D}_0^k \mathbf{u}^{(i+1)}(0) + ... + \mathbf{D}_{N-1}^k \mathbf{u}^{(i+1)}(N - 1) + \mathbf{x}_H(k) \leq \begin{bmatrix} S_{1+} \\ S_{2+} \\ ... \\ S_{n+} \end{bmatrix}, \quad k = 0, 1, ..., N.$$

Consequently, the problem to be solved is to find the control values that satisfy the constraints

$$
\begin{bmatrix} M_{1-} \\ M_{2-} \\ \dots \\ M_{m-} \end{bmatrix} \le \mathbf{u}^{(i+1)}(k) \le \begin{bmatrix} M_{1+} \\ M_{2+} \\ \dots \\ M_{m+} \end{bmatrix}, \quad k = 0, 1, \dots, N,
$$

$$
\begin{bmatrix} S_{1-} \\ S_{2-} \\ \dots \\ S_{n-} \end{bmatrix} \le \mathbf{D}_0^k \mathbf{u}^{(i+1)}(0) + \dots + \mathbf{D}_{N-1}^k \mathbf{u}^{(i+1)}(N-1) + \mathbf{x}_H(k) \le \begin{bmatrix} S_{1+} \\ S_{2+} \\ \dots \\ S_{n+} \end{bmatrix}, \quad k = 0, 1, \dots, N,
$$

$$
\begin{bmatrix} T_{1j} \\ T_{2j} \\ \dots \\ T_{nj} \end{bmatrix} \le \mathbf{D}_0^j \mathbf{u}^{(i+1)}(0) + \dots + \mathbf{D}_{N-1}^j \mathbf{u}^{(i+1)}(N-1) + \mathbf{x}_H(j), j \text{ specified,}
$$

and minimize the function of Nm variables given by

$$
J_D = h(\mathcal{U}^{(i+1)}) + \Delta t \sum_{k=0}^{N-1} g(\mathcal{U}^{(i+1)}).
$$

A summary of the gradient projection algorithm for determining optimal trajectories can be found in [137] and references therein.

C.6 RELATIONSHIP BETWEEN DYNAMIC PROGRAMMING (DP) AND MINIMUM PRINCIPLE

As explained in detail previously, the optimal control problem is to find a $\mathbf{u}^* \in U$ causing the system $\dot{\mathbf{x}}(t) = \mathbf{a}(\mathbf{x}(t), \mathbf{u}(t), t$ to respond, so that the performance measure $J = h(x(t_f), t_f) + \int_{t_0}^{t_f} g(\mathbf{x}(t), \mathbf{u}(t), t) dt$ is minimized. The optimal control and its trajectory must satisfy the Hamilton–Jacobi–Bellman (HJB) equation of a Dynamic Programming (DP) ([26]) formulation

$$
0 = J_t^*(\mathbf{x}(t), t) + \min_{\mathbf{u}(t)} \left\{ g(\mathbf{x}(t), \mathbf{u}(t), t) + \left[J_x^*(\mathbf{x}(t), t) \right]^T \mathbf{a}(\mathbf{x}(t), \mathbf{u}(t), t) \right\}.
$$

If $(\mathbf{x}^*(t), t)$ is a point in the state-time space, then the $\mathbf{u}^*(t)$ corresponding to this point will yield

$$
0 = J_t^*(\mathbf{x}^*(t), t) + g(\mathbf{x}^*(t), \mathbf{u}^*(t), t) + \left[J_x^*(\mathbf{x}^*(t), t) \right]^T \mathbf{a}(\mathbf{x}^*(t), \mathbf{u}^*(t), t).
$$

If t_f is fixed and $x(t_f)$ free, a boundary condition is $J^*(\mathbf{x}^*(t_f),t_f) = h(\mathbf{x}^*(t_f),t_f)$.

By applying Pontryagin's minimum principle to the same problem, in order to obtain necessary conditions for $\mathbf{u}^*,\mathbf{x}^*$ to be optimal control-trajectory, respectively, yields

$$\dot{\mathbf{x}}^*(t) = \frac{\partial \mathcal{H}}{\partial \mathbf{p}}(\mathbf{x}^*,\mathbf{u}^*,\mathbf{p}^*,t), \tag{C.66}$$

$$\dot{\mathbf{p}}^*(t) = -\frac{\partial \mathcal{H}}{\partial \mathbf{x}}(\mathbf{x}^*,\mathbf{u}^*,\mathbf{p}^*,t), \tag{C.67}$$

$$\mathcal{H}(\mathbf{x}^*,\mathbf{u}^*,\mathbf{p}^*,t) \leq \mathcal{H}(\mathbf{x}^*,\mathbf{u},\mathbf{p}^*,t), \tag{C.68}$$

for all admissible $\mathbf{u}(t)$ and for all $t \in [t_0,t_f]$. The boundary conditions for $2n$ first-order state-costate differential equations are

$$\mathbf{x}^*(t) = \mathbf{x}_0, \tag{C.69}$$

$$\mathbf{p}^*(t_f) = \frac{\partial h}{\partial \mathbf{x}}(\mathbf{x}^*(t_f),t_f). \tag{C.70}$$

If $J_x^*(\mathbf{x}^*(t),t) = \mathbf{p}^*(t)$, then the equations of Pontryagin's minimum principle can be derived from the HJB functional equation.

If the values in the neighborhood of $\mathbf{x}^*(t)$ are considered, i.e. $\mathbf{x}(t) = \mathbf{x}^*(t) + \delta\mathbf{x}(t)$, for $\|\delta\mathbf{x}\| < \epsilon$, the scalar function $v = J_t^* + g + J_{x_1}^* a_1 + J_{x_2}^* a_2 + \ldots + J_{x_n}^* a_n$ has a local minimum at point $\mathbf{x}(t) = \mathbf{x}^*(t)$ for fixed $\mathbf{u}^*(t)$ and t, and therefore the gradient of v with respect to $\mathbf{x}$ is $\frac{\partial v}{\partial \mathbf{x}}(\mathbf{x}^*(t),\mathbf{u}^*(t)) = \mathbf{0}$, if $\mathbf{x}(t)$ is not constrained by any boundaries. By taking the aforementioned gradient of v and setting $\psi_i(\mathbf{x}^*(t),t) = J_{x_i}^*(\mathbf{x}^*(t),t)$ for $i = 1,2,\ldots,n$, the ith equation of the gradient can be written as

$$\frac{\partial \psi_i}{\partial t} + \frac{\partial \psi_i}{\partial x_1}a_1 + \frac{\partial \psi_i}{\partial x_2}a_2 + \ldots + \frac{\partial \psi_i}{\partial x_n}a_n = -\psi_1 a_{1x_i} - \psi_2 a_{2x_i} - \ldots - \psi_n a_{nx_i} - g_{x_i},$$

for $i = 1,2,\ldots,n$. Since, $\frac{\partial}{\partial x_i}\left[\frac{\partial J^*}{\partial t}\right] = \frac{\partial}{\partial t}\left[\frac{\partial J^*}{\partial x_i}\right]$, $\frac{\partial}{\partial x_i}\left[\frac{\partial J^*}{\partial x_j}\right] = \frac{\partial}{\partial x_j}\left[\frac{\partial J^*}{\partial x_i}\right]$, and $\frac{dx_i^*(t)}{dt} = a_i(\mathbf{x}^*(t),\mathbf{u}^*(t),t)$ for all $i = 1,2,\ldots,n$, the above equation yields

$$\frac{\partial \psi_i}{\partial t} + \frac{\partial \psi_i}{\partial x_1}\frac{dx_1^*}{dt} + \frac{\partial \psi_i}{\partial x_2}\frac{dx_2^*}{dt} + \ldots + \frac{\partial \psi_i}{\partial x_n}\frac{dx_n^*}{dt} = -\psi_1 a_{1x_i} - \psi_2 a_{2x_i} - \ldots - \psi_n a_{nx_i} - g_{x_i},$$

and after some algebra

$$\frac{d\boldsymbol{\psi}}{dt} = -\left[\frac{\partial \mathbf{a}}{\partial \mathbf{x}}\right]^T - \frac{\partial g}{\partial \mathbf{x}}, \tag{C.71}$$

with the boundary condition $\boldsymbol{\psi}(\mathbf{x}^*(t_f),t_f) = \frac{\partial h}{\partial \mathbf{x}}(\mathbf{x}^*(t_f),t_f)$.

The vector equation derived from the gradient of v can be eventually obtained as

$$\frac{d\boldsymbol{\psi}}{dt}(\mathbf{x}^*(t),t) = -\left[\frac{\partial \mathbf{a}}{\partial \mathbf{x}}(\mathbf{x}^*(t),\mathbf{u}^*(t),t)\right]^T \boldsymbol{\psi}(\mathbf{x}^*(t),t) - \frac{\partial g}{\partial \mathbf{x}}(\mathbf{x}^*(t),\mathbf{u}^*(t),t), \tag{C.72}$$

where the optimal $\mathbf{u}^*(t)$ satisfies

$$g(\mathbf{x}^*(t), \mathbf{u}^*(t), t) + [\boldsymbol{\psi}(\mathbf{x}^*(t), t)]^T \mathbf{a}(\mathbf{x}^*(t), \mathbf{u}^*(t), t)$$
$$= \min_{\mathbf{u}(t)} \left\{ g(\mathbf{x}^*(t), \mathbf{u}^*(t), t) + [\boldsymbol{\psi}(\mathbf{x}^*(t), t)]^T \mathbf{a}(\mathbf{x}^*(t), \mathbf{u}(t), t) \right\}, \tag{C.73}$$

while the state equations $\dot{\mathbf{x}}^*(t) = \mathbf{a}(\mathbf{x}^*(t), \mathbf{u}^*(t), t)$ and boundary conditions $\boldsymbol{\psi}(\mathbf{x}^*(t_f), t_f) = \frac{\partial h}{\partial \mathbf{x}}(\mathbf{x}^*(t_f), t_f)$ must be satisfied as well.

Consequently, $\boldsymbol{\psi}(\mathbf{x}^*(t), t) = \mathbf{p}^*(t)$ satisfy the same differential equations and the same boundary conditions, when the state variables are not constrained by any boundaries. If $\delta J^*(\mathbf{x}^*(t), t, \delta\mathbf{x}(t))$ denotes the first-order approximation to the change of minimum value of the performance measure when the state at t deviates from $\mathbf{x}^*(t)$ by $\delta\mathbf{x}(t)$, then

$$\delta J^*(\mathbf{x}^*(t), t, \delta\mathbf{x}(t)) = \left[J_x^*(\mathbf{x}^*(t), t) \right]^T \delta\mathbf{x}(t) = \mathbf{p}^{*T}(t)\delta\mathbf{x}(t), \tag{C.74}$$

which means that the extremal costate is the sensitivity of the minimum value of the performance measure to changes in the state value.

Using the principle of optimality, the Dynamic Programming multistage decision process can be reduced to a sequence of single-stage decision process

$$J_{N-K, N}^*(\mathbf{x}(N - K))$$
$$= \min_{\mathbf{u}(N-K)} \left\{ g_D(\mathbf{x}(N - K), \mathbf{u}(N - K)) + J_{N-(K-1), N}^*(\mathbf{a}_D(\mathbf{x}(N - K), \mathbf{u}(N - K))) \right\}.$$
$$\tag{C.75}$$

In general, optimal control theory can be considered an extension of the calculus of variations. The method is largely due to the independent work of Pontryagin and Bellman. In our framework, it has been used extensively in Chapter 6 for casting malware diffusion problems in the form of optimal control ones and it could be further used in various extensions studying various attack strategies and obtaining several properties of the corresponding controls associated with the analyzed problems.

Bibliography

REFERENCES

1. Albert R, Barabási A-L. Emergence of scaling in random networks. Science 1999;401:130–1.
2. Abramson N. The ALOHA system final technical report. Boston (MA, USA): Advanced Research Projects Agency; 1974.
3. Afifi N, Chung K-S. Small world wireless mesh networks. In: International conference on innovations in information technology (IIT). 2008. p. 500–4.
4. Akdere M, Cagatay C, Gerdaneri O, Korpeoglu I, Ulusoy O, Cetintemel U. A comparison of epidemic algorithms in wireless sensor networks. Comput Commun 2006;29(13):2450–557.
5. Akyildiz IF, Wang X, Wang W. Wireless mesh networks: a survey. Comput Netw 2005;47:445–87.
6. Albert R, Barabási A-L. Topology of evolving networks: local events and universality. Phys Rev Lett 2000;85(24):5234–7.
7. Albert R, Barabási A-L. Statistical mechanics of complex networks. Rev Modern Phys 2002;74(1):47–97.
8. Anagnostopoulos C, Hadjiefthymiades S, Zervas E. An analytical model for multi-epidemic information dissemination. J Parallel Distrib Comput 2011;71(1):87–104.
9. Anagnostopoulos C, Sekkasb O, Hadjiefthymiades S. An adaptive epidemic information dissemination model for wireless sensor networks. Pervasive Mobile Comput 2012;8(5):751–63.
10. Asadi A, Qing W, Mancuso V. A survey on device-to-device communication in cellular networks. IEEE Commun Surv Tutor 2014;16(4):1801–19.
11. Atkinson KE. An introduction to numerical analysis. John Wiley & Sons; 1989.
12. Attiya H. Distributed computing. 2nd ed. Providence (RI, USA): Wiley; 2004.
13. Aycock J. Computer viruses and malware. Arlington (VA, USA): Springer; 2006.
14. Baddeley A, Barany I, Schneider R, Weil W. Stochastic geometry. Lecture notes in mathematics. Springer-Verlag; 2004.
15. Barabási A-L. Scale-free networks: a decade and beyond. Science 2009;325(5939):412–3.
16. Barabási A-L, Bonabeau E. Scale-free networks. In: Sci Am; 2003. pp. 50–9.
17. Baras JS, Hovareshti P. Efficient and robust communication topologies for distributed decision making in networked systems. In: Proceedings of the 48th IEEE conference on decision and control; 2009. pp. 3751–6.
18. Baras JS, Theodorakopoulos G. Path problems in networks. Synthesis lectures on communication networks. Morgan and Claypool Publishers; 2010.
19. Barrat A, Barthélemy M, Pastor-Satorras R, Vespignani A. The architecture of complex weighted networks. Proc Natl Acad Sci USA 2004;101(11):3747–52.
20. Barrat A, Barthélemy M, Vespignani A. Modeling the evolution of weighted networks. Phys Rev Lett 2004;70(6). doi: 10.1103/PhysRevE.70.066149.
21. Barrat A, Barthélemy M, Vespignani A. Weighted evolving networks: coupling topology and weight dynamics. Phys Rev Lett 2004;92(22):228701(4).
22. Barrat A, Barthelemy M, Vespignani A. Dynamical processes on complex networks. Cambridge University Press; 2008.
23. Barrenetxea G, Berefull-Lozano B, Vetterli M. Lattice networks: capacity limits, optimal routing, and queueing behavior. IEEE/ACM Trans Netw 2006;14(3):492–505.
24. Barthélemy M, Barrat A, Pastor-Satorras R, Vespignani A. Characterization and modeling of the weighted networks. Phys A 2005;346:34–43.
25. Bertsekas D, Gallager R. Data networks. 2nd ed. Prentice-Hall; 1992.
26. Bertsekas DP. Dynamic programming and optimal control, vol. I. 3rd ed. Nashua (NH, USA): Athena Scientific; 2005.

27. Bertsekas DP, Nedic A, Ozdaglar AE. Convex analysis and optimization. Nashua (NH, USA): Athena Scientific; 2003.
28. Bertsimas D, Tsitsiklis J. Simulated annealing. Statist Sci 1993;8(1):10–5.
29. Bettstetter C. On the connectivity of wireless multihop networks with homogeneous and inhomogeneous range assignment. In: Proc. IEEE vehicular technology conf. (VTC). 2002.
30. Bettstetter C. On the minimum node degree and connectivity of a wireless multihop network. In: Proc. ACM intern. symp. on mobile ad hoc networking and computing (MobiHoc). 2002.
31. Bettstetter C. On the connectivity of ad hoc networks. Computer J 2004;47(4):432–47 [special issue].
32. Bettstetter C, Hartmann C. Connectivity of wireless multihop networks in a shadow fading environment. Wirel Netw 2005;11(5):571–9.
33. Bocharov PP, D'Apice C, Pechinkin AV. Queuing theory. De Gruyter; 2013.
34. Bollobas B. Random graphs. Cambridge (UK): Cambridge University Press; 2001.
35. Bollobas B, Riordan O. Percolation. Cambridge (UK): Cambridge University Press; 2006.
36. Bose S. An introduction to queuing systems. Springer; 2001.
37. Le Boudec J-Y, Vojnovic M. The random trip model: stability, stationary regime, and perfect simulation. IEEE/ACM Trans Netw 2006;14(6):1153–66.
38. Le Boudec J-Y, Vojnovic M. Perfect simulation and stationarity of a class of mobility models. In: Proc. of 24th IEEE conference on computer communications (INFOCOM); 2005. p. 2743–54.
39. Boukerche A, editor. Algorithms and protocols for wireless, mobile ad hoc networks. Hoboken (NJ, USA): Wiley-IEEE Press; 2008.
40. Boyce WE, Diprima RC. Elementary differential equations and boundary value problems. 9th ed. Hoboken (NJ, USA): John Wiley & Sons; 2009.
41. Boyd S, Ghosh A, Prabhakar B, Shah D. Randomized gossip algorithms. IEEE Trans Inform Theory 2006;52(6):2508–30.
42. Boyd S, Vandenberghe L. Convex optimization. Cambridge (UK): Cambridge University Press; 2004.
43. Bracewell RN. The fourier transform and its applications. 3rd ed. McGraw-Hill; 2000.
44. Brittain JE. Thevenin's theorem. IEEE Spectrum 1990;27(3):42.
45. Cavalcanti D, Agrawal D, Kelner J, Sadok D. Exploiting the small-world effect to increase connectivity in wireless ad hoc networks. In: Telecommunications and networking - ICT; 2004. pp. 388–93.
46. Chen P-Y, Chen K-C. Information epidemics in complex networks with opportunistic links and dynamic topology. In: Proc. IEEE globecom. 2010.
47. Chen P-Y, Chen K-C. Optimal control of epidemic information dissemination in mobile ad hoc networks. In: Proc. IEEE glebecom; 2011.
48. Cheng S-M, Karyotis V, Chen P-Y, Chen K-C, Papavassiliou S. Diffusion models for information dissemination dynamics in wireless complex communication networks. J Complex Syst 2013;2013:972352.
49. Chiang M, Low SH, Calderbank AR, Doyle JC. Layering as optimization decomposition: a mathematical theory of network architectures. Proc IEEE 2007;95:255–312.
50. Chiang M. Balancing transport and physical layers in wireless multihop networks: jointly optimal congestion control and power control. IEEE J Sel Areas Commun 2005;23(1):104–16.
51. Chitradurga R, Helmy A. Analysis of wired short-cuts in wireless sensor networks. In: IEEE/ACS international conference on pervasive services. 2004. p. 167-77.
52. Christodorescu M, Jha S, Maughan D, Song D, Wang C, editors. Malware detection. USA: Springer Science; 2010.
53. Chung FRK. Spectral graph theory. Rhode Island: American Mathematical Society Province; 1997.

54. Stanford large network dataset collection, Providence, RI, USA. `http://snap.stanford.edu/data/`; 2015 [accessed 22.08.15].

55. Comer DE. Internetworking with TCP/IP (volume one). 6th ed. Pearson Higher Education; 2013.

56. Cormen TH, Leiserson CE, Rivest RL, Stein C. Introduction to algorithms. 3rd ed. Boston (MA, USA): The MIT Press; 2009.

57. Cussler EL. Diffusion: mass transfer in fluid systems. Cambridge (UK): Cambridge University Press; 2009.

58. Daley DJ, Gani J. Epidemic modelling: an introduction. Cambridge (UK): Cambridge University Press; 2001.

59. Daly EM, Haahr M. Social network analysis for information flow in disconnected delay-tolerant manets. IEEE Trans Mobile Comput 2009;8(5):606–21.

60. De P, Liu Y, Das SK. An epidemic theoretic framework for vulnerability analysis of broadcast protocols in wireless sensor networks. IEEE Trans Mobile Comput 2009;8(3):413–25.

61. Diestel R. Graph theory. Graduate texts in mathematics, vol. 173. Heidelberg (Germany): Springer; 2005.

62. Dorogovtsev SN, Mendes JFF, Samukhin AN. Structure of growing networks: exact solution of the barabási-albert model. Phys Rev Lett 2000;85:4633–6.

63. Dshalalow TJH. Frontiers in queueing: models and applications in science and engineering. USA: CRC Press; 1997.

64. Stauffer D, Aharony A. Introduction to percolation theory. 2nd ed. Philadelphia (PA, USA): Taylor & Francis Group; 1994.

65. Duchon P, Hanusse N, Lebhar E, Schabanel N. Could any graph be turned into a small-world?. Theoret Comput Sci Complex Netw 2006;355(1):96–103.

66. Durrett R. Probability: theory and examples. 3rd ed. Duxbury Press (Thomson Brooks/Cole); 2004.

67. Erdös P, Rényi A. On random graphs. Publ Math Debrecen 1959;6:290–7.

68. Eugster PT, Guerraoui R, Kermarrec A-M, Massoulie L. Epidemic information dissemination in distributed systems. IEEE Computer 2004;37(5):60–7.

69. Fagiolo G. Clustering in complex directed networks. Phys Rev Lett 2007;76(2):026107.

70. Feichtinger G, Caulkins JP, Grass D, Tragler G, Behrens DA. Optimal control of nonlinear processes: with applications in drugs, corruption and terror. Springer; 2008.

71. Feller W. An introduction to probability theory and its applications. 3rd ed. New York: John Wiley and Sons; 1968.

72. Filippov AF, Arscott FM. Differential equations with discontinuous righthand sides: control systems, vol. 18. Springer Science & Business Media; 1988.

73. Flett TM. Differential analysis: differentiation, differential equations and differential inequalities. AMC 1980;10:12.

74. Frauenthal JC. Mathematical modeling in epidemiology. New York (NY, USA): Springer-Verlag; 1980.

75. Freeman L. A set of measures of centrality based on betweenness. Sociometry 1977;40(1):35–41.

76. Freeman LC, Borgatti S, White DR. Centrality in valued graphs: a measure of betweenness based on network flow. Social Netw 1991;13:141–54.

77. Friedkin NE. Theoretical foundations for centrality measures. Amer J Sociol 1991;96(6):1478–504.

78. Fronczak A, Fronczak P, Holyst JA. Average path length in random networks. Phys Rev Lett 2004;70(5):056110.

79. Fudenberg D, Tirole J. Game theory. Boston (MA, USA): MIT Press; 1991.

80. Ganesh A, Massoulié L, Towsley D. The effect of network topology on the spread of epidemics. In: Proc. of 24th annual joint conf. of IEEE comp. and comm. societies (INFOCOM). 2005. p. 1455–66.

81. Garabedian PR. Partial differential equations. New York: John Wiley & Sons Inc.; 1964.

82. Garetto M, Gong W, Towsley D. Modeling malware spreading dynamics. In: Proc. of 22nd annual joint conf. of IEEE comp. and comm. societies (INFOCOM). 2003.
83. Geman S, Geman D. Stochastic relaxation, gibbs distributions, and the bayesian restoration of images. IEEE Trans Pattern Anal Mach Intell 1984;6(6):721–41.
84. Gilbert EN. Capacity of a burst-noise channel. Bell Syst Tech J 1960;39:1253–65.
85. Gilbert EN. Random plane networks. J Soc Indus Appl Math 1961;9(4):533–43.
86. Goodman D. The wireless internet: promises and challenges. IEEE Computer 2000;33(7):36–41.
87. Gordis L. Epidemiology. 4th ed. Philadelphia (PA, USA): Saunders (Elsevier); 2009.
88. Grimmett G, Stirzaker D. Probability and random processes. 3rd ed. New York: Oxford University Press; 2001.
89. Grimmett GR. Percolation. 2nd ed. Berlin (Germany): Springer-Verlag; 1999.
90. Gross D, Shortle JF, Thompson JM, Harris CM. Fundamentals of queuing theory. Wiley; 2011.
91. Haas ZJ, Halpern JY, Li L. Gossip-based ad hoc routing. IEEE/ACM Trans Netw 2006;14(3):479–91.
92. Haenggi M. Stochastic geometry for wireless networks. Cambridge (UK): Cambridge University Press; 2012.
93. Haenggi M, Andrews JG, Baccelli F, Dousse O, Franceschetti M. Stochastic geometry and random graphs for the analysis and design of wireless networks. IEEE J Sel Areas Commun 2009;27(7):1029–46.
94. Harary F. Graph theory. Reading (MA, USA): Addison-Wesley; 1969.
95. Harras KA, Almeroth KC, Belding-Royer EM. Delay tolerant mobile networks (dtmns): controlled flooding in sparse mobile networks. In: Proc. IFIP networking. 2005. p. 1180–92.
96. Haykin S. Cognitive radio: brain-empowered wireless communications. IEEE J Sel Areas Commun 2005;23(2):201–20.
97. Heinzelman W, Kulik J, Balakrishnan H. Adaptive protocols for information dissemination in wsn. In: Proc. ACM mobicom. 1999. p. 174–185.
98. Helmy A. Small worlds in wireless networks. IEEE Commun Lett 2003;7(10):490–2.
99. Hethcote HW. The mathematics of infectious diseases. SIAM Rev 2000;42(4):599–653.
100. Hethcote HW, Stech HW, van den Driessche P. Periodicity and stability in epidemic models: a survey. In: Busenberg SN, Cooke KL, editors. Differential equations and applications in ecology epidemics and population problems. New York (NY, USA): Academic Press; 1981.
101. Hillebrand , Friedhelm, editors. GSM and UMTS, the creation of global mobile communications. Arlington (VA, USA): John Wiley and Sons; 2001.
102. Hogg R, Mckean J, Craig A. Introduction to mathematical statistics. Upper Saddle River (NJ, USA): Pearson Prentice Hall; 2005.
103. Holler J, Tsiatsis V, Mulligan C, Karnouskos S, Avesand S, Boyle D. From machine-to-machine to the internet of things: introduction to a new age of intelligence. Arlington (VA, USA): Elsevier; 2014.
104. Holzer S, Pinkolet YA, Smula J, Wattenhofer R. Monitoring churn in wireless networks. J Theoret Comput Sci 2012;453.
105. Housley R, Curran J, Huston G, Conrad D. The internet numbers registry system. IETF. RFC 7020. Boca Raton (FL, USA); 2013.
106. Hovareshti P, Baras JS, Gupta V. Average consensus over small world networks: a probabilistic framework. In: 47th IEEE conference on decision and control. 2008. p. 375–80.
107. Hu H, Myers S, Colizza V, Vespignani A. WiFi networks and malware epidemiology. Proc Natl Acad Sci USA 2009;106(5):1318–23.
108. Hwang YJ, Ko S-W, Lee SI, Cho BI, Kim S-L. Wireless small-world networks with beamforming. In: IEEE region 10 conference TENCON; 2009. pp. 1–6.

109. Mitola III J, Maguire Jr. GQ. Cognitive radio: making software radios more personal. IEEE Pers Commun Mag 1999;6(4):13–8.

110. Institute of Electrical & Electronics Engineers. IEEE 802.11: Wireless LAN medium access control (MAC) and physical layer (PHY) specifications. Arlington (VA, USA): IEEE-SA; 2012.

111. Jiang C-J, Chen C, Chang J-W, Jan R-H, Chiang TC. Construct small worlds in wireless networks using data mules. In: IEEE international conference on sensor networks, ubiquitous and trustworthy computing (SUTC). 2008. p. 28–35.

112. Johnson DH. Origins of the equivalent circuit concept: the current-source equivalent. Proc IEEE 2003;91(5):817–21.

113. Kadanoff LP. More is the same; phase transitions and mean field theories. J Stat Phys 2009;137(5–6):777–9.

114. Kailath T. Linear systems. Dover Books; 1980.

115. Kar K, Krishnamurthy A, Jaggi N. Dynamic node activation in networks of rechargeable sensors. IEEE/ACM Trans Netw 2006;14(1):15–26.

116. Karyotis V. Markov random fields for malware propagation: the case of chain networks. IEEE Commun Lett 2010;14(9):875–7.

117. Karyotis V, Grammatikou M, Papavassiliou S. A closed queueing network model for malware spreading over non-propagative ad hoc networks. In: Proc. of mediterranean ad hoc networking workshop (Med-Hoc-Net); 2007.

118. Karyotis V, Grammatikou M, Papavassiliou S. On the asymptotic behavior of malware-propagative mobile ad hoc networks. In: Proc. of 4th IEEE international conference on mobile ad hoc and sensor networks. 2007.

119. Karyotis V, Kakalis A, Papavassiliou S. Malware-propagative mobile ad hoc networks: asymptotic behavior analysis. J Comput Sci Technol 2008;23(3):389–99.

120. Karyotis V, Papavassiliou S. Topology control in cooperative ad hoc networks. In: Zhang Y, Chen H-H, Guizani M, editors. Cooperative wireless communications. CRC Press, Taylor & Francis Group; 2009. pp. 167–89.

121. Karyotis V, Papavassiliou S. Evaluation of malware spreading in wireless multihop networks with churn. In: Proc. 6th international conference on ad hoc networks (EAI ADHOCNETS). 2014.

122. Karyotis V, Papavassiliou S. Macroscopic malware propagation dynamics for complex networks with churn. IEEE Commun Lett 2015;19(4):577–80.

123. Karyotis V, Papavassiliou S. On the malware spreading over non-propagative wireless ad hoc networks: the attacker's perspective. In: Proc. of 3rd ACM international workshop on QoS and security for wireless and mobile networks (Q2SWinet). 2007.

124. Karyotis V, Papavassiliou S, Grammatikou M, Maglaris V. A novel framework for mobile attack strategy modeling and vulnerability analysis in wireless ad-hoc networks. Int J Security Netw 2006;1(3/4):255–65 4th quarter.

125. Karyotis V, Stai E, Papavassiliou S. Evolutionary dynamics of complex communications networks. Boca Raton (FL, USA): CRC Press - Taylor & Francis Group; 2013.

126. Keefe M. A short history of hacks, worms and cyberterror. Arlington (VA, USA): Computerworld; 2009.

127. Kelly FP. Reversibility and stochastic networks. Cambridge University Press; 2011.

128. Kelly FP, Maulloo AK, Tan DKH. Rate control for communication networks: shadow prices, proportional fairness and stability. J Oper Res Soc 1998;237–52.

129. Kempe D, Kleinberg J, Demers A. Spatial gossip and resource location protocols. J ACM 2004;51(6):943–67.

130. Kephart JO, White SR. Direct-graph epidemiological models of computer viruses. In: Proc. of the 1991 IEEE computer society symposium on research in security and privacy. 1991.

131. Kermarrec A, Ganesh A, Massoulie L. Probabilistic reliable dissemination in large-scale systems. IEEE Trans Parallel Distrib Syst 2003;14(3):248–58.

132. Khouzani MHR, Sarkar S. Dynamic malware attack in energy-constrained mobile wireless networks. In: Proc. 5th symp. inf. theory and applications. 2010.

133. Khouzani MHR, Sarkar S, Altman E. Maximum damage malware attack in mobile wireless networks. In: Proc. IEEE INFOCOM. 2010.

134. Khouzani MHR, Altman E, Sarkar S. Optimal quarantining of wireless malware through power control. In: (Invited Paper), Proc. of the 4th symposium on information theory and applications. 2009.

135. Kim S, Lee C-H, Eun DY. Superdiffusive behavior of mobile nodes and its impact on routing protocol performance. IEEE Trans Mobile Comput 2010;9(2):288–304.

136. Kindermann R, Snell JL. Markov random fields and their applications. Providence (RI, USA): American Mathematical Society; 1980.

137. Kirk DE. Optimal control theory - an introduction. Dover Publications; 1996.

138. Kirk DE. Optimal control theory: an introduction. Prentice Hall; 1970.

139. Kirkpatrick S, Gelatt CD, Vecchi MP. Optimization by simulated annealing. Science 1983;220(4598):671–80.

140. Klain DA, Rota G-C. Introduction to geometric probability. Cambridge (UK): Cambridge University Press; 1997.

141. Kleinberg J. The small-world phenomenon: an algorithmic perspective. In: 32nd ACM symposium on theory of computing; 2000.

142. Kleinrock L. Queuing systems: computer applications (vol. 1 & 2). Wiley; 1976.

143. Knight FB. Essentials of Brownian motion and diffusion. USA: The American Mathematical Society; 1981.

144. Krapivsky PL, Redner S, Leyvraz F. Connectivity of growing random networks. Phys Rev Lett 2000;85(21):4629–32.

145. Krylov NV, Aries AB. Controlled diffusion processes. Berlin-Heidelberg: Springer-Verlag; 2009.

146. Kuhn HW, Szegö GP. Differential games and related topics. North-Holland Pub. Co.; 1971.

147. Kulkarni RV, Almaas E, Stroud D. Exact results and scaling properties of small-world networks. Phys Rev Lett 2000;61(4):4268.

148. Kurose JF, Ross KW. Computer networking: a top-down approach. 6th ed. Boston (MA, USA): Pearson Education; 2012.

149. Landesman M. A brief history of Malware; The First 25 Years. Available online, Arlington, VA, USA, http://antivirus.about.com/od/whatisavirus/a/A-Brief-History-Of-Malware-The-First-25-Years.htm.

150. Landherr A, Friedl B, Heidemann J. A critical review of centrality measures in social networks. Business Inf Syst Eng 2010;2(6):371–85.

151. Larsson T, Patriksson M, Strömberg A-B. Ergodic primal convergence in dual subgradient schemes for convex programming. Math Program 1999;86(2):283–312.

152. Lazar AA. Optimal control of an m/m/m queue. In: Proc. of the ACM computer network performance symposium (SIGMETRICS). 1982. p. 14–20.

153. Lee E. Cyber physical systems: design challenges. University of California, Berkeley Technical Report No. UCB/EECS-2008-8; 2008.

154. Lee WCY. Mobile cellular telecommunications systems. Arlington (VA, USA): McGraw-Hill; 1989.

155. Lewis TG. Network science: theory and applications. Hoboken (NJ, USA): John Wiley & Sons; 2009.

156. Lindgren A, Doria A, Schelen O. Probabilistic routing in intermittently connected networks. In: Proc. SAPIR 2004. 2004. p. 239–54.

157. Lui S, Gopalakrishnan S, Xue L, Qixin W. Cyber-physical systems: a new frontier. In: IEEE international conference on sensor networks, ubiquitous and trustworthy computing; 2008. pp. 1–9.

158. Metropolis N, Rosenbluth AW, Rosenbluth MN, Teller AH, Teller E. Equation of state calculations by fast computing machines. J Chem Phys 1953;21(6):1087–92.

159. Misra V, Gong W, Towsley D. A fluid based analysis of a network of AQM routers supporting TCP flows with an application to RED. In: Proc. of the ACM annual conference of the special interest group on data communication; 2000.

160. Mollison D, editor. Epidemic models: their structure and relation to data. Cambridge (UK): Cambridge University Press; 2008.

161. Molloy M, Reed B. A critical point for random graphs with a given degree sequence. Random Structures Algorithms 1995;6:161–80.

162. Negus KJ, Petrick A. History of wireless local area networks (WLANs) in the unlicensed bands. Arlington (VA, USA): George Mason University Law School Conference, Information Economy Project; 2008.

163. Nelson B, Phillips A, Steuart C. Guide to computer forensics & investigations. 4th ed. USA: Course Technology; 2009.

164. Newman M. The structure and function of complex networks. SIAM Rev 2003;45(2):167–256.

165. Newman M. Networks: an introduction. New York (NY, USA): Oxford University Press; 2010.

166. Obaidat M, Boudriga N. Security for e-systems and computer networks. USA: Cambridge University Press; 2007.

167. Olfati-Saber R. Ultrafast consensus in small-world networks. In: American control conference, vol. 4; 2005. pp. 2371–8.

168. Olsson M, Sultana S, Rommer S, Frid L, Mulligan C. EPC and 4G packet networks. Oxford (UK): Academic Press (Elsevier); 2013.

169. Onnela J-P, Saramäki J, Kertész J, Kaski K. Intensity and coherence of motifs in weighted complex networks. Phys Rev Lett 2005;71(6):065103.

170. Opsahl T, Agneessens F, Skvoretz J. Node centrality in weighted networks: generalizing degree and shortest paths. Social Netw 2010;32:245–51.

171. Opsahl T, Panzarasa P. Clustering in weighted networks. Social Netw 2009;31(2):155–63.

172. Ozkasap O, Genc Z, Atsan E. Epidemic-based reliable and adaptive multicast for mobile ad hoc networks. Comput Netw 2009;53(9):1409–30.

173. Papavassiliou S, Zhou J. A continuum theory-based approach to the modeling of dynamic wireless sensor networks. IEEE Commun Lett 2005;9(4):337–9.

174. Papoulis A, Pillai SU. Probability, random variables and stochastic processes. McGRAW-HILL; 2002.

175. Parkval S, Astely D. The evolution of LTE toward LTE advanced. J Commun 2009;4(3):146–54.

176. Pastor-Satorras R, Vespignani A. Epidemics and immunization in scale-free networks. Berlin (Germany): WILEY-VCH Verlag; 2005.

177. Pastor-Satorras R, Vespignani A. Epidemic spreading in scale-free networks. Phys Rev Lett 2001;86(14):3200–3.

178. Pastor-Satorras R, Vespignani A. Epidemics and immunization in scale-free networks. In: Handbook of graphs and networks: from the genome to the Internet. Wiley-VCH; 2002.

179. Peng S, Yu S, Yang A. Smartphone malware and its propagation modeling: a survey. IEEE Commun Surv Tutor 2014;16(2):925–41.

180. Penrose M. Random geometric graphs. Oxford (UK): Oxford University Press; 2003.

181. Perra N, Fortunato S. Spectral centrality measures in complex networks. Phys Rev Lett 2008;78(3):036107.

182. Raval A, Ray A. Introduction to biological networks. Boca Raton (FL, USA): CRC Press - Taylor & Francis Group; 2013.

183. Rezaei BA, Sarshar N, Roychowdhury VP. Random walks in a dynamic small-world space: robust routing in large-scale sensor networks. In: 60th IEEE vehicular technology conference, vol. 7; 2004. p. 4640–44.

184. Reznik A, Kulkarni SR, Verdú S. A small world approach to heterogeneous networks. Commun Inf Syst 2004;3(4):325–48.

185. Rhee I, Shin M, Hong S, Lee K, Chong S. On the levy-walk nature of human mobility. In: Proc. IEEE INFOCOM. 2008. p. 924–32.

186. Robert CP, Casella G. Monte Carlo statistical methods. 2nd ed. Providence (RI, USA): Springer-Verlag; 2004.

187. Ross S. Stochastic processes. 2nd ed. New York (NY, USA): John Wiley & Sons; 1996.

188. Rugh WJ. Linear system theory. 2nd ed. Prentice Hall; 1996.

189. Ruszczyński AP. Nonlinear optimization, vol. 13. Princeton University Press; 2006.

190. Sahneh FD, Scoglio CM. Optimal information dissemination in epidemic networks. In: Proc. IEEE 51st annual conference on decision and control. 2012. p. 1657–62.

191. Santi P. Topology control in wireless ad hoc and sensor networks. England (UK): John Wiley and Sons; 2005.

192. Saramäki J, Kivelä M, Onnela J-P, Kaski K, Kertész J. Generalizations of the clustering coefficient to weighted complex networks. Phys Rev Lett 2007;75(2):027105.

193. Schmidt V, editor. Stochastic geometry, spatial statistics and random fields: models and algorithms. Springer International Publishing; 2015.

194. Schwartz M. Telecommunication networks: protocols, modeling and analysis. Boston (MA, USA): Addison-Wesley Longman Publishing; 1986.

195. Symantec Enterprise Security. Symantec Internet Security Threat Report. Available online at: `https://www.symantec.com/content/de/de/about/downloads/PressCenter/summary.pdf` (July 2015), USA, 2006.

196. Sharma G, Mazumdar R. Hybrid sensor networks: a small world. In: 6th ACM international symposium on mobile ad hoc networking and computing (MobiHoc'05). 2005. p. 366–77.

197. Shellke SH, Shroff NB, Bagchi S. Modeling and automated containment of worms. IEEE Trans Depend Secure Comput 2008;5(2):71–86.

198. Shi G, Johansson KH. The role of persistent graphs in the agreement seeking of social networks. IEEE J Sel Areas Commun September 2013;31(9):595–606 [special issue]

199. Shiu Y-S, Chang SY, Wu H-C, Huang SC-H, Chen H-H. Physical layer security in wireless networks: a tutorial. IEEE Wirel Commun Mag 2011;18(2):66–74.

200. Shoch J, Hupp J. The 'worm' programs – early experience with a distributed computation. Commun ACM 1982;25(3):172–80.

201. Stai E, Karyotis V, Papavassiliou S. A socially-driven topology improvement framework with applications in content distribution and trust management. J Internet Serv Appl 2011;2(2):113–27.

202. Stai E, Karyotis V, Papavassiliou S. Topology enhancements in wireless multi-hop networks: a top-down approach. IEEE Trans Parallel Distrib Syst 2012;23(7):1344–57.

203. Stai E, Karyotis V, Papavassiliou S. Wireless multi-hop network topology control optimization and trade-off analysis. In: 35th IEEE sarnoff symposium. 2012. p. 1-6.

204. Staniford S, Paxson V, Weaver N. How to own the Internet in your spare time. In: Proc. the 11th USENIX security symposium. 2002. p. 149–67.

205. Stoyan D, Kendall WS, Mecke J. Stochastic geometry and its applications. 2nd ed. Sussex: John Wiley and Sons; 1995.

206. Tan X, Xi W, Baras J. Decentralized coordination of autonomous swarms using parallel gibbs sampling. Automatica 2010;46(12):2068–76.

207. Tanachaiwiwat S, Helmy A. Vaccine: war of the worms in wired and wireless networks. In: The 25th IEEE conference on computer communications (INFOCOM). 2006.

208. Tanenbaum AS, Wetherall DJ. Computer networks. 5th ed. Boston (MA, USA): Pearson Education; 2010.

209. Tanner M. Practical queuing analysis. IBM McGraw-Hill; 1994.

210. Tao T. An introduction to measure theory. USA: American Mathematical Society; 2011.

211. Thomopoulos NT. Fundamentals of queuing systems: statistical methods for analyzing queuing models. In: Springer Statistics for business, economics, operations research/decision theory, management science; 2012.

212. Toh CK. Ad hoc mobile wireless networks. Arlington (VA, USA): Prentice Hall Publishers; 2002.

213. Trubin VA. Strength of a graph and packing of trees and branchings. Cybern Syst Anal 1993;29:379–84.

214. Tse D, Viswanath P. Fundamentals of wireless communications. Cambridge University Press; 2005.

215. Vahdat A, Becker D. Epidemic routing for partially-connected ad hoc networks. Technical Report CS-2000-06. Duke Univ., July 2000.

216. van Mieghem P, Omic J, Kooij R. Virus spread in networks. ACM/IEEE Trans Netw 2009;17(1):1–14.

217. Verma CK, Tamma BR, Manoj BS, Rao R. A realistic small-world model for wireless mesh networks. Commun Inf Syst 2011;15(4):455–7.

218. Vojnovic M, Gupta V, Karagiannis T, Gkantsidis C. Sampling strategies for epidemic-style information dissemination. In: Proc. IEEE INFOCOM. 2008. p. 2351–9.

219. Vynnycky E, White RG. An introduction to infectious disease modelling. Oxford University Press; 2009.

220. Wan P-J, Yi C-W. Asymptotic critical transmission ranges for connectivity in wireless ad hoc networks with bernoulli nodes. In: Proc. of the 6th IEEE wireless communications and networking conference. 2005. p. 2219–24.

221. Wang P, Gonzalez M, Hidalgo C, Barabasi A. Understanding the spreading patterns of mobile phone viruses. Science 2009;324:1071–5.

222. Wang W, Wang B, Hu B, Yan G, Ou Q. General dynamics of topology and traffic on weighted technological networks. Phys Rev Lett 2005;94(18):188702.

223. Wang Y, Wen S, Xiang Y, Zhou W. Modeling the propagation of worms in networks: A survey. IEEE Commun Surv Tutor 2014;16(2):942–60.

224. Watts DJ. Small worlds: the dynamics of networks between order and randomness. Princeton Studies in Complexity; 1999.

225. Watts DJ, Strogatz SH. Collective dynamics of 'small-world' networks. Nature 1998;393:440–2.

226. Weaver N. Measure theory and functional analysis. Hackensack (NJ, USA): World Scientific Publishing; 2013.

227. Weiser M. The computer for the 21st century. Sci Am 1991;265(3):94–104.

228. Winkler G. Image analysis, random fields and markov chain monte carlo methods: a mathematical introduction. 2nd ed. Springer; 2003.

229. Wolff R. Stochastic modeling and the theory of queues. Prentice-Hall; 1989.

230. Yan G, Cuellar L, Eidenbenz S. Bluetooth worm propagation: Mobility pattern matters! In: Proc. ASIACCS. 2007. p. 32–44.

231. Zhang X, Negli G, Kurose J, Towsley D. Performance modeling of epidemic routing. Comput Netw 2007;51(8):2867–91.

232. Zou CC, Gong W, Towsley D, Gao L. The monitoring and early detection of internet worms. IEEE/ACM Trans Netw 2005;13(5):961–74.

233. Zou CC, Towsley D, Gong W. Email virus propagation modeling and analysis. Technical report: TR-CSE-03-04. Apr. 2004.

234. Zou CC, Gong W, Towsley D. Worm propagation modeling and analysis under dynamic quarantine defense. In: Proc. of ACM workshop on rapid malcode. 2003. p. 51–60.

235. Zou CC, Gong W, Towsley D. Code red worm propagation modeling and analysis. In: Proc. of 9th ACM conf. on computer and communications security. 2002.

236. Zou CC, Towsley D, Gong W. Modeling and simulation study of the propagation and defense of Internet e-mail worms. IEEE Trans Depend Secure Comput 2007;4(2):105–18.

Author Index

J

Jaggi, N., 2006
Jan, R.-H., 2008
Jha, S., 2010
Jiang, C.-J., 2008
Johansson, K.H., 2013
Johnson, D.H., 2003

K

Kadanoff, L.P., 2009
Kailath, T., 1980
Kakalis, A., 2008
Kar, K., 2006
Karagiannis, T., 2008
Karnouskos, S., 2014
Karyotis, V., 2006, 2007, 2008, 2009, 2010,
    2011, 2012, 2013, 2014, 2015
Kaski, K., 2005, 2007
Keefe, M., 2009
Kelly, F.P., 1998, 2011
Kelner, J., 2004
Kempe, D., 2004
Kendall, W.S., 1995
Kephart, J.O., 1991
Kermarrec, A., 2003
Kermarrec, A.-M., 2004
Kertész, J., 2005, 2007
Khouzani, M.H.R., 2009, 2010
Kim, S., 2010
Kim, S.-L., 2009
Kindermann, R., 1980
Kirk, D.E., 1970, 1996
Kirkpatrick, S., 1983
Kivelä, M., 2007
Klain, D.A., 1997
Kleinberg, J., 2000, 2004
Kleinrock, L., 1976
Knight, F.B., 1981
Ko, S.-W., 2009
Kooij, R., 2009
Korpeoglu, I., 2006
Krapivsky, P.L., 2000
Krishnamurthy, A., 2006
Krylov, N.V., 2009
Kuhn, H.W., 1971
Kulik, J., 1999
Kulkarni, R.V., 2000
Kulkarni, S.R., 2004
Kurose, J., 2007
Kurose, J.F., 2012

L

Landesman, M.
Landherr, A., 2010
Larsson, T., 1999
Lazar, A.A., 1982
Le Boudec, J.-Y., 2005, 2006
Lebhar, E., 2006
Lee, C.-H., 2010
Lee, E., 2008
Lee, K., 2008
Lee, S.I., 2009
Lee, W.C.Y., 1989
Leiserson, C.E., 2009
Lewis, T.G., 2009
Leyvraz, F., 2000
Li, L., 2006
Lindgren, A., 2004
Liu, Y., 2009
Low, S.H., 2007
Lui, S., 2008

M

Maglaris, V., 2006
Maguire Jr., G.Q., 1999
Mancuso, V., 2014
Manoj, B.S., 2011
Massoulié, L., 2003, 2004, 2005
Maughan, D., 2010
Maulloo, AK, 1998
Mazumdar, R., 2005
Mckean, J., 2005
Mecke, J., 1995
Mendes, J.F.F., 2000
Metropolis, N., 1953
Misra, V., 2000
Mitola III, J., 1999
Mollison, D., 2008
Molloy, M., 1995
Mulligan, C., 2013, 2014
Myers, S., 2009

N

Nedic, A., 2003
Negli, G., 2007
Negus, K.J., 2008
Nelson, B., 2009
Newman, M., 2003, 2010

Subject Index

Printed in the United States
By Bookmasters